Topics in Neuroscience

Managing Editor:

GIANCARLO COMI

Co-Editor:

JACOPO MELDOLESI

Associate Editors:

UGO ECARI

MASSIMO FILIPPI

GIANVITO MARTINO

Springer

*Milano
Berlin
Heidelberg
New York
Barcelona
Hong Kong
London
Paris
Singapore
Tokyo*

G. Martino • L. Adorini (Eds)

From Basic Immunology to Immune-Mediated Demyelination

 Springer

GIANVITO MARTINO
Department of Neuroscience
S. Raffaele Scientific Institute
Milan, Italy

LUCIANO ADORINI
Roche Milano Ricerche
Milan, Italy

The Editors and Authors wish to thank FARMADES-SCHERING GROUP (Italy) for the support and help in the realization and promotion of this volume

---

© Springer-Verlag Italia, Milano 1999

ISBN 88-470-0054-8

Library of Congress Cataloging-in-Publication Data: applied for

This work is subject to copyright. All rights are reserved, whether the whole or part of the material is concerned, specifically the rights of translation, reprinting, re-use of illustrations, recitation, broadcasting, reproduction on microfilms or in other ways, and storage in data banks. Duplication of this publication or parts thereof is only permitted under the provisions of the Italian Copyright Law in its current version, and permission for use must always be obtained from Springer-Verlag. Violations are liable for prosecution under the Italian Copyright Law.

The use of registered names, trademarks, etc. in this publication does not imply, even in the absence of a specific statement, that such names are exempt from the relevant protective laws and regulations and therefore free for general use.

Product liability: the publisher cannot guarantee the accuracy of any information about dosage and application contained in this book. In every individual case the user must check such information by consulting the relevant literature.

Typesetting: Photo Life (Milan)
Printing and binding: Staroffset (Cernusco sul Naviglio, Milan)
Cover design: Simona Colombo

Printed in Italy

SPIN: 10715398

*To Nicolò and Cristina,
who make my life serene*
(G. MARTINO)

*To Silvano, who
wants to become a scientist*
(L. ADORINI)

Preface

Morphological, physiological and pathological evidence demonstrates that nervous and immune systems interact not only in maintaining brain homeostasis, but also in causing neurological diseases. The study of these interactions represents the basis on which neuroimmunology has grown during the years. At present, several neurological diseases are recognized to be caused by a derangement of the immune system in either its regulatory or effector functions.

The main scope of this book, to discuss how an unbalanced immune system may lead to immune-mediated neurological diseases, is achieved in three parts. The first part provides an overview on how the immune system works. This is propaedeutical to understanding interactions between the immune and nervous systems, which are discussed in the second part. The third part of the book focuses on one particular area of neuroimmunology, the immune disorders leading to the damage of central and peripheral myelin.

Given the opportunity to review first the immune system in itself and then how it operates during immune-mediated demyelinating disorders, we have tried to provide the reader with a basis for clearly understanding how interactions between the immune and nervous systems can be protective or pathogenic. This knowledge is a prerequisite for a rationale immune intervention targeting these disorders.

G. Martino
L. Adorini

Table of Contents

Chapter 1 - Protective Versus Harmful Responses and Immune Regulation
F. Di Rosa, V. Barnaba 1

Chapter 2 - Antigen Recognition and Autoimmunity
F. Sinigaglia, J. Hammer 7

Chapter 3 - Assembly and Function of Immunoglobulins During B Cell Development
R. Sitia 16

Chapter 4 - Pathogenetic Mechanisms of Autoimmunity
L. Adorini 26

Chapter 5 - Adhesion Receptors Involved in Leukocyte Functions
E. Bianchi, M. Fabbri, R. Pardi 38

Chapter 6 - The Endothelium of the Brain Microvasculature and the Organization of Intercellular Junctions
M.G. Lampugnani, G. Bazzoni, E. Dejana 47

Chapter 7 - Chemokines and Chemokine Receptors
A. Mantovani, P. Allavena, C. Garlanda, S. Ramponi, C. Paganini,
A. Vecchi, S. Sozzani 58

Chapter 8 - Th1/Th2 Cytokine Network
M.M. D'Elios, G. Del Prete 68

Chapter 9 - Lymphocyte Trafficking in the Central Nervous System
H. Lassmann 83

Chapter 10 - Antigen Presentation in the Central Nervous System
F. Aloisi 89

Chapter 11 - Myelination of the Central Nervous System
G.G. Consalez, V. Avellana-Adalid, C. Alli, A. Baron Van Evercooren 101

Chapter 12 - Genomic Screening in Multiple Sclerosis
P. Momigliano Richiardi 116

Chapter 13 - MHC and Multiple Sclerosis
M.G. Marrosu... 130

Chapter 14 - Cytokine Genes in Multiple Sclerosis
F.L. Sciacca, L.M.E. Grimaldi 137

Chapter 15 - Adhesion Molecules and the and Blood-Brain-Barrier in Multiple Sclerosis
J.J. Archelos, H.-P. Hartung 149

Chapter 16 - Non-Myelin Antigen Autoreactivity in Multiple Sclerosis
G. Ristori, C. Montesperelli, C. Buttinelli, L. Battistini, S. Cannoni,
G. Borsellino, R. Bomprezzi, A. Perna, M. Salvetti..................... 162

Chapter 17 - Myelin Antigen Autoreactivity in Multiple Sclerosis
M. Vergelli... 170

Chapter 18 - Inflammation and Multiple Sclerosis: a Close Interplay
G. Martino, R. Furlan, P.L. Poliani 185

Chapter 19 - Magnetic Resonance and Blood-Brain Barrier Dysfunction in Multiple Sclerosis
M. Rovaris, C. Tortorella, J.C. Sipe, M. Filippi....................... 195

Chapter 20 - Immunotherapies for Multiple Sclerosis
P. Perini, P. Gallo 210

Chapter 21 - Animal Models of Demyelination of the Central Nervous System
A. Uccelli... 233

Chapter 22 - MHC and Non-MHC Genetics of Experimental Autoimmune Encephalomyelitis
T. Olsson, I. Dahlman, E. Wallström 246

Chapter 23 - Genetics of Hereditary Neuropathies
G.L. Mancardi 265

Chapter 24 - Immunopathogenetic Role of Anti-Neural Antibodies in Demyelinating Dysimmune Neuropathies
E. Nobile-Orazio, M. Carpo.. 274

Chapter 25 - Treatment of Inflammatory Demyelinating Polyneuropathy
G. Comi, L. Roveri 287

Subject Index 313

List of Contributors

L. Adorini
Roche Milano Ricerche, Milan, Italy

P. Allavena
Mario Negri Institute of Pharmacological
Research, Milan, Italy

C. Alli
Department of Neuroscience,
San Raffaele Scientific Institute,
Milan, Italy

F. Aloisi
Neurophysiology Unit, Laboratory of
Organ and System Pathophysiology,
Istituto Superiore di Sanità, Rome, Italy

J.J. Archelos
Department of Neurology,
Karl-Franzens-Universität, Graz, Austria

V. Avellana-Adalid
INSERM 134, Cellular, Molecular
and Clinical Neurobiology,
Hôpital de la Salpetrière, Paris, France

V. Barnaba
Andrea Cesalpino Foundation,
I Institute of Internal Medicine,
University of Rome La Sapienza, Rome,
Italy; Pasteur-Cenci Bolognetti Institute,
Rome, Italy

A. Baron Van Evercooren
INSERM 134, Cellular, Molecular
and Clinical Neurobiology,
Hôpital de la Salpetrière, Paris, France

L. Battistini
Neuroimmunology Laboratory,
Santa Lucia Scientific Institute,
Rome, Italy

G. Bazzoni
Laboratory of Vascular Biology,
Mario Negri Instiute of Pharmacological
Research, Milan, Italy

E. Bianchi
Human Immunology Unit, DIBIT-
San Raffaele Scientific Hospital,
Milan, Italy

R. Bomprezzi
Department of Neurological Sciences,
University of Rome La Sapienza,
Rome, Italy

G. Borsellino
Neuroimmunology Laboratory,
Santa Lucia Scientific Institute,
Rome, Italy

C. Buttinelli
Department of Neurological Sciences,
University of Rome La Sapienza,
Rome, Italy

S. Cannoni
Department of Neurological Sciences,
University of Rome La Sapienza,
Rome, Italy

M. Carpo
Giorgio Spagnol Service of Clinical Neuroimmunology, Centro Dino Ferrari, Institute of Clinical Neurology, University of Milan, IRCCS Ospedale Maggiore Policlinico, Milan, Italy

G.G. Consalez
Department of Neuroscience,
San Raffaele Scientific Institute,
Milan, Italy

G. Comi
Department of Clinical Neurophysiology,
San Raffaele Scientific Institute,
Milan, Italy

I. Dahlman
Neuroimmunology Unit,
Center for Molecular Medicine,
Department of Medicine, Karolinska Institutet, Stockholm, Sweden

E. Dejana
Laboratory of Vascular Biology, Mario Negri Institute of Pharmacological Research, Milan, Italy

M.M. D'Elios
Institute of Internal Medicine
and Immunoallergology,
University of Florence, Italy

G. Del Prete
Institute of Internal Medicine
and Immunoallergology, University of Florence, Italy

F. Di Rosa
Andrea Cesalpino Foundation,
I Institute of Internal Medicine,
University of Rome La Sapienza,
Rome, Italy

M. Fabbri
Human Immunology Unit, DIBIT
Scientific Institute San Raffaele,
Milan, Italy

M. Filippi
Neuroimaging Research Unit,
Department of Neuroscience,
San Raffaele Scientific Institute,
Milan, Italy

R. Furlan
Departement of Neuroscience,
San Raffaele Scientific Institute,
Milan, Italy

P. Gallo
Department of Neurological
and Psychiatrical Sciences,
Second Neurological Clinic,
University of Padua, School of Medicine,
Geriatric Hospital, Padua, Italy

C. Garlanda
Mario Negri Institute of
Pharmacological Research, Milan, Italy

L.M.E. Grimaldi
Neuroimmunology Unit,
Department of Neuroscience,
San Raffaele Scientific Institute,
Milan, Italy

J. Hammer
Hoffmann-La-Roche Inc., Preclinical R&D, 3 Nutley, NJ, US

H.-P. Hartung
Department of Neurology,
Karl-Franzens-Universität,
Graz, Austria

H. Lassmann
Institute of Neurology,
University of Vienna, Austria

M.G. Lampugnani
Laboratory of Vascular Biology,
Mario Negri Institute of
Pharmacological Research, Milan, Italy

G.L. Mancardi
Department of Neurological Sciences,
University of Genoa, Italy

List of Contributors

A. Mantovani
Mario Negri Institute of
Pharmacological Research, Milan, Italy;
Department of Biotechnology,
Section of General Pathology,
University of Brescia, Italy

G. Martino
Department of Neuroscience,
San Raffaele Scientific Institute,
Milan, Italy

M.G. Marrosu
Neurophysiopathology Section,
Department of Neuroscience,
University of Cagliari, Cagliari, Italy

P. Momigliano Richiardi
Chair of Human Genetics, Department
of Medical Sciences, University of
Eastern Piedmont "A. Avogadro",
Novara, Italy

C. Montesperelli
Department of Neurological Sciences,
University of Rome La Sapienza, Rome,
Italy

E. Nobile-Orazio
Giorgio Spagnol Service of Clinical
Neuroimmunology, Centro Dino
Ferrari, Institute of Clinical Neurology,
University of Milan, IRCCS Ospedale
Maggiore Policlinico, Milan, Italy

T. Olsson
Neuroimmunology Unit, Center for
Molecular Medicine, Department of
Medicine, Karolinska Institutet,
Stockholm, Sweden

C. Paganini
Mario Negri Institute of Pharmacological
Research, Milan, Italy

R. Pardi
Human Immunology Unit,
DIBIT- San Raffaele Scientific Institute,
Milan, Italy

A. Perna
Department of Neurological Sciences,
University of Rome La Sapienza,
Rome, Italy

P. Perini
Department of Neurological and
Psychiatrical Sciences, Second
Neurological Clinic, University of
Padua, School of Medicine, Geriatric
Hospital, Padua, Italy

P.L. Poliani
Departement of Neuroscience,
San Raffaele Scientific Institute,
Milan, Italy

S. Ramponi
Mario Negri Institute of Pharmacological
Research, Milan, Italy

L. Roveri
Department of Clinical Neurophysiology,
San Raffaele Scientific Institute,
Milan, Italy

G. Ristori
Department of Neurological Sciences,
University of Rome La Sapienza, Rome,
Italy

M. Rovaris
Neuroimaging Research Unit,
Department of Neuroscience,
San Raffaele Scientific Institute,
Milan, Italy

M. Salvetti
Department of Neurological Sciences,
University of Rome La Sapienza,
Rome, Italy

F.L. Sciacca
Neuroimmunology Unit,
Department of Neuroscience,
San Raffaele Scientific Institute,
Milan, Italy

F. Sinigaglia
Roche Milano Ricerche, Milan, Italy

J.C. Sipe
Scripps Clinic, Division of Neurology
and The Scripps Research Institute,
La Jolla, California, USA

R. Sitia
Laboratory of Molecular Immunology,
DIBIT-San Raffaele Scientific Institute,
Milan, Italy

S. Sozzani
Mario Negri Institute of Pharmacological
Research, Milan, Italy

C. Tortorella
Neuroimaging Research Unit,
Department of Neuroscience,
San Raffaele Scientific Institute,
Milan, Italy

A. Uccelli
Neuroimmunology Unit,
Department of Neurological Sciences
and Neurorehabilitation,
University of Genoa, Italy

A. Vecchi
Mario Negri Institute of Pharmacological
Research, Milan, Italy

M. Vergelli
Department of Neurological
and Psychiatric Sciences,
University of Florence, Italy

E. Wallström
Neuroimmunology Unit,
Center for Molecular Medicine,
Department of Medicine, Karolinska
Institutet, Stockholm, Sweden

Chapter 1

Protective Versus Harmful Responses and Immune Regulation

F. Di Rosa[1], V. Barnaba[1,2]

Introduction

The immune system can mount a response against a wide array of different antigens, and the rules governing its functioning are not clear yet. We outline here how the immune response is regulated at the level of (1) induction, (2) balance between different arms of immune defense, and (3) termination. We suggest that it is not easy to draw a line between protective anti-pathogen defense and harmful autoimmune attack, and provide a few examples explaining how anti-viral responses may favor autoimmunity.

Induction of an Immune Response

The first question is what determines whether an antigen will induce an immune response or not. The general view is that antigen recognition by specific membrane receptors expressed by peripheral B and T cells is not sufficient for cell activation, and an additional signal is required. For activation of $CD4^+$ T cells, the second signal is provided by the antigen-presenting cell (APC) and can be given by membrane molecules of the B7 family or by soluble mediators [1]. Once $CD4^+$ T cells have been activated, they can help B cells and $CD8^+$ T cells to respond to the antigen.

The display of an effective second signal by APC is regulated by many factors, such as dose of antigen, route of its administration, and more importantly whether the antigen is given in conjunction with adjuvant or not. The role of the adjuvant can be either to promote inflammation, with subsequent APC activation, or to turn on the APC via direct means. The presence of an infectious agent and the cellular damage caused by it are considered the most potent stimuli to turn on the APC, and the use of adjuvant is aimed at mimicking the effect of a pathogen entry [2, 3].

---

[1] Fondazione Andrea Cesalpino, I Clinica Medica, Università La Sapienza, V. le dell'Università 30 - 00161 Rome, Italy. e-mail: barnaba@uniroma1.it
[2] Istituto Pasteur-Cenci Bolognetti, 00185 Rome, Italy

To avoid harmful response to self-antigens, many mechanisms operate both at the level of primary lymphoid organs (central tolerance) and of secondary and tertiary lymphoid organs (peripheral tolerance). The repertoire of antigen receptors that can be expressed by B and T cells is enormously heterogeneous, and includes potentially harmful self-reactive cells. However, B and T cells specific for self-antigen expressed respectively in bone marrow and thymus do not complete their development successfully, because of central tolerance mechanism [4, 5]. Due to selection during development, mature T cells which populate the periphery can recognize self-peptides at low affinity, or are specific at high affinity for self-antigen not expressed in the primary lymphoid organs. These mature self-reactive cells are potentially harmful and are normally kept under control by different mechanisms of peripheral tolerance [6, 7].

Viral-induced immune Responses

The immune response to antigens carried by viruses is usually protective, and is essential to eradicate the infection. However, during acute viral-induced inflammation, the immune system can develop an anti-self response, in addition to the response to viral antigens [8, 9]. The anti-self response in the course of viral infection can be due to cross-reactivity between viral antigens and self-antigens, as already proposed many years ago [10]. Many examples from experimental models [11, 12] and human diseases [13, 14] support this view [15].

Viruses can also favor the immune response to "cryptic" self epitopes, which are not normally presented by major histocompatibility complex (MHC) molecules [16]. In fact, during viral infection, many new self epitopes can be generated, because of either virus direct effect on cellular protein expression and processing, or virus indirect effect promoting inflammation, with subsequent self-antigen release and processing [17]. Two interesting examples of cryptic antigens which become targets of a self-reactive attack in the presence of a virus are the membrane molecule CD4 and the cytoskeletal protein vinculin, in patients infected by human immunodeficiency virus (HIV) [18, 19].

The response to self-antigens, either cross-reactive to viral antigens or cryptic, or both, is favored because viral infection can induce inflammation and APC activation [9]. The most potent APC are the dendritic cells (DC) which are localized in normal tissues in a quiescent state. Antigen presentation by dormant DC is not an efficient stimulus for T cell activation and it is believed to induce tolerance [20-22]. In the presence of an infectious agent, APC are activated, pick up antigens in the tissue environment, and migrate to lymphoid organs where they stimulate T cells. Viruses can activate DC to prime CD4$^+$ T cells and to directly stimulate CD8$^+$ T cells independently of CD4$^+$ help [23-25]. Moreover, cytokines released because of viral infection may induce bystander T cell activation [26-28].

Th1 and Th2 Cell Activation

The second question concerns the type of response that is induced by the antigen. It has been clearly shown that the pattern of cytokines produced at the beginning of an immune response determines whether T helper 1 (Th1) or Th2 cells will be preferentially activated [29]. In fact, high levels of interleukin (IL)-12 lead to Th1 differentiation, whereas high levels of IL-4 drive the response to Th2 type. The two subsets of T cells, Th1 and Th2, have different functions and help cellular mediated and humoral responses, respectively. Most autoimmune diseases are caused by CD4+ Th1 cells, which produce mainly IL-2, interferon gamma (IFN-γ), and tumor necrosis factor (TNF). When Th1 cells are activated, they express adhesion molecules which enable them to pass the endothelium and gain access into tissues [30]. Self-reactive Th1 cells are restimulated by antigen presented in the tissue and release chemokines and cytokines, recruit other T cells and macrophages, and induce inflammation [31].

Viral Infection and Type of Immune Response

Th1 responses and cytotoxic T lymphocytes are also the main host defense against viruses, especially for stopping viral replication and killing virus-infected cells. Therefore, some viruses evolved different strategies to modulate host immune responses, for example Epstein-Barr virus produces an IL-10-like protein that interferes with the host immune response [32]. However, in most cases the anti-viral response has a strong Th1 polarization. Thus, if an anti-self response is generated in the course of viral infection, a Th1 bias will usually be favored.

Termination of the Immune Response

The third question is about the regulation of the length of the immune response, and whether there is an acute or chronic immune activation. Recently, many investigators have turned their attention to the different mechanisms responsible for termination of the immune response (references in [33]). It has been proposed that effector T and B cells generated during the expansion phase are eliminated at the end of the response while memory cells are spared. The control of cell death and survival in this phase is important both for the homeostasis of the immune system and to limit possible detrimental effects of the response. Interactions between molecules of the Fas/FasL system can promote apoptosis of terminally differentiated effector cells, whereas members of the Bcl protein family have been implicated in the modulation of the cellular response to apoptotic stimuli [34, 35]. In addition, cytotoxic lymphocyte antigen-4 (CTLA-4) ligation on activated T cells can deliver negative signals, possibly via dephosphorylation of activating kinases, and may contribute to the downregulation of the effector arm of the immune response [36]. Recently, it has been proposed that T-T cell presentation by effector T cells in an inflammatory infiltrate may induce downregulation of the immune response, due to lack of CD40-CD40L interaction [37].

Chronic Immune Responses, Viruses and Autoimmune Diseases

The emerging view is that waves of activated effector cells are expanded following antigen stimulation, but they have a limited lifespan. If the antigen is carried by an infectious agent, cytotoxic cells and antibodies generated during the response are important for clearance of the pathogen. The immune response tends to be self-limited and ends a short while after stimulus withdrawal. However, if the pathogen persists, then the immune response will become chronic and may lead to immune pathology. This is the case for many chronic viral infections. Moreover, autoimmunity is probably sustained in most cases by persisting viral infections.

Experimental models mimicking human autoimmune disease have been used to investigate the regulation of the length of the immune response. For example, experimental allergic encephalomyelitis can be induced in susceptible mice by immunization with myelin-derived antigens, leading to activation of self-reactive Th1 cells which migrate to the central nervous system (CNS) and cause inflammation. This process in most cases is self-limited, and mice spontaneously recover due to unknown regulatory mechanisms. Although it has been proposed that a shift from a Th1-type to a Th2-type response induces downregulation of the autoimmune attack to CNS [38, 39], there is no strong supporting experimental evidence. Recent data suggest that this mechanism does not play a major role in turning off the response [40]. An alternative hypothesis is that in the absence of a persisting stimulus the immune response usually ends. Thus, it seems unlikely that events such tissue damage, new self-antigen release, and renewed autoreactive cell stimulation are able to generate a vicious circle responsible for chronic autoimmunity. However, an anti-self response may more easily become a chronic autoimmune disease if sustained by the persistence of an infectious agent, such as a virus.

Acknowledgements: This work was supported by Ministero della Sanità- Istituto Superiore di Sanità (I and II Progetto Sclerosi Multipla).

References

1. Schwartz RH (1990) A cell culture model for T lymphocyte clonal anergy. Science 248: 1349-1356
2. Janeway CJ (1989) Approaching the asymptote? Evolution and revolution in immunology. Cold Spring Harb Symp Quant Biol 1: 1-13
3. Matzinger P (1994) Tolerance, danger, and the extended family. Annu Rev Immunol 12: 991-1045
4. Kappler JW, Roehm N, Marrack P (1987) T cell tolerance by clonal elimination in the thymus Cell 49: 273-280

5. Nemazee DA, Burki K (1989) Clonal deletion of B lymphocytes in a transgenic mouse bearing anti-MHC class I antibody genes Nature 337: 562-566
6. Schonrich G, Kalinke U, Momburg F et al. (1991) Down-regulation of T cell receptors on self-reactive T cells as a novel mechanism for extrathymic tolerance induction. Cell 65: 293-304
7. King C, Sarvetnick N (1997) Organ-specific autoimmunity. Curr Opin Immunol 9: 863-887
8. Aichele P, Bachmann MF, Hengartner H, Zinkernagel RM (1996) Immunopathology or organ-specific autoimmunity as a consequence of virus infection. Immunol Rev 152: 145-156
9. Bhardwaj N (1997) Interactions of viruses with dendritic cells: A double-edged sword. J Exp Med 186: 795-799
10. Fujinami RS, Oldstone MB (1985) Amino acid homology between the encephalitogenic site of myelin basic protein and virus: Mechanism for autoimmunity. Science 230: 1043-1045
11. Ohashi PS, Oehen S, Buerki K et al. (1991) Ablation of "tolerance" and induction of diabetes by virus infection in viral antigen transgenic mice. Cell 65: 305-317
12. Oldstone MB, Nerenberg M, Southern P et al. (1991) Virus infection triggers insulin-dependent diabetes mellitus in a transgenic model: Role of anti-self (virus) immune response. Cell 65: 319-331
13. Atkinson MA, Bowman MA, Campbell L et al. (1994) Cellular immunity to a determinant common to glutamate decarboxylase and coxsackie virus in insulin-dependent diabetes. J Clin Invest 94: 2125-2129
14. Shimoda S, Nakamura M, Ishibashi H et al. (1995) HLA DRB4*0101-restricted immunodominant T cell autoepitope of pyruvate dehydrogenase complex in primary biliary cirrhosis: Evidence of molecular mimicry in human autoimmune disease. J Exp Med 181: 1835-1845
15. Barnaba V, Sinigaglia F (1997) Molecular mimicry and T cell-mediated autoimmune disease. J Exp Med 185: 1529-1531
16. Sercarz EE, Lehmann PV, Ametani A et al. (1993) Dominance and cripticity of T cell antigenic determinants. Annu Rev Immunol 11: 729-766
17. Barnaba V (1996) Viruses, hidden self-epitopes and autoimmunity. Immunol Rev 152: 47-66
18. Salemi S, Caporossi AP, Boffa L et al. (1995) HIVgp120 activates autoreactive CD4-specific T cell responses by unveiling of hidden CD4 peptides during processing. J Exp Med 181: 2253-2257
19. di Marzo Veronese F, Arnott D, Barnaba V et al. (1996) Autoreactive cytotoxic T lymphocytes in human immunodeficiency virus type 1-infected subjects. J Exp Med 183: 2509-2516
20. Sallusto F, Cella M, Danieli C, Lanzavecchia A (1995) Dendritic cells use macropinocytosis and the mannose receptor to concentrate macromolecules in the major histocompatibility complex class II compartment: Downregulation by cytokines and bacterial products. J Exp Med 182: 389-400
21. Cella M, Engering A, Pinet V et al. (1997) Inflammatory stimuli induce accumulation of MHC class II complexes on dendritic cells. Nature 388: 782-787
22. Banchereau J, Steinman RM (1998) Dendritic cells and control of immunity. Nature 392: 245-252
23. Ridge JP, Di Rosa F, Matzinger P (1998) A conditioned dendritic cell can be a temporal bridge between a CD4+ T helper and a T killer cell. Nature 393: 474-478

24. Bennett SRM, Carbone FR, Karamalis F et al. (1998) Help for cytotoxic T cell responses is mediated by CD40 signalling. Nature 393: 478-480
25. Schoenberger SP, Toes REM, van der Voort EIH et al. (1998) T-cell help for cytotoxic T lynphocytes is mediated by CD40-CD40L interactions. Nature 393: 480-483
26. Unutmaz D, Pileri P, Abrignani S (1994) Antigen-independent activation of naive and memory resting T cells by a cytokine combination. J Exp Med 180: 1159-1164
27. Tough DF, Borrow P, Sprent J (1996) Induction of bystander T cell proliferation by viruses and type I interferon in vivo. Science 272: 1947-1950
28. Benoist C, Mathis D (1998) The pathogen connection. Nature 394: 227-228
29. Paul WE, Seder RA (1994) Lymphocyte responses and cytokines. Cell 76: 241-251
30. Austrup F, Vestweber D, Borges E et al. (1997) P- and E-selectin mediate recruitment of T-helper-1 but not T-helper-2 cells into inflamed tissues. Nature 385: 81-83
31. Sallusto F, Lenig D, Mackay CR, Lanzavecchia A (1998) Flexible program of chemokine receptor expression on human polarized T helper 1 and 2 lymphocytes. J Exp Med 187: 875-883
32. Moore KW, Vieira P, Fiorentino DF, Trounstine ML, Khan TA, Mosmann TR (1990) Homology of cytokine synthesis inhibitory factor (IL-10) to the Epstein-Barr virus gene BCRFI. Science 248: 1230-1234
33. Van Parijs L, Abbas AK (1998) Homeostasis and self-tolerance in the immune system: Turning lymphocytes off. Science 280: 243-248
34. Nagata S, Golstein P (1995) The Fas death factor. Science 267: 1449-1456
35. Sprent J (1997) Immunological memory. Curr Opin Immunol 9: 371-379
36. Saito T (1998) Negative regulation of T cell activation. Curr Opin Immunol 10: 313-321
37. De Vita L, Accapezzato D, Mangino G et al. (1998) Defective Th1 and Th2 cytokine synthesis in the T-T cell presentation model for lack of CD40-CD40 ligand interaction. Eur J Immunol 28: 3552-3563
38. Karpus WJ, Gould KE, Swanborg RH (1992) CD4+ suppressor cells of autoimmune encephalomyelitis respond to T cell receptor-associated determinants on effector cells by interleukin-4 secretion. Eur J Immunol 22: 1757-1763
39. Kennedy MK, Torrance DS, Picha KS, Mohler KM (1992) Analysis of cytokine mRNA expression in the central nervous system of mice with experimental autoimmune encephalomyelitis reveals that IL-10 mRNA expression correlates with recovery. J Immunol 149: 2496-505
40. Di Rosa F, Francesconi A, Di Virgilio A et al. (1998) Lack of Th2 cytokine increase during spontaneous remission of experimental allergic encephalomyelitis. Eur J Immunol 28: 3893-3903
41. Di Rosa F, Barnaba V (1998) Persisting viruses and chronic inflammation: understanding their relation to autoimmunity. Immunol Rev 164: 17-27

Chapter 2

Antigen Recognition and Autoimmunity

F. SINIGAGLIA[1], J. HAMMER[2]

Introduction

The genes for the human leukocyte antigen (HLA) class II molecules lie within the major histocompatibility complex (MHC) on chromosome 6. The HLA class II loci are clustered into three regions, known as DR, DQ and DP. Each of these regions contains at least one α gene and one β gene. The gene products form an αβ heterodimer which is expressed as a membrane-bound protein on the cell surface. HLA class II proteins play a central role in T-cell selection and activation. They bind peptide fragments derived from protein antigens and display them on the cell surface for interaction with the antigen-specific receptors of T lymphocytes.

HLA class II loci are extremely polymorphic. Allelic variation between HLA class II molecules of different individuals accounts for the functional differences revealed by HLA typing specificities, allograft reactivity, and, most importantly, differential ability to bind and display antigenic peptides. Allelic variations of HLA class II genes also seem to play a major role in autoimmunity. HLA typing of large groups of patients with various autoimmune diseases revealed that some HLA alleles occurred with a higher frequency in these patients than in the general population. Among the diseases strongly associated with HLA class II are for example, rheumatoid arthritis, insulin-dependent diabetes mellitus and multiple sclerosis.

In the last few years several important breakthroughs and technological advances have made it possible to clarify the role of polymorphism and the molecular events in peptide interaction with MHC class II proteins. Based on this knowledge, the structural basis for MHC-linked susceptibility to autoimmune diseases can now be reassessed with sufficient detail in order to solve long-standing questions in this field.

---

[1] Roche Milano Ricerche, Via Olgettina 58 - 20132 Milan, Italy. e-mail: francesco.sinigaglia@roche.com
[2] Hoffmann-La Roche Inc., Preclinical R&D, 340 Kingsland Street Nutley, NJ 07110 US

Peptide-MHC Molecule Interactions

The determination of the X-ray structure of different human class II molecules [1-3], the selection of large class II-bound peptide repertoires using M13 peptide display [4, 5] and the characterization of class II-eluted, naturally-processed peptides [6-9] have elucidated the structural requirements for peptides binding to class II molecules.

The peptide binding groove of class II molecules is open at both ends [1, 2] thus allowing class II-bound peptides to extend beyond both termini. As a result, these peptides are longer than those bound by class I molecules and exhibit considerable length variation, typically 12-24 residues. Class II molecules form many conserved hydrogen bonds with the peptide main chain [1], forcing class II-bound peptides into a similar conformation despite differences in their primary sequence. The sequence-independent network of hydrogen bonds between conserved MHC residues and the peptide main chain gives rise to a broad but not unlimited peptide binding capacity. Indeed, most natural peptide sequences lack the characteristics necessary to bind to MHC molecules. This is because peptide main chain interactions are not the only mode of MHC binding. Some of the peptide side chains contact residues within the MHC cleft and increase the overall binding affinity and specificity of the associated peptides (anchor residues) [2, 5] while others interfere with residues of the MHC cleft and reduce binding (inhibitory residues) [10, 11]. These sequence-dependent interactions are due to the irregular surface of the MHC cleft. MHC side chains protrude into the cleft and form pockets or ridges, resulting in strong preferences for interaction with particular amino acid side chains.

Notably, most pockets in the MHC groove are shaped by clusters of polymorphic MHC residues and thus have variable chemical and size characteristics in different MHC alleles. For example, a negatively charged side chain in one MHC molecule may preferentially interact with positively charged peptide residues, while a positively charged side chain in another MHC molecule may only bind to negatively charged peptide residues; similarly, a small side chain in one MHC molecule may greatly extend a pocket and allow for the interaction with many different amino acid residues, while a large amino acid side chain in another MHC molecule may completely block the access to this pocket. MHC alleles can therefore be characterized by differences in the precise nature and position of polymorphic pockets, thereby playing a major role in allele-specific peptide binding.

MHC Peptide-Binding Motifs

The interaction of peptide side chains with pockets of the MHC cleft imposes sequence requirements on the binding peptides. These requirements are summarized in peptide-binding motifs. A breakthrough for the analysis of MHC binding motifs was the characterization of large MHC-selected peptide pools which allowed the generalization of rules for peptide binding to MHC molecules. Class

I motifs were analyzed by the pooled peptide sequencing technique, which consists of the elution of endogenous class I peptide pools and their subsequent analysis, involving Edman sequencing (reviewed by Rammensee [12]). The length restriction on peptides bound by class I molecules leads to their natural alignment in the binding groove. Thus, position-dependent preferences for particular amino acid side chains (class I anchors) can be identified by sequencing mixtures of the eluted peptides. Class II motifs were identified by the analysis of large peptide pools selected from M13 bacteriophage peptide display libraries comprising several million random nonamer peptides. Sequence analysis of the DNA encoding the displayed peptides led to the identification of class II anchors (reviewed by Sinigaglia and Hammer [13]).

Pool sequencing and the screening of phage libraries have been used for the identification of several allele-specific class I and class II binding motifs, respectively. These motifs generally consist of 2-4 anchor positions that are at fixed distances from one another.

Prediction of MHC Binding Peptides

The discovery of motifs through pool sequencing and phage library screening would imply simple rules for peptide-MHC interactions. Several lines of evidence indicate, however, that these motifs are only the tip of the iceberg and that the rules for MHC-peptide binding are much more complex [14] since: (i) inhibitory

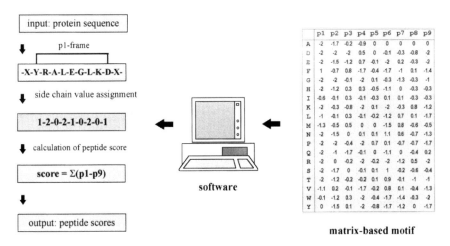

Fig. 1. Principle of matrix-based epitope prediction. The position-specific and amino acid-specific matrix values, e.g. binding data or amino acid frequencies, are assigned to each residue of a given peptide frame. Subsequently, values are mathematically processed, resulting in a peptide "score". The correlation between peptide score and binding affinity is the prerequisite for an effective HLA class II ligand prediction. (From [15] with permission)

residues are as important for MHC-peptide binding as are anchor residues, and binding studies have revealed both position-specific and allele-specific properties of such inhibitory residues; (ii) apart from the main anchor positions identified by pool sequencing and phage library screening, additional "secondary" anchors were found to be important for MHC-peptide binding, thus rendering difficult the prediction of MHC binding sequences from "simple" MHC motifs. As a consequence, new quantitative approaches were developed that allowed for a more complete characterization of MHC-peptide binding specificity. The effects of all natural amino acids at each peptide position were quantified using multiple-peptide synthesis technology, together with new high-flux in vitro binding assays [11]. The resulting quantitative motifs are "fingerprints" of the MHC clefts since they represent the specificity of the various pockets forming the polymorphic HLA class II cleft. A small database of pocket-specificity profiles was sufficient to generate *in silico* a large number of HLA class II matrices, covering the majority of human HLA class II peptide binding specificity. These virtual matrices were incorporated in a software named TEPITOPE, capable of predicting HLA class II ligands (Fig. 1) [15].

HLA and Disease Associations

Autoimmune diseases appear to occur when a specific immune response is mounted against self. Although it is not known what triggers the autoimmune response, both environmental and genetic factors are important. Genes in the HLA-complex appear to account for the strongest genetic predisposition. For example, a genome-wide scan for diabetes susceptibility genes has recently demonstrated that several genes contribute to the disease process, but that genes of the HLA complex are the most important [16]. The HLA complex consists of more than 100 different HLA [17]. Sequencing of these HLA genes indicated that disease-associated class II molecules often share unique amino acid residues in the peptide binding cleft (reviewed in [18-20]). These residues could affect predisposition to autoimmune disease by several mechanisms that are not mutually exclusive, including for example shaping of the T-cell receptor repertoire and altering the specificity of the peptide binding site [21-23].

The effect of disease-associated class II polymorphic residues in determining peptide binding specificity has been addressed by detailed HLA class II peptide binding studies. Although these studies alone cannot definitively establish the role of MHC class II alleles in determining resistance or susceptibility to autoimmune disease, they nonetheless provide important information toward a clearer understanding of the autoimmune disease process. Identification of a correlation between binding specificity and disease association, for example, strongly supports the hypothesis that selective binding of autoantigenic peptides is the major mechanism underlying HLA-disease association. Furthermore, the knowledge of HLA class II peptide binding rules could lead to the identification of autoantigenic peptides selectively presented by the disease-associated HLA class II molecules.

Rheumatoid Arthritis

Susceptibility to rheumatoid arthritis (RA) is specifically associated with the HLA-DR locus [24]. More than 80% of Caucasian RA patients express DRB1*0401, DRB1*0404, DRB1*0408 or DRB1*0101 [25]. In Japanese patients, the frequency of another DR4 subtype, encoded by DRB1*0405, increases [26]. Interestingly, the RA-associated DR molecules all carry a short stretch of amino acids at DRβ 67-74 ("shared epitope") that is highly conserved, even though DRβ 67-74 covers an otherwise very polymorphic region of the DR cleft [27-29]. The fact that DRB1*0402, a closely related molecule not associated with RA, differs from some RA-linked molecules only in this shared epitope region, suggests that this part of the molecule plays a critical role in disease association.

In an attempt to analyze the role of DRβ 67-74 on HLA-DR peptide binding specificity, we compared the binding characteristics of RA-associated DR4 subtypes with non-associated DR molecules [30]. Striking differences, especially in the specificity of pocket 4, were identified between DR4 subtypes which are associated with RA and those which are not. For example, peptides with negatively charged residues at position 4 bound to RA-associated DRB1*0401 or DRB1*0404 molecules, but not to the non-associated DRB1*0402 molecule; the reverse was true for peptides with positively charged residues at position 4. Site-directed mutagenesis demonstrated that positively (DRB1*0401) or negatively (DRB1*0402) charged residues at DRβ71 were responsible for most of these effects [30]. Altogether, these results demonstrated a striking correlation between binding specificity and disease association, thus supporting the hypothesis that selective binding of autoantigenic peptides is the mechanism underlying HLA association in RA.

Although the target self antigen that initiates the autoimmune process is unknown, a number of candidate autoantigens have been implicated in the pathogenesis of RA. Some were derived from normal self proteins such as type II collagen or other joint-associated proteins [31, 32], and others were from microorganisms [33]. Recently, a panel of DRB1*0401-restricted mouse T cell hybridomas specific for bovine type II collagen were generated from DRB1*0401 transgenic mice [34]. The vast majority recognized a single determinant corresponding to the conserved residues 390-402 in human. This determinant was indeed predicted by TEPITOPE to bind selectively to RA-associated HLA-DR molecules [15], thus supporting the value of this approach for the identification of potential autoantigens.

Multiple Sclerosis

Multiple sclerosis (MS) is a chronic inflammatory disease of the human central nervous system (CNS) characterized by demyelination and focal infiltrates of macrophages, plasma cells and T cells in the CNS. Susceptibility to MS is specifically associated with the HLA-DR locus [24]. Approximately 50%-70% of MS

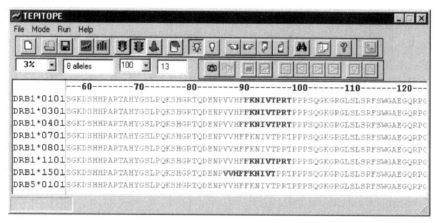

Fig. 2. Prediction of a promiscuous peptide in MBP with TEPITOPE. MBP 87-99 represents an immunodominant region of MBP [41, 44]. DRB1*1501, in contrast to most other DR molecules, is predicted to bind MBP region 87-99 in a slightly different binding frame. (From [15] with permission)

patients carry the DRB1*1501 allele versus 20%-30% of normal individuals. Besides DRB1*1501, other HLA class II alleles are over-represented in certain ethnic groups, such as DRB1*0401 in southern Italians and in Arabs [35, 36].

Myelin basic protein (MBP) and proteolipid protein (PLP), putative autoantigens involved in the pathogenesis of MS, can induce experimental autoimmune encephalitis (EAE) in mice and rats [37-39]. In humans, MBP or PLP-reactive HLA-DR-restricted T cells have been isolated from both MS patients and healthy controls [40-42]. MBP-specific T cells in MS patients were activated while those from controls were in a resting state [43]. Activated T cells reactive to MBP or PLP might therefore be involved in the pathogenesis of MS.

Computer analysis of MBP with TEPITOPE identified a promiscuous sequence (MBP 87-99) likely to be bound by DRB1*1501; this MBP region represents a slightly different binding frame than that recognized by most other DR molecules (Fig. 2) [15]. A number of investigators demonstrated that MBP 87-99 indeed represents the immunodominant region of MBP in the context of the MS-associated DR allele DRB1*1501 [44, 45], and that DRB1*1501 binds a peptide frame that is different from the one recognized by most other DR alleles [46].

Epitope scanning with overlapping 20-residue-long peptides of PLP led to the identification of an immunodominant epitope (PLP 175-192) for HLA-DR4 individuals [47]. Once again, this region could be predicted by TEPITOPE. HLA-DR4-IE chimeric class II transgenic mice were immunized with MBP 87-106 and PLP 175-192 to investigate the development of the HLA-DR associated disease [48]. PLP 175-192 provoked a strong proliferative response of lymph node T cells, and caused inflammatory lesions in the white matter of the CNS as well as symptoms of experimental allergic encephalomyelitis. Immunization with MBP 87-106 elicited a weak proliferative T cell response and caused mild EAE. The amino acid sequences of both PLP 175-192 and MBP 87-106 are identical in humans and in mice, thus representing potential autoantigens in both species. Notably, mice lacking the DRB1*0401 transgene did not develop EAE after immunization with either PLP 175-192 or MBP 87-106, indicating that a human MHC class II binding site alone can confer susceptibility to an experimentally induced murine autoimmune disease.

Conclusions

A striking characteristic of autoimmune diseases is the increased frequency of certain HLA class II alleles in affected individuals. Although there is much to learn about the pathogenesis of autoimmunity, it is generally believed that disease-associated HLA class II molecules have the capacity to bind and present autoantigenic peptides to T cells. A nearly global coverage of HLA-DR peptide binding specificity, combined with newly emerging bioinformatic tools, have led to effective ways of studying the role of disease-associated class II residues and to the identification of peptides selective for disease-associated HLA molecules. The detailed knowledge of HLA class II peptide interaction may therefore represent a step forward in understanding autoimmune disease processes.

References

1. Brown JH, Jardetzky TS, Gorga JC et al. (1993) 3-Dimensional structure of the human class-II histocompatibility antigen HLA-DR1. Nature 364: 33-39
2. Stern LJ, Brown JH, Jardetzky TS et al. (1994) Crystal structure of the human class II MHC protein HLA-DR1 complexed with an influenza virus peptide. Nature 368: 215-221
3. Ghosh P, Amaya M, Mellins E, Wiley DC (1995) The structure of an intermediate in class II MHC maturation: CLIP bound to HLA-DR3. Nature 378: 457-462
4. Hammer J, Takacs B, Sinigaglia F (1992) Identification of a motif for HLA-DR1 binding peptides using M13 display libraries. J Exp Med 176: 1007-1013
5. Hammer J, Valsasnini P, Tolba K et al. (1993) Promiscuous and allele-specific anchors in HLA-DR binding peptides. Cell 74: 197-203
6. Rudensky AY, Preston-Hurlburt P, Hong SC et al. (1991) Sequence analysis of peptides bound to MHC class II molecules. Nature 353: 622-627
7. Hunt DF, Michel H, Dickinson TA et al. (1992) Peptides presented to the immune system by the murine class II major histocompatibility complex molecule I-A^d. Science 256: 1817-1820
8. Chicz RM, Urban RG, Gorga JC et al. (1993) Specificity and promiscuity among naturally processed peptides bound to HLA-DR Alleles. J Exp Med 178: 27-47
9. Rammensee H-G, Friede T, Stevanovic S (1995) MHC ligands and peptide motifs: First listing. Immunogenetics 41: 178-228
10. Boehncke W-H, Takeshita T, Pendleton CD et al. (1993) The importance of dominant negative effects of amino acid side chain substitution in peptide-MHC molecule interactions and T cell recognition. J Immunol 150: 331-341

11. Hammer J, Bono E, Gallazzi F et al. (1994) Precise prediction of major histocompatibility complex class II peptide interaction based on peptide side chain scanning. J Exp Med 180: 2353-2358
12. Rammensee H-G (1993) Peptides naturally presented by MHC class I molecules. Annu Rev Immunol 11: 213-244
13. Sinigaglia F, Hammer J (1994) Defining rules for the peptide-MHC Class II interaction. Curr Opin Immunol 6: 52-56
14. Hammer J (1995) New methods to predict MHC-binding sequences within protein antigens. Curr Opin Immunol 7: 263-269
15. Hammer J, Sturniolo T, Sinigaglia F (1997) HLA class II peptide binding specificity and autoimmunity. Adv Immunol 66: 67-100
16. Davies JL, Kawaguchi Y, Bennett ST et al. (1994) A genome-wide search for human type 1 diabetes susceptibility genes. Nature 371: 130-136
17. Campbell RD, Trowsdale J (1994) Map of the human MHC. Immunol Today 14: 349-352
18. Nepom GT, Erlich HA (1991) MHC class II molecules and autoimmunity. Annu Rev Immunol 9: 493-520
19. Todd JA, Acha-Orbea H, Bell JI et al. (1988) A molecular basis for MHC class II-associated autoimmunity. Science 240: 1003-1009
20. Thorsby E (1995) HLA-associated disease susceptibility - Which genes are primarily involved? Immunologist 3: 51-58
21. Kappler J, Roehm N, Marrack P (1987) T cell tolerance by clonal elimination in the thymus. Cell 49: 273-280
22. Roy S, Scherer MT, Briner TJ et al. (1989) Murine MHC polymorphism and T cell specificities. Science 244: 572-575
23. Teh HS, Kisielow P, Scott B et al. (1988) Thymic major histocompatibility complex antigens and the ab T-cell receptor determine the CD4/CD8 phenotype of T cells. Nature 335: 229-233
24. Tiwari JL, Terasaki PI (1985) HLA and disease associations. Springer, Berlin Heidelberg New York
25. Winchester R (1994) The molecular basis of susceptibility to rheumatoid arthritis. Adv Immunol 56: 289-466
26. Otha N, Nishimura YK, Tanimoto T et al. (1982) Association between HLA and Japanese patients with rheumatoid arthritis. Hum Immunol 5: 123-132
27. Gregersen PK, Shen M, Song QL et al. (1986) Molecular diversity of HLA-DR4 haplotaypes. Proc Natl Acad Sci USA 83: 2642-2646
28. Nepom GT, Byers P, Seyfried C et al. (1989) HLA genes associated with rheumatoid arthritis. Identification of susceptibility alleles using specific oligonucleotides probes. Arthritis Rheum 32: 15-21
29. Wordsworth BP, Lanchbury JS, Sakkas LI et al. (1989) HLA-DR4 subtype frequencies in rheumatoid arthritis indicate that DRB1 is the major susceptibility locus within the HLA class II region. Proc Natl Acad Sci USA 86: 10049-10053
30. Hammer J, Gallazzi F, Bono E et al. (1995) Peptide binding specificity of HLA-DR4 molecules: Correlation with rheumatoid arthritis association. J Exp Med 181: 1847-1855
31. Holmdal R, Andersson M, Goldschmidt K et al. (1990) Type II collagen autoimmunity in animals and provocations leading to arthritis. Immunol Rev 118: 193-232
32. Mikecz K, Glant TT, Poole AR (1987) Immunity to cartilage proteoglycans in BALB/c mice with progressive polyarthritis and ankylosing spondylitis induced by injection of human cartilage proteoglycan. Arthritis Rheum 30: 306-318

33. Gaston JSH, Life PF, Jenner PJ, Colston MJ, Bacon PA (1990) Recognition of a Mycobacteria-specific epitope in the 65-kD heat-shock protein by synovial fluid-derived T cell clones. J Exp Med 171: 831-841
34. Fugger L, Rothbard J, Sonderstrup-McDevitt G (1996) Specificity of an HLA-DRB1*0401 restricted T cell response to type II collagen. Eur J Immunol 26: 928-933
35. Marrosu MG, Muntoni F, Murru MR et al. (1988) Sardinian multiple sclerosis is associated with HLA-DR4: A serologic and molecular analysis. Neurology 348: 1749-1753
36. Yacub BA, Daif AK (1988) Multiple sclerosis in Saudi Arabia. Neurology 328: 621-625
37. Pettinelli CB, Fritz RB, Chou CHJ, McFarlin DEJ (1982) Encephalitogenic activity of guinea pig myelin basic protein in the SJL mouse. Immunology 129: 1209-1211
38. Fritz RB, Chou CHJ, McFarlin DE (1983) Relapsing murine experimental allergic encephalomyelitis induced by myelin basic protein. J Immunol 130: 1024-1026
39. Enddoh M, Tabira T, Kunishita T et al. (1986) A proteolipid apoprotein is an encephalitogen of acute and relapsing autoimmune encephalomyelitis in mice. J Immunol 137: 3832-3835
40. Martin R, Jaraquemada D, Flerlage M et al. (1990) Fine specificity and HLA restriction of myelin basic protein-specific cytotoxic T cell lines from multiple sclerosis patients and healthy individuals. J Immunol 145: 540-548
41. Ota K, Matsui M, Milford E et al. (1990) T-cell recognition of an immuno-dominant myelin basic protein epitope in multiple sclerosis. Nature 346: 183-187
42. Pette M, Fujita K, Wilkinson D et al. (1990) Myelin autoreactivity in multiple sclerosis: Recognition of myelin basic protein in the context of HLA-DR2 products by T lymphocytes of multiple-sclerosis patients and healthy donors. Proc Natl Acad Sci USA 87: 7968-7972
43. Zhang J, Markovic-Plese S, Lacet B et al. (1994) Increased frequency of interleukin 2-responsive T cell specific for myelin basic protein and proteolipid protein in peripheral blood and cerebrospinal fluid of patients with multiple sclerosis. J Exp Med 179: 973-984
44. Wucherpfennig KW, Sette A, Southwood S et al. (1994) Structural requirements for binding of an immunodominant myelin basic protein peptide to DR2 isotypes and for its recognition by human T cell clones. J Exp Med 179: 279-290
45. Valli A, Sette A, Kappos L et al. (1993) Binding of myelin basic protein peptides to human histocompatibility leukocyte antigen class II molecules and their recognition by T cells from multiple sclerosis patients. J Clin Invest 91: 616-628
46. Vogt AB, Kropshofer H, Kalbacher H et al. (1994) Ligand motifs of HLA-DRB5*0101 and DRB1*1501 molecules delineated from self-peptides. J Immunol 153: 1665-1673
47. Markovic-Plese S, Fukaura H, Zhang J et al. (1995) T cell recognition of immunodominant and cryptic proteolipid protein epitopes in humans. J Immunol 155: 982-992
48. Ito K, Bian H-J, Molina M et al. (1996) HLA-DR4-IE chimeric class II transgenic, murine class II-deficient mice susceptible to experimental allergic encephalomyelitis. J Exp Med 183: 2635-2644

Chapter 3

Assembly and Function of Immunoglobulins During B Cell Development

R. SITIA

A Cell Biologist's View of the Problem of Ig Production

In all multicellular organisms, individual cells must continuously exchange information. In general, they do so by secreting ligand molecules into the extracellular space, and expressing suitable receptors on their surfaces. The fidelity of intercellular communication thus depends on the specificity of the interactions between ligands and receptors, which is in turn determined by the three-dimensional structure of the molecules involved. As many of these molecules are complex oligomeric structures, correct execution of the folding and assembly pathways becomes of paramount importance for the social life of all cells. It is therefore not surprising that cells have evolved sophisticated quality control mechanisms to ensure that the molecules they release into the external world have attained their proper structure. Over the last decade, a vast series of investigations have been conducted to dissect the molecular mechanisms underlying the quality control of newly synthesized proteins destined for export. In the same manner that immunoglobulin (Ig) genes have provided a rich source of information for molecular biologists, the products of these genes, the antibodies, have become a favorite model for studying the basic principles of protein synthesis, assembly and secretion. This was largely due to the availability of powerful model systems, such as cell hybridomas or transfectomas producing various forms and types of immunoglobulins in large amounts.

The Endoplasmic Reticulum as the Folding Compartment for Proteins Destined to the Extracellular Space

The three-dimensional structure of proteins is dictated by their amino acid sequence [1]. Nonetheless, it has become clear that folding, assembly and oligomerization are catalyzed in vivo by a family of conserved enzymes and

---

Laboratory of Molecular Immunology, DIBIT-San Raffaele Scientific Institute, Via Olgettina 58 - 20132 Milan, Italy. e-mail: sitia.roberto@hsr.it

"chaperone" molecules [2]. Proteins destined to the extracellular space begin their journey through the cell in the endoplasmic reticulum (ER), where they are translocated cotranslationally. A complex system based on the recognition of a signal sequence (or leader peptide [3-5]) on nascent proteins ensures that ribosomes that are synthesizing secretory or membrane proteins arrest translation until they dock to the cytosolic face of the ER membrane. Correct docking opens a membrane translocation complex (the so-called translocon) and reactivates polypeptide elongation [6, 7]. The nascent protein is somehow forced through the translocon, emerging in the unfolded state within the ER lumen. Here, it finds a series of devoted proteins that assist its folding and assembly as well as several post-translational modifications such as glycosylation and disulfide bond formation (see [8, 9] for reviews). While proteins undergo structural maturation within the ER, they are constantly being inspected by strict quality control mechanisms [10] that somehow permit to exit the ER only those polypeptides or polypeptide complexes that have completed their folding and assembly. Intermediates are retained within the ER. It is possible that ER retention serves not only to prevent the transport of immature molecules (an event that might be detrimental to intercellular communication and other extracellular functions) but also to facilitate their maturation. For instance, the rate of oligomerization might be increased if retention increases the local concentration of individual subunits. Whatever its teleology, there remains little doubt that selective retention is one of the key elements in the control of intracellular protein transport.

If the subunits of a heterodimeric protein are synthesized in equal amounts, it may be expected that – given enough time – the proper dimers will be formed and transported to the Golgi. However, unbalanced synthesis is very frequent and results in the accumulation of the component made in excess. This might create serious problems in living cells, should quality control mechanisms not be tightly coupled to a proteolytic system. Studies on Igs and the T cell receptor have been instrumental to reveal that the ER is equipped not only with folding and retention machineries, but also with a selective protein degradation system, independent from the endolysosomal one, which maintains homeostasis and serves some physiological functions [11-13]. It has been recently shown that cytosolic proteasomes are responsible for the degradation of many membrane and secretory ER proteins [14-16]. This implies that proteins marked for degradation are retrotranslocated across the ER membrane to be dispatched to the cytosol where proteasomes reside. Unexpectedly, the same components that mediate the entry of proteins into the ER (namely Sec61p) seem to be responsible for the retrograde translocation as well [17, 18], raising the question of what determines vectoriality of transport across the ER membrane. Ubiquitination [19] as well as proteasomes themselves [20] may act as pulling forces in mediating the extraction of proteins from the ER.

Quality Control of Ig Production

All classes of antibodies share the basic H2L2 structure, and are produced in two forms during B cell differentiation: as integral membrane proteins on the surface of B lymphocytes where they act as antigen receptors, or as soluble effector molecules secreted by plasma cells. This dual topology of antibodies represented a crucial problem for immunologists and cell biologists, until it was demonstrated that membrane and secreted Igs contain different heavy (H) chains, produced by alternate RNA processing. Heavy chains present in antigen receptors (Hm) have a stretch of hydrophobic amino acids in their C-terminal end which act as a stop transfer sequence allowing insertion into the lipid bilayer. In contrast, secreted H chains (Hs) are endowed with carboxy-terminal hydrophilic peptides that allow the complete translocation into the ER lumen, where folding and assembly take place (see [21] and references therein).

Ig Assembly in B and Plasma Cells

The folding, assembly and polymerization pathways of newly synthesized Igs (see [21-23] for extensive reviews) can be summarized as follows:
- H and L chains are synthesized on distinct polysomes.
- Both H and L chains possess N-terminal leader sequences and are cotranslationally translocated across the ER membrane.
- Folding begins cotranslationally and seems to occur independently for H and L, proceeding vectorially from the amino terminus to the carboxy terminus.
- With few important exceptions [24], folding precedes assembly.
- HL assembly and HH dimerization also initiate before translation is completed, and the sequence of these events depends on the isotype and on the size of the L chain pool.
- Plasma cells generally produce L chains in excess (the Bence Jones protein in myelomas).
- Free L chains can be secreted, while free H are retained and degraded intracellularly.
- Polymerization of IgM and IgA is slow and restricted to plasmacytoid cells.

The Ig domain, a homology unit of about 110 amino acids composed of antiparallel ß-sheets connected by loops, is the building block of antibodies and of proteins of the Ig superfamily. The Ig domain has a strong tendency to dimerize. This property is central in determining the quaternary structure of antibodies, and implies that in general subunit folding precedes assembly.

Domain folding can be readily monitored by the formation of the conserved disulfide bridge that links the two cysteines: this intradomain bond is formed as soon as translation of the domain is completed. Vectorial folding is thought to be important for preventing aberrant intrachain disulfide bond formation [25]. However, it has been shown that proper folding of L chains can occur also posttranslationally [26]. Another feature of the rapid, cotranslational domain folding could be preventing the retrograde movement of the growing chain from the ER lumen into the cytosol [27]. Other post-translational modifications, such as glycosylation and interactions with other proteins within the ER lumen, might be crucial for tethering the newly made protein in this compartment [28].

Retention of Free H Chains

In the absence of L chains, H chains are not secreted. Myeloma variants which lose H chains but still produce and secrete L chains are easily isolated. These variants correspond to the frequent clinical finding of myelomas producing L chains in vast excess (the Bence Jones proteins). In contrast, the isolation of cells that lose L chain synthesis is a rare event, and often yields unstable clones which retain the H chain intracellularly [29]. Myeloma patients with monoclonal peaks consisting of free H chains are extremely rare: interestingly, the monoclonal H chains of these "heavy chain disease" patients are characterized by extensive deletions, which always encompass the CH1 domain [30]. The reason underlying this situation became clear when John Kearney and coworkers demonstrated that – in the absence of L chains – the CH1 domain binds rather tightly to BiP, a chaperone molecule related to hsp70 present in large quantities in the ER of all cells [31]. L chains can displace BiP from H, to form stable HL complexes [32]. Therefore, BiP seems to play a crucial role in Ig assembly, preventing the aggregation of free H chains and facilitating their interactions with L [33].

Secretion of Free L Chains

The presence of Bence-Jones proteins in the blood and urine of myeloma patients implies that free L chains can escape quality control to be secreted even in the absence of H. This is largely due to the fact that – unlike the CH1 domain – the two L domains can form homotypic dimers, attaining a conformation that negotiates intracellular transport. Indeed, Bence Jones proteins are often L2 homodimers, either covalent or non-covalent [21]. As homodimers are less stable than HL pairs, allowing the secretion of free L chains seems to be an elegant solution to optimize Ig assembly reducing the risks of intracellular accumulation. Plasma cells synthesize L chains in excess: retention is sufficient to generate an ER pool sufficient to complement all H chains, but weak enough to prevent engulfment problems.

Normal L chains also transiently interact with BiP. However, the half-life of the L-BiP complex is only a few minutes. In contrast, non-secreted L mutants interact stably with BiP, until they dissociate to be degraded by the ER-associated degradation pathway [34].

Molecular Mechanisms of ER Retention and Quality Control

As to the mechanisms of retention, a unifying concept is that structurally immature proteins expose reactive surfaces or motifs which prevent their mobility through the secretory pathway. The vast array of chaperones in the ER is thought to provide an interactive matrix that coordinates attempts to fold with retention (and eventually retrotranslocation to the cytosol for proteasomal degradation). Folding and assembly mask reactive surfaces allowing exit from the ER. Three molecular mechanisms of ER retention and quality control can be described:

(a) *Hydrophobic surfaces.* Hydrophobic patches reduce the diffusibility of unassembled or unfolded proteins by favoring interactions with BiP and/or other proteins of the ER matrix. A paradigm of this model is the retention of Ig H chains in the absence of L chains.

(b) *Thiol-mediated retention.* Work on IgM and IgA, two polymeric immunoglobulins that are secreted by plasma cells but retained and degraded by B lymphocytes, has allowed my colleagues and I to disclose another way – thiol-mediated retention – by which the quality control machineries discriminate between assembled molecules and unassembled precursors [35, 36]. In oligomeric proteins stabilized by interchain disulfide bonds, unassembled subunits are likely to expose reactive thiols. We have provided evidence suggesting that a single thiol group can confer ER retention to a polypetide [37]. This model, originally described for polymeric immunoglobulins, has been recently confirmed for other secretory [38], transmembrane [39] and glycophosphatidylinositol (GPI)-anchored surface proteins [40]. Interestingly, the strength of an unpaired cysteine as an ER retention element is modulated by its surrounding amino acid context: the vicinity of acidic residues weakens retention and allows the secretion of some unpolymerized molecules [36, 41]. In monomeric Ig L chains secreted in the absence of H, Cys213, normally utilized to form disulfide bonds with the CH1 domain, remains unpaired. In this case, transport to the Golgi seems to be negotiated by oxidation of Cys213, through the formation of a disulfide bond with a free cysteine residue [41].

Not only is thiol-mediated retention regulated at the level of substrate; cellular factors are also involved. Thus, B cells do not secrete IgM because they are incapable of forming covalent polymers, and as a result thiol-mediated retention via the μ-chain Cys575 predominates. In contrast, IgM are polymerized and secreted by plasma cells [35]. However, B cells do not seem to lack any structural components for polymerization [42], as secretion of covalent hexamers (the molecular form of IgM secreted by cells that – like B lymphocytes – lack J chains [43, 44]) can be induced by transient alterations of the intracellular redox state. It is possible that redox regulation of protein transport takes place for a wide class of molecules, particularly those containing a cysteine-rich motif [45, 46].

(c) *The calnexin cycle.* Many secretory and membrane proteins contain N-linked sugars. In addition to their role in conferring stability and solubility to the polypeptide backbones, these glycans serve a major function in quality control as well. Thus, it has been shown that calnexin and calreticulin, two abundant ER-resident proteins, bind monoglucosylated proteins: these are transient intermediates in the normal glycan trimming reactions in the ER [47]. Interestingly, there is an enzymatic activity that selectively adds glucose moieties to unfolded, but not to properly folded glycoproteins [48]. These cycles of glucose removal and readditions can serve as a timer to maintain folding and assembly intermediates in the optimal environment for their maturation, preventing premature degradation and secretion.

Retention and Allelic Exclusion: Signalling from the ER?

Since the rearrangement of the Ig H locus precedes that at the Ig L locus, pro-B and pre-B cells synthesize H chains in the absence of L. In agreement with the laws of quality control, pro-B and pre-B cells largely retain H chains intracellularly, even if assembled with the so-called surrogate L chains (SLC). SLC are monomorphic molecules composed of the products of two genes: VpreB and $\lambda 5$ [49]. The former resembles a VL domain and associates with the latter (structurally homologous to a CL domain) to produce a structure that mimicks L chains. The SLC can pair with μ chains (mainly of the membrane form, μm) to form a pre-B cell receptor (PBCR). Ample genetic evidence indicates that the PBCR is crucial for allelic exclusion (blocking further rearrangements at the Ig H locus and concomitantly activating the rearrangement of the K and λ loci). Similarly to that described for the mature B cell receptors, signals generated by the PBCR depend on Ig-α and Ig-β, two transmembrane molecules endowed with tyrosine motifs [50, 51].

The SLC can be viewed as a means to select productive rearrangements. Only if capable to interact with SLC would newly rearranged μm chains dispatch the proper signals, allowing progression of the rearrangement programs. However, only minute amounts of the PBCR are expressed on the cell surface, as most of them are retained and degraded in the ER [52]. In addition, despite the efforts of many laboratories, ligands for PBCRs have not yet been identified. Such a ligand should be able to recognize all PBCRs, independently from the nature of the VH domain generated by VDJ recombination. The question arises as to whether signals are generated by PBCRs at the cell surface or, alternatively, by ER-retained receptors. As a similar series of ordered rearrangements occurs in T lymphocytes, intracellular activation of pre-T cell receptors could be important for T cell development as well. However, Mike Owen and coworkers have recently obtained evidence indicating that pre-T cell receptors must exit the ER to generate the signals necessary for developmental progression [53].

There is ample evidence indicating that signals can indeed originate from the ER. For instance, a number of mutated tyrosine kinase membrane receptors (including the Ret and Ron protooncogenes) seem to generate signals in the absence of ligands even if the mutation causes retention in the ER (see [39] and references therein). In addition, cells respond to the accumulation of unfolded or misfolded proteins in the ER by increasing the transcription of a set of genes encoding ER chaperones, enzymes and membrane proteins [54]. In yeast, the presence of unfolded proteins is sensed by the product of the Ire1 gene, a kinase exclusively localized in the ER. Recently, Ire1p homologs have been identified in mammalian cells; signals originating from Ire1p are important for stress responses, and can also induce cell death [55].

It is therefore tempting to speculate that association with SLC stabilizes newly generated μm chains in the ER, preventing their rapid degradation. The increase in the pool of intracellular molecules might facilitate interactions with Ig-α and Ig-β and generate the required signals. This model could help explaining the findings of Rajewsky and coworkers [56] who showed that B cell receptor (BCR) expression is necessary for B cell survival, in the absence of antigen [57]. During the process of assembly, collisions among receptors may be favored in the ER, and perhaps stabilized by the formation of reversible intermolecular disulfide bonds [41, 42]. These transient interactions might generate the low-level signalling that is necessary for survival. In contrast, antigen-mediated cross-linking of BCRs at the cell surface would lead to full-fledged signalling followed by activation.

References

1. Anfinsen CB (1973) Principles that govern the folding of protein chains. Science 181: 223-230
2. Gething MJ, Sambrook J (1992) Protein folding in the cell. Nature 355: 33-45
3. Milstein C, Brownlee GG, Harrison TM, Mathews MB (1972) A possible precursor for immunoglobulin light chains. Nature 205: 1171-1173
4. Blobel G, Dobberstein B (1975) Transfer of proteins across membranes. I. Presence of proteolytically processed and unprocessed nascent immunoglobulin light chains on membrane-bound ribosomes of murine myeloma. J Cell Biol 67: 835-851
5. Von Heijne G (1990) Protein targeting signals. Curr Opin Cell Biol 2: 604-608
6. Rapoport TA, Jungnickel B, Kutay U (1996) Annu Rev Biochem 65: 271-303
7. Andrews DW, Johnson AE (1996) The translocon: More than a hole in the ER membrane? Trends Biochem Sci 21: 365-369
8. Helenius A, Marquardt T, Braakman I (1992) The endoplasmic reticulum as a protein-folding compartment. Trends Cell Biol 2: 227-231
9. Gaut JR, Hendershot LM (1993) The modification and assembly of proteins in the endoplasmic reticulum. Curr Opin Cell Biol 5: 589-595
10. Hammond C, Helenius A (1995) Quality control in the secretory pathway. Curr Biol 7: 523-529
11. Klausner RD, Sitia R (1990) Protein degradation in the endoplasmic reticulum. Cell 62: 611-614
12. Bonifacino JS, Lippincott-Schwartz J (1991) Degradation of proteins within the endoplasmic reticulum. Curr Opin Cell Biol 3: 592-600
13. Fra AM, Sitia R (1993) The endoplasmic reticulum as a site of protein degradation. Plenum, New York, pp 143-168 (Subcellular Biochemistry, vol 21)
14. Lord JM (1996) Go outside and see the proteasome. Protein degradation. Curr Biol 6: 1067-1069

15. Kopito RR (1997) ER quality control: the cytoplasmic connection. Cell 88: 427-430
16. Cresswell P, Hughes EA (1997) Protein degradation: The ins and outs of the matter. Curr Biol 7: 552-555
17. Wiertz EJ, Tortorella D, Bogyo M et al. (1996) Sec61-mediated transfer of a membrane protein from the endoplasmic reticulum to the proteasome for destruction. Nature 384: 432-438
18. Pilon M, Schekman R, Romisch K (1997) Sec61p mediates export of a misfolded secretory protein from the endoplasmic reticulum to the cytosol for degradation. EMBO J 16: 4540-4548
19. de Virgilio M, Weninger H, Ivessa NE (1998) Ubiquitination is required for the retro-translocation of a short-lived luminal endoplasmic reticulum glycoprotein to the cytosol for degradation by the proteasome. J Biol Chem 273: 9734-9743
20. Mayer TU, Braun T, Jentsch S (1998) Role of the proteasome in membrane extraction of a short-lived ER-transmembrane protein. EMBO J 17: 3251-3257
21. Sitia R, Cattaneo A (1995) Synthesis and assembly of antibodies in natural and artificial environments. In: Zanetti M, Capra JD (eds) The antibodies, vol 1. Harwood Academic, Luxembourg, pp 127-168
22. Melnick J, Argon Y (1995) Molecular chaperones and the biosynthesis of antigen receptors. Immunol Today 16: 243-250
23. Carayannopoulos L, Capra JD (1993) Immunoglobulins. Structure and function. In: Paul WE (ed) Fundamental immunology, 3rd edn. Raven, New York, pp 283-314
24. Kaloff CR, Haas IG (1995) Coordination of immunoglobulin chain folding and immunoglobulin chain assembly is essential for the formation of functional IgG. Immunity 2: 1-20
25. Bergman LW, Kuehl WM (1979) Formation of intermolecular disulphide bonds on nascent immunoglobulin polypeptides. J Biol Chem 254: 5690-5694
26. Valetti C, Sitia R (1994) The differential effects of dithiothreitol and 2-mercaptoethanol on the secretion of partially and completely assembled immunoglobulins suggest that thiol-mediated retention does not take place in or beyond the Golgi. Mol Biol Cell 5: 1311-1324
27. Ooi CE, Weiss J (1992) Bidirectional movement of a nascent polypeptide across microsomal membranes reveals requirements for vectorial translocation of proteins. Cell 71: 87-96
28. Nicchitta CV, Blobel G (1993) Lumenal proteins of the mammalian endoplasmic reticulum are required to complete protein translocation. Cell 73: 989-998
29. Kohler G (1980) Immunoglobulin chain loss in hybridoma lines. Proc Natl Acad Sci USA 77: 2197-2199
30. Seligmann ME, Preudhomme JL, Danon F, Brouet JC (1979) Heavy chain disease: current findings and concepts. Immunol Rev 48: 145-167
31. Bole DG, Hendershot LM, Kearney JF (1986) Posttranslational association of immunoglobulin heavy chain binding protein with nascent heavy chains in non-secreting and secreting hybridomas. J Cell Biol 102: 1558-1566
32. Hendershot LM (1990) Immunoglobulin heavy chain and binding protein complexes are dissociated in vivo by light chain addition. J Cell Biol 111: 829-837
33. Haas IG (1991) BiP – A heat shock protein involved in immunoglobulin chain assembly. Curr Top Microb Immunol 167: 71-82
34. Knittler MR, Haas IG (1992) Interaction of Bip with newly synthesized immunoglobulin light chain molecules: Cycles of sequential binding and release. EMBO J 11: 1573-1581

35. Sitia R, Neuberger MS, Alberini CM et al. (1990) Developmental regulation of IgM secretion: The role of the carboxy-terminal cysteine. Cell 60: 781-790
36. Guenzi S, Fra AM, Sparvoli A et al. (1994) The efficiency of cysteine-mediated intracellular retention determines the differential fate of secretory IgA and IgM in B and plasma cells. Eur J Immunol 24: 2477-2482
37. Isidoro C, Maggioni C, Demoz M et al. (1996) Exposed thiols confer localization in the endoplasmic reticulum by retention rather than retrieval. J Biol Chem 271: 26138-26142
38. Kerem A, Kronman C, Bar-Nun S et al. (1993) Interrelations between assembly and secretion of recombinant human acetylcholinesterase. J Biol Chem 268: 180-184
39. Collesi C, Santoro MM, Gaudino G, Comoglio PM (1996) A splicing variant of the RON transcript induces constitutive tyrosine kinase activity and an invasive phenotype. Mol Cell Biol 16: 5518-5526
40. Wilbourn B, Nesbeth DN, Wainwright LJ, Field MC (1998) Proteasome and thiol involvement in quality control of glycophosphatidylinositol anchor addition. Biochem J 332: 111-118
41. Reddy P, Sparvoli A, Fagioli C, Fassina G, Sitia R (1996) Formation of reversible disulfide bonds with the protein matrix of the endoplasmic reticulum correlates with the retention of unassembled Ig-light chains. EMBO J 15: 2077-2085
42. Carelli S, Ceriotti A, Sitia R (1997) A stringent thiol-mediated retention in B lymphocytes and Xenopus oocytes correlates with inefficient IgM polymerization. Eur J Immunol 27: 1283-1291
43. Cattaneo A, Neuberger MS (1987) Polymeric IgM is secreted by transfectants of non-lymphoid cells in the absence of immunoglobulin J chain. EMBO J 6: 2753-2758
44. Brewer JW, Randall TD, Parkhouse RME, Corley RB (1994) IgM hexamers? Immunol Today 15: 165-168
45. Olson TS, Lane MD (1989) A common mechanism for posttranslational activation of plasma membrane receptors? FASEB J 3: 1618-1624
46. Bauskin AR, Alkalay I, Ben-Neriah Y (1991) Redox regulation of a protein tyrosine kinase in the endoplasmic reticulum. Cell 66: 685-696
47. Helenius A, Trombetta ES, Hebert DN, Simons JF (1997) Calnexin, calreticulin and the folding of glycoproteins. Trends Cell Biol 7: 193-200
48. Fernandez F, D'Alessio C, Fanchiotti S, Parodi AJ (1998) A misfolded protein conformation is not a sufficient condition for in vivo reglucosylation by the UDP-Glc: glycoprotein glucosyltransferase. EMBO J 17: 5877-5886
49. Melchers F, Karasuyama H, Haasner D et al. (1993) The surrogate light chain in B cell development. Immunol Today 14: 60-68
50. Venkitaraman AR, Williams GT, Dariavach P, Neuberger MS (1991) The B-cell antigen receptor of the five immunoglobulin classes. Nature 352: 777-781
51. Papavasilou F, Jankovic M, Gong S, Nussenzweig MC (1997) Control of immunoglobulin rearrangements in developing B cells. Curr Opin Immunol 9: 233-238
52. Bornemann KD, Brewer J, Perez S et al. (1997) Secretion of soluble pre-B cell receptors by pre-B cells. J Immunol 158: 2551-2557
53. O'Shea CC, Thornell AP, Rosewell IR et al. (1997) Exit of the pre-TCR from the ER/cis-Golgi is necessary for signaling differentiation, proliferation, and allelic exclusion in immature thymocytes. Immunity 5: 591-599
54. Shamu CE, Cox JS, Walter P (1994) The unfolded-protein response pathway in yeast. Trends Cell Biol 4: 56-60
55. Wang XW, Harding HP, Zhang Y et al. (1998) Cloning of mammalian Ire1 reveals diversity in the ER stress response. EMBO J 17: 5708-5717

56. Lam KP, Kuhn R, Rajewsky K (1997) In vivo ablation of surface immunoglobulin on mature B cells by inducible gene targeting results in rapid cell death. Cell 90: 1073-1083
57. Neuberger MS (1997) Antigen receptor signaling gives lymphocytes a long life. Cell 90: 971-973

Chapter 4

Pathogenetic Mechanisms of Autoimmunity

L. Adorini

Introduction

The function of the immune system is to preserve the biological identity of the individual (*self*). This requires the capacity to distinguish *self* from *nonself*, a discrimination primarily carried out by CD4+ T cells which have evolved to a sophisticated level of complexity in higher vertebrates. The basic strategy has been to generate a vast repertoire of antigen-specific receptors, distribute them clonally in different lymphocytes, and then eliminate cells capable of recognizing self antigens. This strategy at the same time renders possible the differentiation of lymphocytes potentially capable of recognizing nonself antigens. However complex the mechanisms utilized, self-nonself discrimination is teleologically simple: it positively selects T cells potentially capable of recognizing nonself while eliminating, physically or functionally, those responding to self-antigens.

The primary site for self-nonself discrimination is within the thymus, where the T cell repertoire is shaped by rescuing thymocytes from programmed cell death (positive selection) and by deleting cells expressing T cell receptors (TCR) with high affinity for self (negative selection). Both positive and negative selection of developing T cells occur via TCR-mediated recognition of complexes between self-peptides and self-MHC molecules. Positive selection is probably promoted by low avidity interactions, whereas negative selection depends on relatively high avidity interactions with specific self-ligands. Although presentation of self-peptides in sufficient amounts leads to deletion of developing T cells, the mechanism of negative selection is not absolute, as demonstrated by the presence of peripheral T cells capable of responding to self-antigens. In fact, negative thymic selection cannot occur for antigens expressed only on extra-thymic tissues, nor for those available at concentrations too low to be effectively presented by thymic antigen-presenting cells (APC). In these cases, autoreactive T cells can be exported to the periphery and peripheral tolerance mechanisms must exist to control their reactivity.

Tolerance to self is learned during development, rather than being genetically programmed. The structure of an antigenic molecule does not determine its capacity to be recognized as self or nonself. Key factors in self-nonself discrimi-

---

Roche Milano Ricerche, Via Olgettina 58 - 20132 Milan, Italy. e-mail: luciano.adorini@roche.com

nation include the time when lymphocytes are first confronted with antigenic epitopes, the site of the encounter, the nature of the cells presenting the antigenic epitopes, and the expression of costimulatory molecules by these cells. Since the control of self-nonself discrimination is so critical, the variety of strategies utilized by the immune system to induce and maintain tolerance is perhaps not surprising. Tolerance is truly redundant, and deletional and non-deletional mechanisms operate both in central and in peripheral lymphoid organs. Following central clonal deletion, whereby thymocytes expressing TCR specific for complexes of self-antigens and MHC molecules are physically eliminated by programmed cell death, T cells specific for antigens not expressed on thymic APC are exported to the periphery. It is now clear that several post-thymic mechanisms contribute to the induction and maintenance of tolerance in mature, peripheral T lymphocytes. The failure of self-tolerance can result in autoimmune diseases.

Autoimmune diseases can be quite diverse, but some characteristics are remarkably preserved: (1) association of disease with particular MHC alleles, (2) multifactorial origin, (3) polygenic control of susceptibility, and (4) chronic course, characterized by remission and relapse. This last feature suggests a balance between positive and negative regulatory factors which control susceptibility and progression of disease.

Th1 and Th2 Cells

Over ten years have passed since the original description of two subsets of differentiated $CD4^+$ T lymphocytes in the mouse, T helper 1 (Th1) and Th2 cells [1]. This subdivision was initially met with some skepticism because it did not seem to hold true in man. Further analysis of T cell clones from allergic patients clearly defined antigen-specific Th1 and Th2 cells also in humans [2] and, dispelling doubts [3], the Th1/Th2 paradigm rapidly became a key concept in immunoregulation.

The subdivision of both mouse and human $CD4^+$ as well as $CD8^+$ [4] T cells into three major subsets, Th1, Th2, and Th0, is based on their pattern of cytokine production. Th1 cells are characterized by secretion of interferon-γ (IFN-γ), interleukin (IL)-2, and tumor necrosis factor beta (TNF-β), and they promote cell-mediated immunity able to eliminate intracellular pathogens. Conversely, Th2 cells selectively produce IL-4 and IL-5, and are involved in the development of humoral immunity protecting against extracellular pathogens. Th0 cells, which could either represent precursors of Th1/Th2 cells or a terminally differentiated subset, are not restricted in their lymphokine production. The development of Th1 and Th2 cells is influenced by several factors, but three are most important: ligand-T cell receptor (TCR) interactions, genetic polymorphism and cytokines (Table 1). Among the cytokines, decisive roles are played by IL-12 and IL-4, guiding T cell responses towards the Th1 or Th2 phenotype, respectively [5, 6].

It is clear that polarized Th1 and Th2 cells represent extremes in a spectrum. Within this spectrum, discrete subsets of differentiated T cell secreting a mixture of Th1 and Th2 cytokines may exist, for example mouse T cells secreting IFN-γ

Table 1. Factors affecting Th1/Th2 development

Cytokines (IL-4, IL-10, IL-12, IL-18, IFN-γ)
Avidity of MHC-peptide/TCR interaction
  APC type
  Mode of antigen administration
  Dose of antigen
  Affinity of peptide-MHC binding
  Costimulation
Genetic background

APC, antigen-presenting cell; *TCR*, T cell receptor.

and IL-10 [7]. Differentiated CD4$^+$ cells characterized by unique cytokine secretion, such as Tr1 cells, have also been described. Tr1 cells produce high levels of IL-10, low levels of IL-2 and IFN-γ and no IL-4, and are able to suppress antigen-specific T cell responses [8]. It has been suggested that the polarization of the immune response may be achieved by altering the probabilities of IFN-γ, IL-4 and other cytokine gene expression at the population level, rather than by selective expansion of distinct T cell subsets [9, 10]. However, molecular mechanisms to explain the polarization of Th1 and Th2 subsets, based on the differential expression of the receptors for IFN-γ and IL-12, do exist. The ability of IFN-γ to inhibit the proliferation of Th2 but not of Th1 cells may be related to the lack of IFN-γR β-chain expression in Th1 cells [11]. However, IFN-γR β-chain loss also occurs in IFN-γ-treated Th2 cells, and therefore does not appear to represent a Th1 cell-specific differentiation event [12]. Conversely, developmental commitment to the Th2 lineage results from rapid loss of IL-12 signaling in Th2 cells [13]. The inability of Th2 cells to respond to IL-12 appears to be due to selective downregulation of IL-12R β2-subunit [14, 15]. Inhibition of Th1 and induction of Th2 in vivo is also related to downregulation of the IL-12R β2-subunit expression [16]. These findings are therefore consistent with a general model in which selective modulation of IL-12 signaling plays an important role in the acquisition of polarized Th cell phenotypes (Fig. 1).

The strength of peptide/class II-TCR interactions, which depends on the overall avidity of antigen-presenting cell (APC)-T cell interactions, also controls the profile of cytokine secretion by T cells. As demonstrated by using different antigen doses in vitro and in vivo as well as by altered peptide ligands, lower avidity interactions appear to favor Th2 cell development [17]. The reciprocal regulation between Th cell subsets is another driving force polarizing CD4$^+$ T cells into differentiated Th1 or Th2 cells. IL-12 promotes the development of Th1 cells [18-20] and inhibits IL-4-induced IgE synthesis [21]. IFN-γ amplifies the IL-12-dependent development of Th1 cells [22] and inhibits Th2 cell proliferation [23]. Conversely, IL-4 and IL-10 inhibit lymphokine production by Th1 clones [24]. In addition, IL-10 [25], IL-4 and IL-13 [26] suppress the development of Th1 cells through downregulation of IL-12 production by monocytes (Fig. 1).

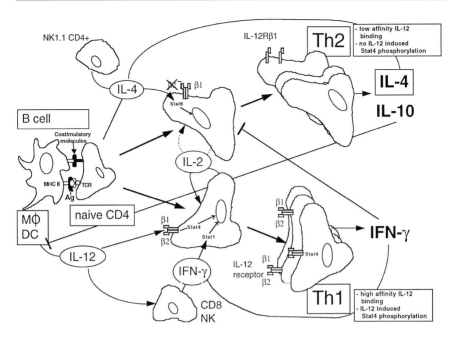

Fig. 1. Role of IL-12 in the regulation of th1/Th2 cell development. Activation of naive T cells throuh triggering of the antigen receptor is sufficient for the initial expression of functional IL-12 receptors. Depending on the cytokines present in the microenvironment, T cells will progressively develop into IL-12-responsive Th1 or IL-12-non-responsive Th2 cells. IL-12 and IL-4 act through Stat4 and Stat6, respectively, to deliver specific differentiation signals. Due to loss of IL-12Rβ2 chain expression, Th2 cells extinguish IL-12 responsiveness. The principal sources of IL-12 are activated macrophages and dendritic cells, whereas IL-4 is initially produced by antigen-stimulated CD4$^+$ T cells and CD4$^+$NK1.1$^+$-type T cells. IL-4 produced by mast cells and basophils may also contribute to Th2 cell development

Th1 and Th2 Cells in Autoimmune Diseases

Th1 cells are considered to be involved in the induction of experimental autoimmune diseases [27-30]. Evidence for this is based on adoptive transfer experiments demonstrating that CD4$^+$ cells producing Th1-type lymphokines can transfer disease, both in experimental allergic encephalomyelitis (EAE) [31] and in insulin-dependent diabetes mellitus (IDDM) [32-34] models. However, cytokine regulation is complex, for example TNF-α and IL-10 have opposite effects on IDDM depending on the developmental stage of the immune system [35, 36]. This could also explain why, in some cases, beta cell destruction in IDDM has been associated with Th2 rather than Th1 cells [37, 38].

The reciprocal regulation between T cell subsets predicts a role for Th2 cells in inhibition of autoimmune diseases. Evidence for this is provided by the reduced

IDDM incidence following IL-4 [39] or IL-10 [40] administration to nonobese diabetic (NOD) mice. A role for Th2 cells regulating the onset of IDDM is also suggested by their capacity to inhibit the spontaneous onset of diabetes in rats [41] and by the correlation between protection from IDDM and IL-4 production in double-transgenic mice on BALB/c background [42]. Furthermore, administration of IL-4 to mice with EAE ameliorates the disease [43], and in a model of streptococcal cell-wall induced arthritis it suppresses the chronic destructive phase [44]. Regulatory T cells that suppress the development of EAE produce Th2-type cytokines [45], and recovery from EAE is associated with increased Th2 cytokines in the central nervous system (CNS) [46]. These findings suggest that activation of Th2 cells may prevent EAE.

To study the role of Th1 and Th2 cells in IDDM, my colleagues and I targeted endogenous IL-12 in NOD mice by administration of the IL-12 antagonist $(p40)_2$ [47]. Administration of $(p40)_2$ from 3 weeks of age, before the onset of insulitis, resulted in the deviation of pancreas-infiltrating $CD4^+$ but not $CD8^+$ cells to the type 2 phenotype as well as in the reduction of spontaneous and cyclophosphamide-accelerated IDDM. After treating NOD mice with $(p40)_2$ from 9 weeks of age, when insulitis is well established, few Th2 and a reduced percentage of Th1 cells were found in the pancreas. This was associated with a slightly decreased incidence of spontaneous IDDM but, at variance with another report [48], no protection from cyclophosphamide-accelerated IDDM. $(p40)_2$ can inhibit in vitro the default Th1 development of naive TCR transgenic $CD4^+$ cells to the Th2 pathway but does not modify the cytokine profile of polarized Th1 cells, although it prevents further recruitment of $CD4^+$ cells into the Th1 subset. When polarized Th1 cells infiltrate the pancreas, targeting endogenous IL-12 has a marginal effect on IDDM incidence. This implies that inhibition of IL-12 may not inhibit pathogenic differentiated Th1 cells in chronic progressive diseases such as IDDM, whereas it could be beneficial in remitting/relapsing diseases such as EAE or some forms of multiple sclerosis (MS). In conclusion, the immune deviation to Th2 is maximal when IL-12 is targeted before the onset of insulitis, and is associated with protection from IDDM.

Collectively, these results indicate that the extent of immune deviation to Th2 is related to the degree of protection from IDDM, as predicted from the Th1/Th2 paradigm. However, they do not reveal whether Th2 cells are directly responsible for protection from IDDM or whether the immune deviation away from Th1 in itself accounts for the decreased IDDM incidence.

When injected into neonatal NOD mice, Th2 cells transgenic for a TCR derived from a clone able to transfer IDDM invaded the islets but did not provoke disease neither did they provide substantial protection [34]. Similar results were also obtained by adoptive transfer of non-transgenic Th1 and Th2 cell lines into neonatal mice [49]. Therefore, these data do not support the concept that Th2 cells afford protection from IDDM at least in the effector phase of the disease. Rather, they are in accord with the observation that transgenic expression by islet cells of IL-10 [36, 50], an inhibitory lymphokine of Th1 cells, actually promotes insulitis and IDDM, instead of inhibiting them. Collectively, these results

Table 2. Autoimmune diseases mediated by Th1 cells

Rheumatoid arthritis
Insulin-dependent diabetes mellitus
Experimental allergic encephalomyelitis/Multiple sclerosis
Autoimmune thyroiditis
Inflammatory bowel disease
Uveoretinitis
Myasthenia gravis

point to a critical role of Th1 cells in the induction of autoimmune diseases, whereas the influence of Th2 cells is still unclear.

Th1 cells also appear to be involved in human organ-specific autoimmune diseases. CD4$^+$ T cell clones isolated from lymphocytic infiltrates of Hashimoto's thyroiditis or Graves' disease exhibit a clear-cut Th1 phenotype [51]. In addition, most T cell clones derived from peripheral blood or cerebrospinal fluid of multiple sclerosis patients show a Th1 lymphokine profile [52].

Involvement of Th1 cells has also been suggested in other human autoimmune diseases. Insulitis in IDDM patients has been shown to comprise a large number of IFN-γ-producing lymphocytes [53]. T cell clones derived from the synovial membrane of rheumatoid arthritis (RA) patients also display a Th1 phenotype as they produce, upon activation, large amounts of IFN-γ and no or very little IL-4 [54]. Another study has shown that most CD4$^+$ and CD8$^+$ clones recovered from synovial fluid of RA patients display a Th1 phenotype [55]. Interestingly, in situ hybridization for T cell cytokine expression demonstrates a Th1-like pattern in most synovial samples from RA patients, whereas samples from patients with reactive arthritis, a disorder with similar synovial pathology but driven by persisting exogenous antigen, express a Th0 phenotype [56].

The situation is less clear in most systemic autoimmune disorders. In general, heterogeneous cytokine profiles are found in the serum or target organs of patients with systemic autoimmunity, such as systemic lupus, Sjögren's syndrome, and primary vasculitis [57]. A list of autoimmune diseases mediated by Th1 cells is shown in Table 2.

Prospects for Treatment of Human Autoimmune Diseases

The considerable interest raised by cytokine-based immunotherapy is already being applied to clinical settings, and particularly in RA patients [58]. Manipulation of the Th1/Th2 balance in RA patients by oral administration of type II collagen has been attempted but preliminary results [58] indicate some benefit only at the lowest dose tested. As the understanding of Th1/Th2 regulation in human autoimmune diseases broadens, we will certainly witness the application of more articulate strategies able to selectively target cytokine production by Th1 or Th2 cells or to modify the Th1/Th2 balance in clinical situations.

At present, consistent results from clinical trials have reported targeting the proinflammatory cytokines IL-1 and TNF-α in RA patients, and results of TNF-α-directed therapies are particularly promising. Animal studies have clearly documented the important role of TNF-α in RA. Mice transgenic for the human TNF-α gene produce high levels of this cytokine and develop arthritis beginning at 4 weeks of age [59]. The disease is mediated by the transgenic molecule, as demonstrated by prevention of arthritis following administration of monoclonal antibodies to human TNF-α. In addition, in a model of type II collagen-induced arthritis, administration of anti-mouse TNF-α – even after disease onset – significantly reduced inflammation and tissue destruction [60].

Based on these results, chimeric anti-TNF-α monoclonal antibody was administered to RA patients [61]. Treatment with anti-TNF-α was safe and well-tolerated, and led to significant clinical and laboratory improvements. After the first administration of anti-TNF-α antibody, remissions lasted, on the average, about three months, but reinjection of the antibody induced a significant anti-globulin response in most patients, reducing considerably the efficacy of the treatment. An alternative approach, using the soluble TNF receptor p55 chain fused to the constant region of human IgG1 heavy chain (sTNFR-IgG1), has been demonstrated to be about 10-fold more effective than anti-TNF-α antibody at neutralizing the activity of endogenous TNF, as assessed in a model of listeriosis [62] or in chronic relapsing EAE [63]. This fusion protein appears to achieve the same clinical effects as anti-TNF-α antibody administration without strong induction of neutralizing antibodies.

Even if clinical results of anti-TNF-α therapy in RA patients are promising, the role of TNF-α in IDDM models is still puzzling. The fact that anti-TNF-α antibody treatment initiated before 3 weeks of age prevents insulitis and IDDM clearly suggests that TNF-α may be an essential mediator for the generation or activation of autoreactive lymphocytes [35]. Intriguingly, administration of TNF-α to adult NOD mice also prevented IDDM, but the mechanism is still unclear [35]. TNF-α appears to have distinct effect on the diabetogenic process depending upon the developmental stage of the immune system and of the target organ, perhaps in a manner analogous to IL-10.

These results stress the importance of the time window of cytokine or anti-cytokine treatment to obtain the desired effect. If this concept cannot be translated to clinical practice, conditions to recreate a situation favoring the protective effects of the anti-cytokine treatment should be optimized. Besides the time window, another important factor to be taken into consideration is the non-MHC-linked polymorphism controlling Th1 and Th2 cell induction [42, 64, 65]. However, soluble antigen administration inhibits Th1 cell development in any mouse strain tested [66], indicating that this approach may be effective at the population level.

Table 3. Targeting Th1 cells

Cytokine-based approaches
Inhibition of IL-12 and/or IFN-γ
Anti-cytokine mAb or cytokine antagonists
Anti-cytokine receptor mAb
Soluble cytokine receptors
Inhibitors of cytokine production
Administration of IL-4 and/or IL-10
Autoantigen-based approaches
Direct
Systemic administration of soluble autoantigen
Indirect (induction of Th2 cells/diversion away from Th1 cells)
Oral administration or inhalation of autoantigen
Presentation of autoantigen by B cells
Administration of peptide analogues of autoantigen
Administration of autoantigen together with antagonists or inhibitors of IL-12
Administration of autoantigen together with IL-4

mAb, monoclonal antibody.

Conclusions

The results reviewed highlight the critical role of Th1 cells in the induction of several organ-specific autoimmune diseases and suggest a possible influence of Th2 cells in controlling, directly or indirectly, disease induction or progression. The efficacy of cytokine-specific treatments in chronic inflammatory disease states raises hopes for immunosuppressive strategies selectively targeting Th1- or Th2-type T cells. In particular, antagonists of IL-12 and inducers of IL-4 or IL-10 offer the possibility to selectively manipulate Th1 and Th2 cell induction, with potential for the treatment of autoimmune diseases (Table 3).

References

1. Mosmann TR, Cherwinski H, Bond MW et al. (1986) Two types of murine helper T cell clone. I Definition according to profile of lymphokine activities and secreted proteins. J Immunol 136: 2348-2357
2. Del Prete G, De Carli M, Mastromauro C et al. (1991) Purified protein derivative of *Mycobacterium tuberculosis* and excretory-secretory antigen(s) of *Toxocara canis* expand in vitro human T cells with stable and opposite (type 1 T helper or type 2 T helper) profile of cytokine production. J Clin Invest 88: 346-350
3. Romagnani S (1991) Human TH1 and TH2: doubt no more. Immunol Today 12: 256-257
4. Erard F, Wild M-T, Garcia-Sanz JA, Le Gros G (1993) Switch of CD8 T cells to non-cytolytic CD8⁻CD4⁻ cells that make Th2 cytokines and help B cells. Science 260: 1802-1805

5. Gately MK, Renzetti LM, Magram J et al. (1998) The interleukin-12/interleukin-12 receptor system: Role in normal and pathologic immune responses. Annu Rev Immunol 16: 495-521
6. Paul WE, Seder RA (1994) Lymphocytes responses and cytokines. Cell 76: 241-251
7. Nanda NK, Sercarz EE, Hsu D-H, Kronenberg M (1994) A unique pattern of lymphokine synthesis is a characteristic of certain antigen-specific suppressor T cell clones. Int Immunol 6: 731-737
8. Groux H, O'Garra A, Bigler M et al. (1997) A $CD4^+$ T-cell subset inhibits antigen-specific T-cell responses and prevents colitis. Nature 389: 737-742
9. Kelso A, Groves P, Troutt AB, Francis K (1995) Evidence for the stochastic acquisition of cytokine profile by $CD4^+$ T cells activated in a T helper type 2-like response in vivo. Eur J Immunol 25: 1168-1175
10. Kelso A (1995) Th1 and Th2 subsets: Paradigms lost? Immunol Today 16: 374-379
11. Pernis A, Gupta S, Gollob KJ et al. (1995) Lack of interferon γ receptor β chain and the prevention of interferon γ signaling in Th1 cells. Science 269: 245-247
12. Bach EA, Szabo S, Dighe AS et al. (1995) Ligand-induced autoregulation of IFN-γ receptor β chain expression in T helper cell subsets. Science 270: 1215-1218
13. Szabo SJ, Jacobson AG, Gubler U, Murphy KM (1995) Developmental commitment to the Th2 lineage by extinction of IL-12 signaling. Immunity 2: 665-675
14. Rogge L, Barberis-Maino L, Biffi M et al. (1997) Selective expression of an interleukin-12 receptor component by human T helper 1 cells. J Exp Med 185: 825-831
15. Szabo SJ, Dighe AS, Gubler U, Murphy KM (1997) Regulation of the interleukin (IL)-12R β2 subunit expression in developing T helper 1 (Th1) and Th2 cells. J Exp Med 185: 817-824
16. Galbiati F, Rogge L, Guéry J-C et al. (1998) Regulation of the interleukin (IL)-12 receptor β2 subunit by soluble antigen and IL-12 in vivo. Eur J Immunol 28: 209-220
17. Abbas AK, Murphy KM, Sher A (1997) Functional diversity of helper T lymphocytes. Nature 383: 787-793
18. Hsieh C-S, Macatonia SE, Tripp CS et al. (1993) Development of Th1 $CD4^+$ T cells through IL-12 produced by *Listeria*-induced macrophages. Science 260: 547-549
19. Manetti R, Parronchi P, Giudizi MG et al. (1993) Natural killer cell stimulatory factor (interleukin 12, IL-12) induces T helper type 1 (Th1)-specific immune responses and inhibits the development of IL-4-producing Th cells. J Exp Med 177: 1199-1204
20. Afonso LCC, Scharton TM, Vieira LQ et al. (1994) The adjuvant effect of interleukin-12 in a vaccine against *Leishmania major*. Science 263: 235-237
21. Kiniwa M, Gately M, Gubler U et al. (1992) Recombinant interleukin-12 suppresses the synthesis of IgE by interleukin-4 stimulated human lymphocytes. J Clin Invest 90: 262-266
22. Schmitt E, Hoehn P, Huels C et al. (1994) T helper type 1 development of naive $CD4^+$ T cells requires the coordinate action of interleukin-12 and interferon-γ and is inhibited by transforming growth factor-β. Eur J Immunol 24: 793-798
23. Gajewski TF, Fitch FW (1988) Anti-proliferative effect of IFN-gamma in immune regulation. I. IFN-gamma inhibits the proliferation of Th2 but not Th1 murine helper T lymphocyte clones. J Immunol 140: 4245-4253
24. Moore K, O'Garra A, de Waal Malefyt R et al. (1993) Interleukin-10. Annu Rev Immunol 11: 165-190
25. D'Andrea A, Aste-Amezaga M, Valiante NM et al. (1993) Interleukin 10 (IL-10) inhibits human lymphocyte interferon-γ production by suppressing natural killer cell stimulatory factor/IL-12 synthesis in accessory cells. J Exp Med 178: 1041-1048

26. de Waal Malefyt R, Figdor CG, Huijbens R et al. (1993) Effects of IL-13 on phenotype, cytokine production, and cytotoxic function of human monocytes. Comparison with IL-4 and modulation by IFN-γ or IL-10. J Immunol 151: 6370-6381
27. Powrie F, Coffmann RL (1993) Cytokine regulation of T cell function: potential for therapeutic intervention. Immunol Today 14: 270-274
28. O'Garra A, Murphy K (1993) T-cell subsets in autoimmunity. Curr Opin Immunol 5: 880-886
29. Liblau RS, Singer SM, McDevitt HO (1995) Th1 and Th2 CD4$^+$ T cells in the pathogenesis of organ-specific autoimmune diseases. Immunol Today 16: 34-38
30. Trembleau S, Germann T, Gately MK, Adorini L (1995) The role of IL-12 in the induction of organ-specific autoimmune diseases. Immunol Today 16: 383-386
31. Ando DG, Clayton J, Kong D et al. (1989) Encephalitogenic T cells in the B10.PL model of experimental allergic encephalomyelitis (EAE) are of the Th1 lymphokine subtype. Cell Immunol 124: 132-143
32. Haskins K, McDuffie M (1990) Acceleration of diabetes in young NOD mice with a CD4$^+$ islet specific T cell clone. Science 249: 1433-1436
33. Bergman B, Haskins K (1994) Islet-specific T-cell clones from the NOD mouse respond to beta-granule antigen. Diabetes 43: 197-203
34. Katz JD, Benoist C, Mathis D (1995) T helper cell subsets in insulin-dependent diabetes. Science 268: 1185-1188
35. Yang X-D, Tisch R, Singer SM et al. (1994) Effect of tumor necrosis factor α on insulin-dependent diabetes mellitus in NOD mice. I. The early development of autoimmunity and the diabetogenic process. J Exp Med 180: 995-1004
36. Wogesen L, Lee M-S, Sarvetnick N (1994) Production of interleukin 10 by islet cells accelerates immune-mediated destruction of β cells in nonobese diabetic mice. J Exp Med 179: 1379-1384
37. Anderson JT, Cornelius JG, Jarpe AJ et al. (1993) Insulin-dependent diabetes in the NOD mouse model. II. Beta cell destruction in autoimmune diabetes is a Th2 and not a Th1 mediated event. Autoimmunity 15: 113-122
38. Akhtar I, Gold JP, Pan L-Y et al. (1995) CD4$^+$ β islet cell-reactive T cell clones that suppress autoimmune diabetes in nonobese diabetic mice. J Exp Med 182: 87-97
39. Rapoport MJ, Jaramillo A, Zipris D et al. (1993) Interleukin 4 reverses T cell proliferative unresponsiveness and prevents the onset of diabetes in nonobese diabetic mice. J Exp Med 178: 87-99
40. Pennline KJ, Roquegaffney E, Monahan M (1994) Recombinant human IL-10 prevents the onset of diabetes in the nonobese diabetic mouse. Clin Immunol Immunopathol 71: 169-175
41. Fowell D, Mason D (1993) Evidence that the T cell repertoire of normal rats contains cells with the potential to cause diabetes. Characterization of the CD4$^+$ T cell subset that inhibits this autoimmune potential. J Exp Med 177: 627-636
42. Scott B, Liblau R, Degermann S et al. (1994) A role for non-MHC genetic polymorphism in susceptibility to spontaneous autoimmunity. Immunity 1: 1-20
43. Racke MK, Bonomo A, Scott DE et al. (1994) Cytokine-induced immune deviation as a therapy for inflammatory autoimmune disease. J Exp Med 180: 1961-1966
44. Allen JB, Wong HL, Costa GL et al. (1993) Suppression of monocyte function and differential regulation of IL-1 and IL-1ra by IL-4 contribute to resolution of experimental arthritis. J Immunol 151: 4344-4351
45. van der Veen RC, Stohlman SA (1993) Encephalitogenic Th1 cells are inhibited by Th2 cells with related peptide specificity: relative roles of interleukin (IL)-4 and IL-10. J Neuroimmunol 48: 213-220

46. Khoury SJ, Hancock WW, Weiner HL (1992) Oral tolerance to myelin basic protein and natural recovery from experimental autoimmune encephalomyelitis are associated with downregulation of inflammatory cytokines and differential upregulation of transforming growth factor β, interleukin 4, and prostaglandin E expression in the brain. J Exp Med 176: 1355-1364
47. Trembleau S, Penna G, Gregori S et al. (1997) Deviation of pancreas-infiltrating cells to Th2 by interleukin-12 antagonist administration inhibits autoimmune diabetes. Eur J Immunol 27: 2230-2239
48. Rothe H, O'Hara RM, Martin S, Kolb H (1997) Suppression of cyclophosphamide induced diabetes development and pancreatic Th1 reactivity in NOD mice treated with the interleukin (IL)-12 antagonist IL-12(p40)$_2$. Diabetologia 40: 641-646
49. Healey D, Ozegbe P, Arden S et al. (1995) In vivo activity and in vitro specificity of CD4$^+$ Th1 and Th2 cells derived from the spleens of diabetic NOD mice. J Clin Invest 95: 2979-2985
50. Moritani M, Yoshimoto K, Tashiro F et al. (1994) Transgenic expression of IL-10 in pancreatic islet A cells accelerates autoimmune insulitis and diabetes in non-obese diabetic mice. Int Immunol 6: 1927-1936
51. De Carli M, D'Elios M, Mariotti S et al. (1993) Cytolytic T cells with Th1-like cytokine profile predominate in retroorbital lymphocytic infiltrates of Graves' ophthalmopathy. J Clin Endocrinol Metab 77: 1120-1124
52. Brod SA, Benjamin D, Hafler DA (1991) Restricted T cell expression of IL-2, IFN-γ mRNA in human inflammatory disease. J Immunol 147: 810-815
53. Foulis AK, McGill M, Farquahrson MA (1991) Insulitis in type I (insulin-dependent) diabetes mellitus in man. Macrophages, lymphocytes and interferon-γ-containing cells. J Pathol 165: 97-103
54. Miltenburg AM, van Laar JM, de Kuiper R et al. (1992) T cells cloned from human rheumatoid synovial membrane functionally represent the Th1 subset. Scand J Immunol 35: 603-610
55. De Carli M, D'Elios MM, Zancuoghi G et al. (1994) Human TH1 and TH2 cells: functional properties, regulation of development and role in autoimmunity. Autoimmunity 18: 301-308
56. Simon AK, Seipelt E, Sieper J (1994) Divergent T-cell cytokine patterns in inflammatory arthritis. Proc Natl Acad Sci USA 91: 8562-8566
57. Romagnani S (1994) Lymphokine production by human T cells in disease states. Annu Rev Immunol 12: 227-257
58. Kingsley G, Lanchbury J, Panayi G (1996) Immunotherapy in rheumatic disease: An idea whose time has come - or gone? Immunol Today 17: 9-12
59. Keffer J, Probert L, Cazlaris H et al. (1991) Transgenic mice expressing human tumor necrosis factor - a predictive genetic model of arthritis. EMBO J 13: 4025-4031
60. Williams RO, Feldmann M, Maini RN (1992) Anti-tumor necrosis factor ameliorates joint disease in murine collagen-induced arthritis. Proc Natl Acad Sci USA 89: 9784-9788
61. Elliott MJ, Maini RN, Feldmann M et al. (1993) Treatment of rheumatoid arthritis with chimeric monoclonal antibodies to tumor necrosis factor. Arthritis Rheum 36: 1681-1690
62. Haak-Frendscho M, Marsters SA, Mordenti J et al. (1994) Inhibition of TNF by a TNF receptor immunoadhesin. Comparison to an anti-TNF monoclonal antibody. J Immunol 152: 1347-1353
63. Baker D, Butler D, Scallon BJ et al. (1994) Control of established experimental allergic encephalomyelitis by inhibition of tumor necrosis factor (TNF) activity within the central nervous system using monoclonal antibodies and TNF receptor-immunoglobulin fusion proteins. Eur J Immunol 24: 2040-2048

64. Reiner SL, Locksley RM (1995) The regulation of immunity to *Leishmania major*. Annu Rev Immunol 13: 151-177
65. Guéry J-C, Galbiati F, Smiroldo S, Adorini L (1997) Non MHC-linked Th2 cell development induced by soluble protein administration predicts susceptibility to *Leishmania major* infection. J Immunol 159: 2147-2153
66. Guéry J-C, Galbiati F, Smiroldo S, Adorini L (1996) Selective development of Th2 cells induced by continuous administration of low dose soluble proteins to normal and β2-microglobulin-deficient BALB/c mice. J Exp Med 183: 485-497

Chapter 5

Adhesion Receptors Involved in Leukocyte Functions

E. Bianchi[1], M. Fabbri[1], R. Pardi[1,2]

Introduction

Leukocytes are the only nucleated somatic cells that spend a significant portion of their life span displaying a nonadherent, circulating phenotype, being transported by the bloodstream and interstitial fluids towards their sites of action. However, most effector functions of leukocytes depend on and are induced by firm adhesion to other cells or to the extracellular matrix. A logical explanation for this apparent paradox is that while most functional responses of immunocompetent cells are triggered by intimate contacts with other cells, the effective search for potentially harmful environmental agents in a multicellular organism (i.e. immune surveillance) requires continuous patrolling of the body across anatomical barriers. Leukocytes have evolved a highly dynamic and finely regulated adhesive behaviour in order to comply with these contrasting requirements.

To accomplish functions requiring such high motility, leukocytes are equipped with a vast array of membrane receptors capable of mediating fluid shear-resistant and mechanical stress-resistant adhesion. Among the expressed receptors are members of the integrin and immunoglobulin superfamilies, CD44 and the selectins. Interestingly, most of these receptors are constantly engaged by their ligands, thus mediating constitutive adhesion, when expressed in differentiated cells other than leukocytes. This suggests that intracellular mechanisms have uniquely evolved in cells of the immune system to maintain adhesion receptors in a "switched off" configuration during their transit in the bloodstream and extracellular fluids. Recent studies have begun to shed light on the molecular mechanisms that are responsible for such a dynamic behaviour of leukocyte adhesion receptors [1, 2].

Selectin-Mediated Adhesion

Transient adhesion, or tethering, and rolling of leukocytes along the endoluminal sides of capillaries is mediated by selectins [3]. These are receptors containing a

---

[1] Human Immunology Unit, DIBIT-San Raffaele Scientific Institute, via Olgettina 58 - 20132 Milan, Italy.
[2] University of Milan School of Medicine, Milan, Italy e-mail: pardi.ruggero@hsr.it

calcium-dependent (C-type) lectin domain that allows them to interact with complex carbohydrates on apposing plasma membranes. The intrinsic affinity of selectins for their ligands is rather low (K_D=100 nM) but the on-off rates of the binding are extremely fast [4, 5]. The resulting interactions with ligands are transient and insufficient to promote firm cell-cell adhesion. In essence, leukocytes "stumble" across complex carbohydrates expressed on the surface of endothelial cells when they marginate from the center of a blood vessel towards the vessel wall, where the relative flow rate is lower. Besides the intrinsically controlled nature of selectin-mediated adhesion, the recently characterized, activation-dependent proteolytic cleavage of the extracellular domain of L-selectin [6] provides an additional level of regulation: once activated, leukocytes shed their membrane-expressed selectins and become transiently incompetent to recognize selectin ligands. Since activation is a prerequisite for the onset of integrin-mediated adhesion (see next paragraph), it is possible that selectins and integrins alternate their functions in a finely regulated temporal sequence. This allows the vectorial progression of leukocytes from the bloodstream into peripheral or lymphoid tissues.

Integrin-Mediated Adhesion

Firm intercellular or cell-matrix adhesion of leukocytes is largely mediated by integrins of the β_1 and β_2 subfamilies, the latter being selectively expressed by cells of lymphoid and myeloid origin. Inherited defects in β_2 integrin expression lead to a severe form of immunodeficiency, the leukocyte adhesion deficiency syndrome (LAD). The instrumental role played by these receptors in maintaining the functional integrity of the immune system is clearly demostrated by their association with LAD [7]. As stated previously, integrins expressed by circulating cells such as leukocytes or platelets, whether they belong to the β_1, β_2 or β_3 subfamilies, are incompetent to engage specific ligands unless activation signals are delivered to the expressing cell [8-11]. The ill-defined allosteric transition of integrins from a low avidity to a high avidity state for the ligand has originally been described as an activation-dependent process in the platelet integrin $\alpha IIb/\beta_3$ [12, 13], and subsequently demonstrated for other members of the integrin family, including β_2 integrins [8, 14] and β_1 integrins expressed by mature lymphocytes [15, 16]. A recent, detailed analysis of the relative affinity of $\alpha L/\beta_2$ (LFA-1) for its ligand ICAM-1 in soluble form indicated that a fraction of surface LFA-1 molecules converts from a low, baseline affinity (K_D= 100 µM) to a state with a 200-fold higher affinity in activated T lymphocytes [17].

The prototypic stimulus for T cell adhesion is represented by the recognition of antigen in a cell-bound form. On the other hand chemokines, the chemotactic cytokines, are likely candidates for being the most relevant physiologic activators of antigen-independent leukocyte adhesion [18]. Chemoattractant receptors are seven transmembrane domain molecules homologous to trimeric G protein coupled receptors [19]. As such, upon ligand binding, they catalyze the exchange of

GDP for GTP by the G_1 protein alpha subunit, resulting in dissociation of the beta and gamma subunits and activation of a number of second messenger pathways, including those implicated in integrin function upregulation [20, 21]. A step-wise model for leukocyte transmigration across anatomical barriers such as the vascular endothelium has recently been proposed [22]. The model combines detailed knowledge of the kinetics and requirements for selectin-mediated and integrin-mediated cell adhesion.

Several molecular events, whose causal relationships and temporal sequence are as yet poorly defined, parallel the energy-dependent conversion of inactive leukocyte integrins to a ligand-binding conformation, downstream from the activating stimulus. These include protein kinase-C activation and translocation [23, 24], phosphorylation on serine residues of both subunits of the heterodimer [25, 26] and acquisition of neo-epitopes by the extracellular domains of the α and β subunits [27-31]. In addition, energy-independent avidity shifts in integrins have been shown to occur as a consequence of changes in extracellular divalent cation concentration [28, 32], release of small unsaturated lipids by activated cells [33], or treatment with selected anti-α or anti-β subunit monoclonal antibodies that "lock" the molecule in an activated state [29, 31, 34]. The activated state of leukocyte or platelet integrins is clearly accompanied by a conformational change (allosteric transition) in the extracellular domains of the heterodimer, as defined by the acquisition of neo-epitopes which are characteristic of the ligand-binding conformation of the receptor [27, 30, 31, 35]. Additionally, reversible association of the adhesion receptor to the actin-based cytoskeleton, likely to be mediated by a complex of peripheral proteins which includes talin, α-actinin and vinculin, has been shown by our group [24] and by others [36-38] to parallel the activation-induced avidity shift of leukocyte and platelet integrins.

Two models can be proposed to causally relate these phenomena. One model predicts that cell activation directly promotes a conformational change in the receptor, followed by ligand recognition, receptor clustering, and cytoskeletal reorganization. Alternatively, it can be postulated that activation-induced physical association of the receptor to cytoskeletal elements precedes, and possibly promotes, conformational changes in ligand binding sites located in the receptor's extracellular domains, with subsequent engagement of the ligand. According to this model, receptor clustering could lead to accumulation of G-actin-binding proteins (such as α-actinin) underneath the plasma membrane at the site of ligand engagement, followed by a focal burst of ATP-dependent actin polymerization originating within the cluster of ligand-bound receptors. This would parallel the phenotypic conversion of leukocytes from a nonadherent to an adherent state. Several independent lines of evidence favour the latter model, and suggest the occurrence of ligand-independent sorting of integrins to specialized sub-domains of the plasma membrane [39], where they complex to cytoskeletal proteins in constitutively adherent cells. This was elegantly shown for β_1 and β_3 integrins by the construction of chimaeric proteins with unrelated extracellular domains, fused to integrin transmembrane and cytoplasmic domains [40].

In circulating cells such as T lymphocytes, in which adhesion receptors are normally disconnected from the cell cytoskeleton, we demonstrated that triggering via multiple activation receptors was sufficient to promote physical association of a fraction of surface LFA-1 to the actin-based cytoskeleton, in the absence of ligand engagement by the adhesion receptor [24]. A recent study by Tominaga et al. [41] demonstrated that phorbol ester-induced activation of LFA-1 could be abrogated by selectively inhibiting the activity of rhoA, a ras-like GTP-binding protein involved in growth factor-promoted cytoskeletal organization. Finally, α-actinin from activated but not resting polymorphonuclear cell extracts was shown by Pavalko and LaRoche [42] to interact with a peptide corresponding to the β_2 integrin cytoplasmic domain. Collectively, these findings indicate that cytoskeletal rearrangement, and ordered plasma membrane-cytoskeletal connections [43], lie upstream of the engagement of specific ligands by adhesion receptors such as integrins, and that some of these mechanisms are constitutively operating in adherent cells like fibroblasts, while requiring triggering by external stimuli in circulating cells such as leukocytes and platelets.

The dynamic organization of the cytoskeleton plays an instrumental role in determining the onset, overall strength, and persistence of cell adhesion. Cytoskeletal proteins form a multimolecular chain underneath the plasma membrane that connects structures exposed to the cell exterior to microfilaments, microtubules or intermediate filaments with high plasticity. To mediate efficient cell adhesion, spreading and locomotion, an integral membrane protein must interact, in a constitutive or regulated fashion, with cytoskeletal components. Integrins are the prototype of membrane receptors endowed with the ability to interact with the cell cytoskeleton. Defined regions in their cytoplasmic domains physically associate with one or more components of the cytoskeletal network, including α-actinin, talin and, possibly, vinculin [44-46]. These proteins in turn mediate the association of the adhesion receptor to actin and control the kinetics of assembly and the qualitatative aspects (enlongation, severing and branching) of ATP-dependent actin polymerization. Pharmacological inhibition of cytoskeletal rearrangement completely abrogates leukocyte adhesion and locomotion mediated by integrin receptors [47].

In addition to its structural contribution to cell adhesion, the cytoskeleton provides a framework whereby signalling molecules come into close association with their effectors underneath the plasma membrane. Within specialized areas of tight membrane-substrate interaction, there is recruitment of regulatory proteins such as kinases ($pp60^{src}$, $pp125^{FAK}$, $pp58^{fgr}$, PKC, MAPK), proteolytic enzymes (calpain II), trimeric or low molecular weight GTP-binding proteins, and a number of phosphoproteins whose functional role, i.e. structural versus regulatory, has yet to be established [48]. The common feature shared by the aforementioned molecular intermediates of transmembrane signalling is their ability to permanently or transiently associate with the cytoskeleton. The organization of multiple cytoskeletal components at discrete sites of the plasma membrane where cell-cell or cell-matrix adhesion take place is therefore one of the clues to understanding the role of adhesion receptors in signal transduction.

Transendothelial Migration of Leukocytes

How and where do leukocytes traverse the endothelium under physiologic conditions? Most evidence indicates that leukocyte transmigration takes place at the endothelial cell (EC) junction. As cell migration does not occur without adhesion, it is important to know which receptor-ligand interactions support leukocyte-EC adhesion at the junctional level. In principle, adjacent ECs loosening their intercellular contact could expose subendothelial matrix to direct recognition by leukocyte β_1 and β_3 integrins. However, this seems unlikely to occur under physiological conditions, given the catastrophic consequences that such exposure would have on the activation of intravascular cascades of prothrombotic and procoagulant mediators. A more likely possibility is that EC junctional molecules are engaged in the transmigration process, allowing gradual and transient loosening of junctions with minimal leakage of the EC barrier. Indeed, only a small and transient leakage of solutes can be demonstrated at perivascular sites where leukocyte transmigration takes place [49].

Some of the candidate molecules expressed by ECs that could support leukocyte migration, such as the ICAMs and vascular cell adhesion molecule-1 (VCAM-1), are involved in supporting firm adhesion to the luminal aspect of ECs. Other molecules (CD31 and cadherins) are particularly attractive candidates because of their constitutive junctional distribution. Most of these molecules, independently of their polarized distribution, have been shown to support activation-dependent rather than spontaneous adhesion of leukocytes; this is an essential requirement for a process involving tightly regulated adhesion and de-adhesion steps [50].

On the basis of several independent reports [51], it can be hypothesized that junctional molecules (e.g. CD31 and cadherins) promote haptotactic (i.e. along a gradient of solid-phase-bound molecules) migration of leukocytes in the absence of a chemotactic gradient at the junctional level. Concomitantly, more diffusely distributed ligands might support vectorial migration driven by a junctional chemotactic gradient established in the subendothelial space. Evidence supporting this hypothesis comes from the observation that mutant CD31 molecules displaying a diffuse apical, rather than junctional, distribution support homophilic adhesion of leukocytes but lose the ability to promote their transmigration, unless a chemotactic gradient is established in the subendothelial space [49]. CD31-mediated migration may be regulated through an activation-dependent change in the affinity of the leukocyte-expressed molecule, which would displace existing homophilic interactions at the EC junctional level. This would provide a fail-safe mechanism in which the leukocyte proceeds through a "molecular zipper" made of CD31 molecules on apposing ECs. Further progression of leukocytes through the EC junction towards the subendothelial matrix could be effected by CD31-mediated transactivation of β_1 integrin-dependent adhesion, in what has been defined as an "adhesion cascade" [52].

While EC-bound leukocytes may detach and re-enter the circulation, the distinct feature of transmigrating leukocytes is their acquired vectorial motility, which entails a fine balance between adhesion and de-adhesion events. How do the extracellular triggers for this behaviour act at the intracellular level? Recent insights into this process suggest that intracellular regulators of actin cytoskeleton rearrangement are central to the complex process of regulated adhesion and migration [53]. The small GTPase RhoA appears to be a critical switch for the "inside-out" regulation of leukocyte adhesion. Indeed, RhoA positively regulates adhesion receptor function in response to activation induced by phorbol ester [54] or chemokines [55], and it may also be a target of protein kinase A-mediated feed-back inhibition of integrin-dependent adhesion [56, 57]. On the basis of its ability to effect contractility-dependent formation of actin stress fibers, and to organize integrins in focal adhesions in several cell types, RhoA may be a critical transducer controlling the dynamic interaction between adhesion receptors and connecting peripheral proteins of the actin-based cytoskeleton.

Rac1, another member of the Rho family of small GTPases, induces cytoskeletal changes such as lamellipodia formation and membrane ruffling, which are typical of motile cells. Rac1 appears to effect the activation of phosphatidylinositol (PI) 3-kinase [53], whose involvement in regulating cell motility has been unequivocally demonstrated in other cellular systems [58]. The recent demonstration that a lymphoid cell migration promoting gene, T cell invasion-associated molecule-1 (TIAM-1), is a specific exchange factor for Rac1 (promoting conversion from its inactive, GDP-bound state to the active, GTP-bound state) [59], supports the idea that Rac1 activation may be critically involved in regulating leukocyte migration.

Conclusions

We have presented several examples relevant to the emerging concept that adhesion receptors are finely regulated molecules whose functions go from providing dynamic adhesion to transducing informational signals related to many cellular activities, including growth and gene expression. Major challenges for the future include understanding the signalling pathways involved in both the inside-out and the outside-in aspects of adhesion receptor function. With regard to cells of the immune system, it will be interesting to dissect these events in various cell subsets to determine whether signalling by adhesion receptors is influenced by the state of differentiation of a given cell subset or by its lineage of origin. Clearly, adhesion receptors provide more than just a "velcro" covering the plasma membrane and controlling the adhesiveness of a cell: they effect essential functions of immunocompetent cells, such as topographic memory and antigen-driven proliferation.

References

1. Dustin ML Springer TA (1991) Role of lymphocyte adhesion receptors in transient interactions and cell locomotion. Annu Rev Immunol 9: 27-66

2. Pardi R, Inverardi L, Bender JR (1992) Regulatory mechanisms in leukocyte adhesion: flexible receptors for sophisticated travelers. Immunol Today 13: 224-230
3. Drickamer K, Taylor ME (1993) Biology of animal lectins. Annu Rev Cell Biol 9: 237-264
4. Springer TA (1994) Traffic signals for lymphocyte recirculation and leukocyte emigration: The multistep paradigm. Cell 76: 301-314
5. Williams A (1991) Out of equilibrium. Nature 352: 473-474
6. Kahn J, Ingraham RH, Shirley F et al. (1994) Membrane proximal cleavage of L-selectin: Identification of the cleavage site and a 6-kD transmembrane peptide fragment of L-selectin. J Cell Biol 125: 461-470
7. Anderson DC, Springer TA (1987) Leukocyte adhesion deficiency: An inherited defect in the Mac-1, LFA-1 and p150, 95 glycoproteins. Annu Rev Immunol 38: 175-212
8. Dustin LM, Springer TA (1989) T cell receptor cross-linking transiently stimulates adhesiveness through LFA-1. Nature 341: 619-624
9. Ginsberg MH, Du X, Plow EF (1992) Inside-out integrin signalling. Curr Opin Cell Biol 4: 766-771
10. Hemler EM (1990) VLA proteins in the integrin family: Structures, functions, and their role on leukocytes. Annu Rev Immunol 8: 365-400
11. Springer TA (1990) Adhesion receptors of the immune system. Nature 346: 425-434
12. Frelinger A, Du X, Plow EF, Ginsberg MH (1991) Monoclonal antibodies to ligand-occupied conformers of integrin $\alpha II\beta/\beta 3$ alters receptor affinity, specificity and function. J Biol Chem 266: 17106-17111
13. Marguerie GA, Plow EF, Edgington TS (1979) Human platelets possess an inducible and saturable receptor specific for fibrinogen. J Biol Chem 254: 5357-5363
14. Wright SD, Meyer BC (1986) Phorbol esters cause sequential activation and deactivation of complement receptors on polymorphonuclear leukocytes. J Immunol 36: 1759-1764
15. Danilov YN, Juliano RL (1989) Phorbol ester modulation of integrin-mediated cell adhesion: a post-receptor event. J Cell Biol 108: 1925-1933
16. Shimizu Y, Van Seventer GA, Horgan KJ, Shaw S (1990) Regulated expression and binding of three VLA (β_1) integrin receptors on T cells. Nature 345: 250-253
17. Lollo BA, Chan KWH, Hanson EM et al. (1993) Direct evidence for two affinity states for lymphocyte function-associated antigen 1 on activated T cells. J Biol Chem 268: 21693-21700
18. Sozzani S, Molino M, Locat M et al. (1993) Receptor-activated calcium influx in human monocytes exposed to monocyte chemotactic protein-1 and related cytokines. J Immunol 150: 1544-1553
19. Baggiolini M, Dewald B, Moser B (1994) Interleukin-8 and related chemotactic cytokines – CXC and CC chemokines. Adv Immunol 55: 97-179
20. Detmers PA, Lo SK, Olsen-Ebret E et al. (1990) Neutrophil activating protein 1/IL-8 stimulates the binding activity of the leukocyte adhesion receptor CD11b/CD18 on human neutrophils. J Exp Med 171: 1155-1162
21. Woldemar Carr M, Roth SJ, Luther E et al. (1994) Monocyte chemoattractant protein 1 acts as a T-lymphocyte chemoattractant. Proc Natl Acad Sci USA 91: 3652-3656
22. Butcher EC (1991) Leukocyte-endothelial cell recognition: three (or more) steps to specificity and diversity. Cell 67: 1033-1037
23. Downey GP, Chan CK, Lea P et al. (1992) Phorbol ester-induced actin assembly in neutrophils: role of protein kinase C. J Cell Biol 116: 695-706
24. Pardi R, Inverardi L, Rugarli C, Bender JR (1992) Antigen-receptor complex stimulation triggers protein kinase C-dependent CD11a/CD18-cytoskeleton association in T lymphocytes. J Cell Biol 116: 1211-1220

25. Hibbs ML, Jakes S, Stacker SA et al. (1991) The cytoplasmic domain of the integrin lymphocyte function-associated antigen 1 β subunit: sites required for binding to intercellular adhesion molecule 1 and the phorbol ester-stimulated phosphorylation site. J Exp Med 174: 1227-1238
26. Valmu LM, Autero P, Siljander M et al. (1991) Phosphorylation of the β-subunit of CD11/CD18 integrins by protein kinase C correlates with leukocyte adhesion. Eur J Immunol 21: 2857-2862
27. Diamond MS, Springer TA (1993) A subpopulation of Mac-1 (CD11b/CD18) molecules mediates neutrophil adhesion to ICAM-1 and fibrinogen. J Cell Biol 120: 545-556
28. Dransfield Icabanas C, Craig A, Hogg N (1992) Divalent cation regulation of the function of the leukocyte integrin LFA-1. J Cell Biol 116: 219-226
29. Altieri DC (1991) Occupancy of CD11b/CD18 (Mac-1) divalent ion binding site(s) induces leukocyte adhesion. J Immunol 147: 1891-1898
30. Landis RC, Bennett RI, Hogg N (1993) A novel LFA-1 activation epitope maps to the I domain. J Cell Biol 120: 1519-1527
31. Van Kooyk Y, Weder P, Hogervorst F et al. (1991) Activation of LFA-1 through Ca^{2+}-dependent epitope stimulates lymphocyte adhesion. J Cell Biol 112: 345-354
32. Gailit J, Ruoslathi E (1988) Regulation of the fibronectin receptor affinity by divalent cations. J Biol Chem 263: 12927-12932
33. Hermanowski-Vostaka A, van Strijp JAG, Swiggard WJ, Wright SD (1992) Integrin modulating factor-1: A lipid that alters the function of leukocyte integrins. Cell 68: 341-352
34. Chan BMC, Hemler ME (1993) Multiple functional forms of integrin VLA-2 can be derived from a single α2 cDNA clone: Interconversion of forms induced by anti-β1 antibody. J Cell Biol 120: 537-543
35. O'Toole ET, Katagiri Y, Faull RJ et al. (1994) Integrin cytoplasmic domains mediate inside-out signal transduction. J Cell Biol 124: 1047-1059
36. Bertagnolli ME, Bekerle MC (1993) Evidence for the selective association of a subpopulation of GPIIb-IIIa with the actin cytoskeletons of thrombin-activated platelets. J Cell Biol 121: 1329-1342
37. Larson RS, Hibbs ML, Springer TA (1990) The leukocyte integrin LFA-1 reconstituted by cDNA transfection in a nonhematopoietic cell line is functionally active and not transiently regulated. Cell Regulation 1: 359-367
38. Luna EJ, Hitt AL (1993) Cytoskeleton-plasma membrane interactions. Science 258: 955-964
39. Briesewitz R, Kern A, Marcantonio EE (1993) Ligand-dependent and -independent integrin focal contact localization: The role of the chain cytoplasmic domain. Mol Biol Cell 4: 593-604
40. LaFlamme SE, Akiyama SK, Yamada KM (1992) Regulation of fibronectin receptor distribution. J Cell Biol 117: 437-448
41. Tominaga T, Sugie K, Hirata M et al. (1993) Inhibition of PMA-induced, LFA-1-dependent lymphocyte aggregation by ADP ribosylation of the small molecular weight GTP binding protein, rho. J Cell Biol 120:1529-1537
42. Pavalko FM, LaRoche SM (1993) Activation of human neutrophils induces an interaction between the integrin $β_2$ subunit (CD18) and the actin binding protein α-actinin. J Immunol 151: 3795-3807
43. Geiger B (1989) Cytoskeleton-associated cell contacts. Curr Opin Cell Biol 1:103-109
44. Sastry SK, Horwitz AF (1993) Integrin cytoplasmic domains: Mediators of cytoskeletal linkages and extra- and intracellular initiated transmembrane signalling. Curr Opin Cell Biol 5: 819-831

45. Reszka AA, Hayashi Y, Horwitz AF (1992) Identification of aminoacid sequences in the integrin β_1 cytoplasmic domain implicated in cytoskeletal association. J Cell Biol 117: 1321-1330
46. Pardi R, Bossi G, Inverardi L et al. (1995) Conserved regions in the α and β subunit cytoplasmic domains of the leukocyte integrin αL/β2 are involved in ER retention, dimerization and cytoskeletal association. J Immunol 155: 1252-1263
47. Hynes RO (1992) Integrins: versatility, modulation, and signaling in cell adhesion. Cell 69: 11-15
48. Schwartz MA (1992) Transmembrane signalling by integrins. Trends Cell Biol 2: 304-307
49. Anderson AO, Shaw S (1993) T cell adhesion to endothelium: the FRC conduit system and other anatomic and molecular features which facilitate the adhesion cascade in lymph node. Semin Immunol 5: 271-282
50. Zocchi MR, Ferrero E, Leone BE et al. (1996) CD31/PECAM-1-driven chemokine-independent transmigration of human T lymphocytes. Eur J Immunol 26: 759-767
51. Tanaka Y, Albelda SM, Horgan KJ et al. (1992) CD31 expressed on distinctive T cell subsets is a preferential amplifier of beta 1 integrin-mediated adhesion. J Exp Med 176: 245-253
52. Ridley AJ (1994) Signal transduction through the GTP-binding proteins Rac and Rho. J Cell Sci Suppl 18: 127-131
53. Tominaga T, Sugie K, Hirata M, et al. (1993) Inhibition of PMA-induced, LFA-1-dependent lymphocyte aggregation by ADP ribosylation of the small molecular weight GTP binding protein, rho. J Cell Biol 120: 1529-1537
54. Laudanna C, Campbell JJ, Butcher EC (1996) Role of Rho in chemoattractant-activated leukocyte adhesion through integrins. Science 271: 981-983
55. Lang P, Gesbert F, Delespine-Carmagnat M et al. (1996) Protein kinase A phosphorylation of RhoA mediates the morphological and functional effects of cyclic AMP in cytotoxic lymphocytes. EMBO J 15: 510-519
56. Rovere P, Inverardi L, Bender JR, Pardi R (1996) Feedback modulation of ligand-engaged alpha L/beta 2 leukocyte integrin (LFA-1) by cyclic AMP-dependent protein kinase. J Immunol 156: 2273-2279
57. Ponzetto C, Bardelli A, Zhen Z et al. (1994) A multifunctional docking site mediates signaling and transformation by the hepatocyte growth factor/scatter factor receptor family. Cell 77: 261-271
58. Michiels F, Habets GG, Stam JC et al. (1995) A role for Rac in Tiam1-induced membrane ruffling and invasion. Nature 375: 338-340
59. Goetzl EJ, Banda MJ, Leppert D (1996) Matrix metalloproteinases in immunity. J Immunol 156: 1-4

Chapter 6

The Endothelium of the Brain Microvasculature and the Organization of Intercellular Junctions

M.G. Lampugnani, G. Bazzoni, E. Dejana

The Endothelium of the Brain Microvasculature

The endothelium is considered to be a sparse organ system due to its vast extension and its ability to exert a complex array of specialized functions [1, 2]. A unique characteristic of endothelial cells (EC) is that, although they present many common functional and morphological features, they also display remarkable heterogeneity in different organs. Here, we will focus on the endothelium of the brain microvasculature, which represents the interface between blood and the central nervous system. We will also examine how its functional properties may be mediated by intercellular junctions. Due to its unique location, the microvascular endothelium of the brain has specific protective properties which strictly regulate the infiltration of plasma components and circulating cells into the brain. The blood-brain barrier normally permits the passage of only small hydrophobic molecules, a limited number of specifically transported nutrients such as glucose and amino acids, and some transcytosed molecules such as transferrin (reviewed in [3-5]).

In general, the barrier activity is due to well-developed junctions between EC, selective intracellular transport systems, and low pinocytotic activity. Among the specialized domains at cell-cell contacts (see following section), tight junctions (TJ) are particularly well developed in brain vessels and are mostly found associated to the P-phase of the cell membrane, indirectly indicating a close linkage to the cytoskeleton [6]. The presence of such a developed system of TJ is probably responsible for the cell polarization found in brain capillaries [5]. Cell polarization helps to direct the transport of solutes from the apical membrane to the basal membrane and viceversa. Brain EC synthesize the multi-drug resistance protein P-glycoprotein (or mdr1a), which actively transports a variety of small molecular weight molecules out of the brain into the circulation [7, 8], thus protecting nervous tissue from accumulation of undesired toxic molecules. Brain endothelium also has specific transport systems directing the flow of substances from the blood to the brain, such as the glucose transporter Glut-1 or the various amino-acid transporters [5, 9].

Laboratory of Vascular Biology, Mario Negri Institute of Pharmacological Research, Via Eritrea 62 - 20157 Milan, Italy. e-mail: dejana@irfmn.mnegri.it

Brain microvasculature derives from meningeal vessels and invades the brain by angiogenesis [3]. These EC acquire the characteristics of blood-brain barrier, most likely through contact with the neuroectoderm during embryogenesis. Astrocytes contribute to the establishment of barrier properties in EC [4, 10]. However, recent evidence shows that some specific markers of the brain microvasculature are already expressed by these cells as early as day 10.5 of gestation when astrocytes are not yet detectable [8, 11]. This suggests that some forms of "early" endothelial committment exist even in absence of the interaction with astrocytes [12].

Interestingly, the vessels which invade brain tumors, such as glioblastoma multiforme, do not have blood-brain barrier properties and this results in brain edema. In this type of tumor, the vascular defect seems to be mostly due to the formation of channels through interendothelial junctions [13]. This observation suggests that, for maintenance of specialized barrier properties, EC require a continuous interaction with normal nervous cells. When EC invade a newly formed tumor such as a glioma or glioblastoma, they come into contact with tumor cells which produce growth factors and, in particular, vascular endothelial growth factor (VEGF), which may be responsible not only for vascular proliferation, but also for the altered permeability properties of the newly formed vessels [14-16].

Having highlighted the relevance of interendothelial junctions in blood-brain barrier function, we will now analyze the components of these structures at both molecular and functional levels.

Interendothelial Junctions

Adhesion of EC to one another is mediated by various surface receptors that belong to several families of ubiquitously expressed cell adhesion molecules, such as cadherins, integrins, immunoglobulins, and proteoglycans. Besides merely acting as attachment sites, most adhesive receptors interact with cytoskeletal and cytoplasmic molecules, thus contributing to the regulation of cell morphology and signalling [17]. Most of the molecules at cell-cell contacts are organized in interendothelial junctional structures, such as adherens junctions (AJ) [18] and TJ [19].

The general molecular organization (with adhesive transmembrane components associated with regulatory cytoplasmic proteins) and the reciprocal relationships of cell-cell adhesive systems in the endothelium are similar to those of other cell types [20]. AJ and TJ differ on the basis of their components, localization in the width of the lateral membrane, and distribution along the vascular tree. TJ are located apically to AJ and present a restricted distribution, being particularly enriched in endothelia that strictly control permeability, such as arterial endothelium and brain capillaries, as already mentioned. No endothelial-specific components of TJ have been identified up to now [20, 21]. In contrast, AJ present an endothelial specific transmembrane component, vascular endothelial (VE-) cadherin (cadherin-5/CD144), which is exclusively expressed by EC [22] since the earliest stages of endothelial differentiation in the embryonic blood islands [23].

VE-cadherin is the major transmembrane component of AJ, which are ubiquitous along the vascular tree [24].

Outside these two complexes, a third cellular adhesive protein called PECAM-1 (platelet endothelial cell adhesion molecule-1/CD31) [25, 26] has been well characterized. PECAM-1 is also expressed on circulating monocytes, neutrophils, and platelets, and regulates leukocyte extravasation [27]. PECAM-1 can transfer and receive signals possibly through binding to the tyrosine phosphatase SHP-2, which may participate in the signalling cascade of different growth factors.

While it is reasonable to envisage that molecules at EC cell-cell contacts control the organization of new vessels and stabilize the integrity of the endothelium, experimental data are still fragmentary. However, information obtained using blocking antibodies against VE-cadherin and PECAM-1 in either in vitro or in vivo models suggests that these molecules are required in the organization of new vessels and the maintenance of endothelial integrity [28-30].

Adherens Junctions

The endothelial VE-cadherin, like the other members of the cadherin family, sustains cell-cell recognition and adhesion by homophilically binding an identical cadherin molecule on an adjoining cell. Intracellularly it is linked to cytoplasmic molecules, collectively known as catenins (α- and β-catenin, plakoglobin, and p120) [31, 32], some of which connect the transmembrane protein to the actin cytoskeleton. These molecules are important mediators of VE-cadherin activity. VE-cadherin inhibits cell migration [33] and proliferation [34], and restricts paracellular permeability [33]. In contrast to the wild type form, a truncated VE-cadherin lacking the region of the cytoplasmic domain which binds catenins is unable to regulate these functions, even if it retains normal adhesive activity [35]. Thus, the biological activity of VE-cadherin is strictly dependent upon its capacity to bind catenins or actin cytoskeleton. The influence of VE-cadherin on the cell functions mentioned above is compatible with a regulatory role in new vessel formation [33]. More direct indications come from the use of monoclonal antibodies to VE-cadherin. These antibodies inhibited the organization of vascular-like structures in in vitro models of angiogenesis using EC of both human and mouse origin (M.G. Lampugnani et al., unpublished observations). In addition, abolishing the function of the VE-cadherin gene in embryonic stem cells through homologous recombination did not affect the appearance of differentiated EC in embryoid bodies. However, these EC were unable to develop a proper vascular-like network, compared to wild type embryoid bodies, and remain in clusters [36].

Other indirect evidence for a role of VE-cadherin in the process of vessel morphogenesis is the following. First, VE-cadherin can be the target of an endothelial morphogen like VEGF, which induces tyrosine phosphorylation of VE-cadherin [37] and may modulate VE-cadherin activity [32]. Second, VE-cadherin at cell-cell contacts is strongly reduced in human malignant angiosarcomas and hemangioendotheliomas, where the vascular network is highly abnormal [38]. Third, cytotrophoblast invasion of uterine vessels is associated with loss of E-cadherin and acquired expression of VE-cadherin [39].

Several cadherins have been described in some types of EC in culture [40]. Some of these cadherins, such as N-cadherin [41-43], P-cadherin [42] and E-cadherin [44], are not endothelium-specific and have been mostly characterized as neural and muscular (N-cadherin) [45], or epithelial (P- and E-cadherin) [46]. EC express high amounts of N-cadherin, which is however preferentially found diffuse on the cell membrane [43]. This suggests that it can play a role in the anchorage of EC to other N-cadherin-expressing cells such as smooth muscle cells, astrocytes, and pericytes.

The role of cadherins in the organization of new vessels has not yet been completely defined. E-cadherin null mutation produced a lethal phenotype before implantation at the morula stage [47], much earlier than the organization of the vascular system [2]. P-cadherin deficient mice were viable and fertile [48], and they did not show any obvious vascular defect. On the other hand, suppression of N-cadherin gene, which resulted in postimplantation lethality and abnormal development of the heart tube, was also associated with defective blood vessels in the yolk sac [49], suggesting a role for N-cadherin in vasculogenesis. It remains to be defined whether the observed defect is exclusively due to the absence of N-cadherin in EC or to the lack of association with pericytes or smooth muscle cells [50].

As mentioned before, cadherins require the contribution of cytoplasmic partners, the catenins, to transmit signals to the cell. The type of catenin associated with VE-cadherin in an endothelial layer in dynamic situations (i.e. during migration or organization of a subconfluent culture) is different from that of a resting and established cell layer [31]. Null mutations of α-catenin, β-catenin and plakoglobin resulted in embryonic lethality, which occured at a preimplantation stage for α-catenin [51], but postimplantation for both β-catenin [52] and plakoglobin [53]. The absence of β-catenin induced lethality at a stage (6.5-7.5 days of embryogenesis) which preceded blood vessel formation. Lethality of plakoglobin null embryo, which was delayed to 12-16 days, was due to impaired myocardial architecture [53]. No vascular defect has been described, except edema in a few embryos surviving for 16-18 days. This effect, however, could be secondary to preexisting heart failure.

Therefore, while many intruiguing suggestions point to AJ components as possible regulators of vessel formation, conclusive experimental data are still missing. The development of conditional mutants or endothelial-targeted mutants may help to define more specifically the role of some of the molecules discussed above in the organization of new vessels.

Tight Junctions

At variance with AJ, which represent the basic and ubiquitous type of organized structures at interendothelial contacts, TJ or occludens junctions are particularly well developed in those endothelia that strictly control the exchanges between blood and tissues (typically at the blood-brain barrier and in the large arteries) [19]. At electron microscopy, TJ present an apparent fusion of the outer leaflets of the plasma membrane of the two contiguous cells, suggesting that they seal the intercellular space giving rise to an adhesive belt located towards the apical surface of the cell layer.

Morphologically, endothelial and epithelial TJ are organized in a similar manner. However, at variance with epithelia, and possibly due to the far more flat aspect of the endothelium, endothelial TJ present a less strictly apical localization and are found spatially intermingled with AJ [19, 54]. More recently, several molecular components of TJ, as well as their reciprocal relationships, have been defined. Some of these constituents are common to endothelium and epithelium, even if subtle differences may exist.

Transmembrane constituents of both epithelial and endothelial TJ are occludin [55] and junctional adhesion molecule (JAM) [21]. Occludin is a 65 kDa protein with four putative membrane-spanning domains and both amino- and carboxy-terminal regions localized in the cytoplasm. Occludin might bind homophilically an identical molecule present on an adjoining cell, possibly through the second extracellular domain [56]. In parallel to the high concentration of TJ in brain endothelium, occludin staining is particularly intense in brain capillaries, while weaker staining is found in heart, muscle, and intestinal endothelium [55]. JAM is another integral membrane protein (35 kDa) with an extracellular domain composed of two Ig-like loops, a single transmembrane domain, and a short cytoplasmic tail. It is ubiquitously expressed in endothelial and epithelial cells (where it colocalizes at intercellular contacts together with TJ proteins), as well as on circulating cells. Due to its location at interendothelial junctions, JAM might play an active role both in leukocyte extravasation and paracellular permeability in inflamed tissues [21]. Finally, claudin-1 and -2 are TJ components that have been recently identified from the junctional fraction of chicken liver. These 22 kDa protein are homologous and, like occludin, have four transmembrane domains. However, they are not related to occludin. Claudins are detected in several tissues but so far no data are available on their cellular distribution [57].

Occludin may need association with cytoskeletal proteins to be localized at TJ [58]. One of these mediators could be zonula occludens-1 (ZO-1), a 225 kDa protein [58-60] that binds to the carboxy-terminal region of occludin [61]. Binding of occludin to ZO-1 is necessary, but apparently not sufficient for targeting occludin to TJ, since some deletion mutants of occludin that do bind ZO-1 are not concentrated at TJ [56]. Therefore other factors may be required for the correct junctional distribution of ZO-1. For instance, ZO-1 might connect occludin to the actin cytoskeleton through its association with spectrin [61], and may also bind actin directly using its carboxy-terminal domain [62].

ZO-1 binds occludin at its amino-terminal region [62]. ZO-1 in turn interacts through one of its three PDZ (PSD-95, discs-large, ZO-1) domains with ZO-2 (zonula occludens-2, 160 kDa, [63]), and both proteins form part of the cytoplasmic undercoat of TJ. ZO-1 and ZO-2 are members of the MAGUK (membrane-associated guanylate kinase) family, as they bear a domain homologous to guanylate kinase, although devoid of enzymatic activity [64]. Cytoplasmic proteins belonging to the MAGUK family in general express organizing and targeting activity towards cell membrane proteins [64]. Besides the guanylate kinase homology region, they contain multiple copies of an 80 amino acid repeat, the PDZ/DHR (discs-large homology region) domain, which is involved in protein-protein interactions. While ZO-2 is exclusively found at TJ both in endothelia and epithelia, ZO-1 shows a far less specific distribution. Indeed it is located at cell-cell contacts independently of the presence of TJ and can also be found in cell types that never develop TJ, such as fibroblasts and cardiac muscle cells [61, 65].

Targeting of ZO-1 to the plasma membrane can be regulated by its binding to cytoplasmic components of AJ, typically catenins (α-, β- and plakoglobin, [66]). In cells which do not develop TJ, ZO-1 is found associated to AJ [61]. However, when TJ form, ZO-1 is segregated with them, suggesting that ZO-1 may represent a cross-talking element between AJ and TJ.

Interestingly, endothelial cells express a specific ZO-1 isotype, called ZO-1a [54]. It derives from alternative RNA splicing and lacks an 80 amino acid region (in the proline-rich carboxy-terminal half of the molecule). No alteration of the molecular relationships of this isoform with other TJ components (for example ZO-2) has been reported although it was suggested that ZO-1a characterizes more dynamic junctions [54].

Two other cytoplasmic components of TJ common to epithelia and endothelia are cingulin (140-108 kDa [67]) and 7H6 antigen (175-155 kDa [68]), the expression of this last molecule being restricted to brain capillaries in vivo. Cingulin is located more peripherally to the plasma membrane than ZO-1 and ZO-2, and its molecular interaction with other constituents of TJ remains to be defined. Notably, epithelial cells express molecules at TJ which are absent in EC such as symplekin [69].

The functional effects of TJ have been classically indicated in the control of paracellular permeability and cell polarity, functions that most epithelia and many endothelia have to accomplish [19]. Increasing evidence indicates that occludin directly contributes to paracellular barrier function [70]. Interestingly, when occludin is displaced from its junctional localization, the cytoplasmic components of TJ (ZO-1, ZO-2 and cingulin) remain in place at the plasma membrane even if both permeability and electrical resistance are severely affected [56]. As observed in tissue sections (see preceding section), also cultured EC from the brain microvasculature express higher levels of occludin in comparison to aortic endothelium and this parallels a more effective barrier function [71].

As far as cytoplasmic components are concerned, the far more studied in terms of functional effects is ZO-1. In cultured EC the amount of ZO-1 is upregulated by cell confluency [72] and downregulated by the vasoactive agent histamine [73]. Tyrosine phosphorylation of ZO-1 correlates with increased paracellular permeability both in epithelial cells and EC [74]. Induction of another cytoplasmic component of TJ, 7H6, with dbcAMP or retinoic acid, enhances the barrier function of EC in culture [75].

Similarly to AJ, there is indirect evidence that TJ can act as cell-to-cell signalling organelles. Non-classic signalling pathways have been outlined for components of TJ. Symplekin was found to be concentrated in the nucleus [69]. In epithelial cells ZO-1 shows a nuclear localization which is inversely related to the maturity of cell-cell contacts [76]. Also, ZO-1 distribution is stimulated at sites of wounding in epithelial cells in vitro and along the outer tip of the villus in tissue sections [76], thus suggesting a possible involvement of TJ components in morphogenetic processes.

Other Structures

In contrast to epithelial cells, EC do not have classic desmosomes, even if they might express similar but possibly simplified desmosomal structures. EC synthesize desmoplakin which is a specific component of desmosomes. Desmoplakin codistributes with VE-cadherin, plakoglobin and vimentin [77], and the association of these molecules might constitute a desmosomal-like structure (also called complexus adherens [78]). Other adhesive molecules, not directly associated to AJ or TJ, have been found to be concentrated at endothelial cell-to-cell contacts. These include PECAM-1 (see the section on Interendothelial Junctions), S-endo-1/Muc 18, CD34, and endoglin (reviewed in [24]).

Conclusions

Intercellular junctions of brain microvascular EC may be active players in the induction and maintenance of the blood-brain barrier function exerted by these cells. The vast amount of evidence collected in recent years about the molecular organization and the functional properties of endothelial AJ and TJ might help to fully elucidate the role of junctions in the brain microvasculature. A striking feature that has emerged so far is the redundacy of adhesive molecules and associated proteins at interendothelial junctions. One obvious explanation for such redundancy is that several adhesive structure might simply cooperate to maintain and stabilize intercellular adhesion. However, it is far more attractive to consider that these structures may have specific biological activities and activate unique signaling pathways.

References

1. Augustin HG, Kozian DH, Johnson RC (1994) Differentiation of endothelial cells: Analysis of the constitutive and activated endothelial cell phenotypes. Bioessays 16: 901-906
2. Risau W (1995) Differentiation of endothelium. FASEB J 9: 926-933
3. Risau W, Wolburg H (1990) Development of the blood-brain barrier. Trends Neurosci 13: 174-178

4. Staddon JM, Rubin LL (1996) Cell adhesion, cell junctions and the blood-brain barrier. Curr Opin Neurobiol 6: 622-627
5. Joo F (1996) Endothelial cells of the brain and other organ systems: Some similarities and differences. Prog Neurobiol 48: 255-273
6. Wolburg H, Neuhaus J, Kniesel U et al. (1994) Modulation of tight junction structure in blood-brain barrier endothelial cells. Effects of tissue culture, second messengers and cocultured astrocytes. J Cell Sci 107: 1347-1357
7. Schinkel AH, Smit JJ, van Tellingen O et al. (1994) Disruption of the mouse mdr1a P-glycoprotein gene leads to a deficiency in the blood-brain barrier and to increased sensitivity to drugs. Cell 77: 491-502
8. Qin Y, Sato TN (1995) Mouse multidrug resistance 1a/3 gene is the earliest known endothelial cell differentiation marker during blood-brain barrier development. Dev Dyn 202: 172-180
9. Maher F, Vannucci SJ, Simpson IA (1994) Glucose transporter proteins in brain. FASEB J 8: 1003-1011.
10. Janzer RC, Raff MC (1987) Astrocytes induce blood-brain barrier properties in endothelial cells. Nature 325: 253-257
11. Achen MG, Clauss M, Schnurch H, Risau W (1995) The non-receptor tyrosine kinase Lyn is localised in the developing murine blood-brain barrier. Differentiation 59: 15-24
12. Holash HA, Noden DM, Stewart PA (1993) Re-evaluating the role of astrocytes in blood-brain barrier induction. Dev Dyn 197: 14-25
13. Coomber BL, Stewart PA, Hayakawa K et al. (1987) Quantitative morphology of human glioblastoma multiforme microvessels: structural basis of blood-brain barrier defect. J Neuroncol 5: 299-307
14. Ferrara N (1993) Vascular endothelial growth factor. Trends Cardiovasc Med 3: 244-250
15. Millauer B, Shawver LK, Plate KH et al. (1994) Glioblastoma growth inhibited in vivo by a dominant-negative Flk-1 mutant. Nature 367: 576-579
16. Breier G, Risau W (1996) Angiogenesis in the developing brain and in brain tumours. Trends Exp Med 6: 362-376
17. Gumbiner BM (1996) Cell adhesion: The molecular basis of tissue architecture and morphogenesis. Cell 84: 345-357
18. Aberle H, Schwartz H, Kemler R (1996) Cadherin-catenin complex: protein interactions and their implications for cadherin function. J Cell Biochem 61: 514-523
19. Anderson JM, Van Itallie CM (1995) Tight junctions and the molecular basis for regulation of paracellular permeability. Am J Pathol 269: G465-G475
20. Lampugnani MG, Dejana E (1997) Interendothelial junctions: structure, signalling and functional roles. Curr Opin Cell Biol 9: 674-682
21. Martin-Padura I, Lostaglio S, Schneemann M et al. (1998) Junctional adhesion molecule, a novel member of the immunoglobulin superfamily that distributes at intercellular junctions and modulates monocyte transmigration. J Cell Biol 142: 117-127
22. Lampugnani MG, Resnati M, Raiteri M, et al.(1992) A novel endothelial-specific membrane protein is a marker of cell-cell contacts. J Cell Biol 118: 1511-1522
23. Breier G, Breviario F, Caveda L et al. (1996) Molecular cloning and expression of murine vascular endothelial-cadherin in early stage development of cardiovascular system. Blood 87: 630-642
24. Dejana E (1996) Endothelial adherens junctions: Implications in the control of vascular permeability and angiogenesis. J Clin Invest 98: 1949-1953
25. Newman PJ, Berndt MC, Gorski J et al. (1990) PECAM-1 (CD31) cloning and relation to adhesion molecules of the immunoglobulin gene superfamily. Science 247: 1219-1222

26. Simmons DL, Walker C, Power C, Pigott R (1990) Molecular cloning of CD31, a putative intercellular adhesion molecule closely related to carcinoembryonic antigen. J Exp Med 171: 2147-2152
27. Muller WA, Weigl SA, Deng X, Phillips DM (1993) PECAM-1 is required for transendothelial migration of leukocytes. J Exp Med 178: 449-460
28. DeLisser HM, Christofidou-Solomidou M, Strieter RM et al. (1997) Involvement of endothelial PECAM-1/CD31 in angiogenesis. Am J Pathol 151: 671-677
29. Matsumura T, Wolff K, Petzelbauer P (1997) Endothelial cell tube formation depends on cadherin 5 and CD31 interactions with filamentous actin. J Immunol 158: 3408-3416
30. Bach TL, Barsigian C, Chalupowicz DG et al. (1998) VE-cadherin mediates endothelial cell capillary tube formation in fibrin and collagen gels. Exp Cell Res 238: 324-334
31. Lampugnani MG, Corada M, Caveda L et al. (1995) The molecular organization of endothelial cell to cell junctions: Differential association of plakoglobin, beta-catenin, and alpha-catenin with vascular endothelial cadherin (VE-cadherin). J Cell Biol 129: 203-217
32. Lampugnani MG, Corada M, Andriopoulou P et al. (1997) Cell confluence regulates tyrosine phosphorylation of adherens junction components in endothelial cells. J Cell Sci 110: 2065-2077
33. Breviario F, Caveda L, Corada M et al. (1995) Functional properties of human vascular endothelial cadherin (7B4/cadherin-5), an endothelium-specific cadherin. Arterioscler Thromb Vasc Biol 15: 1229-1239
34. Caveda L, Martin-Padura I, Navarro P et al. (1996) Inhibition of cultured cell growth by vascular endothelial cadherin (cadherin-5/VE-cadherin). J Clin Invest 98: 886-893
35. Navarro P, Caveda L, Breviario F et al. (1995) Catenin-dependent and -independent functions of vascular endothelial cadherin. J Biol Chem 270: 30965-30972
36. Vittet D, Buchou T, Schweitzer A et al. (1997) Targeted null-mutation in the vascular endothelial-cadherin gene impairs the organization of vascular-like structures in embryoid bodies. Proc Natl Acad Sci USA 94: 6273-6278
37. Esser S, Lampugnani MG, Corada M et al. (1998) Vascular endothelial growth factor induces VE-cadherin tyrosine phosphorylation in endothelial cells. J Cell Sci 111: 1853-1865
38. Martin-Padura I, De Castellarnau C, Uccini S et al. (1995) Expression of VE (vascular endothelial)-cadherin and other endothelial-specific markers in haemangiomas. J Pathol 175: 51-57
39. Zhou Y, Fisher SJ, Janatpour M et al. (1997) Human cytotrophoblasts adopt a vascular phenotype as they differentiate. A strategy for successful endovascular invasion? J Clin Invest 99: 2139-2151
40. Heimark RL, Degner M, Schwartz SM (1990) Identification of a Ca2(+)-dependent cell-cell adhesion molecule in endothelial cells. J Cell Biol 110: 1745-1756
41. Alexander JS, Blaschuk OW, Haselton FR (1993) An N-cadherin-like protein contributes to solute barrier maintenance in cultured endothelium. J Cell Physiol 156: 610-618
42. Liaw CW, Cannon C, Power MD et al. (1990) Identification and cloning of two species of cadherins in bovine endothelial cells. EMBO J 9: 2701-2708
43. Salomon D, Ayalon O, Patel King R et al. (1992) Extrajunctional distribution of N-cadherin in cultured human endothelial cells. J Cell Sci 102: 7-17
44. Rubin LL (1992) Endothelial cells: Adhesion and tight junctions. Curr Opin Cell Biol 4: 830-833
45. Hatta K, Takagi S, Fujisawa H, Takeichi M (1987) Spatial and temporal expression pattern of N-cadherin cell adhesion molecules correlated with morphogenetic processes of chicken embryos. Dev Biol 120: 215-227

46. Nose A, Takeichi M (1986) A novel cadherin cell adhesion molecule: its expression patterns associated with implantation and organogenesis of mouse embryos. J Cell Biol 103: 2649-2658
47. Larue L, Ohsugi M, Hirchenhain J, Kemler R (1994) E-cadherin null mutant embryos fail to form a trophectoderm epithelium. Proc Natl Acad Sci USA 91: 8263-8267
48. Radice GL, Ferreira-Cornwell MC, Robinson SD et al. (1997) Precocious mammary gland development in P-cadherin-deficient mice. J Cell Biol 139: 1025-1032
49. Radice GL, Rayburn H, Matsunami H et al. (1997) Developmental defects in mouse embryos lacking N-cadherin. Dev Biol 181: 64-78
50. Folkman J, D'Amore PA (1996) Blood vessel formation: What is its molecular basis? Cell 87: 1153-1155
51. Torres M, Stoykova A, Huber O et al. (1997) An alpha-E-catenin gene trap mutation defines its function in preimplantation development. Proc Natl Acad Sci USA 94: 901-906
52. Haegel H, Larue L, Ohsugi M et al. (1995) Lack of beta-catenin affects mouse development at gastrulation. Development 121: 3529-3537
53. Ruiz P, Brinkmann V, Ledermann B et al. (1996) Targeted mutation of plakoglobin in mice reveals essential functions of desmosomes in the embryonic heart. J Cell Biol 135: 215-225
54. Balda MS, Anderson JM (1993) Two classes of tight junctions are revealed by ZO-1 isoforms. Am J Physiol 264: C918-C924
55. Furuse M, Hirase T, Itoh M et al. (1993) Occludin: a novel integral membrane protein localizing at tight junctions. J Cell Biol 123: 1777-1788
56. Wong V, Gumbiner BM (1997) A synthetic peptide corresponding to the extracellular domain of occludin perturbs the tight junction permeability barrier. J Cell Biol 136: 399-409
57. Furuse M, Fujita K, Hiiragi T et al. (1998) Claudin-1 and -2: novel integral membrane proteins localizing at tight junctions with no sequence similarity to occludin. J Cell Biol 141:1539-1550
58. Furuse M, Itoh M, Hirase T et al. (1994) Direct association of occludin with ZO-1 and its possible involvement in the localization of occludin at tight junction. J Cell Biol 127: 1617-1626
59. Stevenson BR, Siciliano JD, Mooseker MS, Goodenough DA (1986) Identification of ZO-1: a high molecular weight polypeptide associated with the tight junction (zonula occludens) in a variety of epithelia. J Cell Biol 103: 755-766
60. Anderson JM, Stevenson BR, Jesaitis LA et al. (1988) Characterization of ZO-1, a protein component of the tight junction from mouse liver and Madin-Darby canine kidney cells. J Cell Biol 106: 1141-1149
61. Itoh M, Yonemura S, Nagafuchi A, Tsukita S (1991) A 220-kD undercoat-constitutive protein: its specific localization at cadherin-based cell-cell adhesion sites. J Cell Biol 115: 1449-1462
62. Fanning A, Jameson BT, Anderson JM (1996) Molecular interactions among the tight junction proteins ZO-1, ZO-2 and occludin. Mol Biol Cell 7: 3530 (abstract)
63. Gumbiner B, Lowenkopf T, Apatira D (1991) Identification of a 160-kDa polypeptide that binds to the tight junction protein ZO-1. Proc Natl Acad Sci USA 88: 3460-3464
64. Kim SK (1995) Tight junctions, membrane-associated guanylate kinases and cell signaling. Curr Opin Cell Biol 7: 641-649
65. Howarth AG, Hughes MR, Stevenson BR (1992) Detection of the tight junction-associated protein ZO-1 in astrocytes and other nonepithelial cell types. Am J Pathol 262: C461-C469

66. Rajasekaran AK, Hojo M, Huima T, Rodriguez Boulan E (1996) Catenins and zonula occludens-1 form a complex during early stages in the assembly of tight junctions. J Cell Biol 132: 451-463
67. Citi S, Sabanay H, Jakes R et al. (1988) Cingulin, a new peripheral component of tight junctions. Nature 333: 272-276
68. Zhong Y, Saitoh T, Minase T et al. (1993) Monoclonal antibody 7H6 reacts with a novel tight junction-associated protein distinct from ZO-1, cingulin and ZO-2. J Cell Biol 120: 477-483
69. Keon BH, Schafer SS, Kuhn C et al. (1996) Symplekin, a novel type of tight junction plaque protein. J Cell Biol 134: 1003-1018
70. McCarthy KM, Skare IB, Stankewich MC et al. (1996) Occludin is a functional component of the tight junction. J Cell Sci 109: 2287-2298
71. Staddon JM, Saitou M, Furuse M et al. (1996) Occludin in endothelial cells. Mol Biol Cell 7: 3520 (abstract)
72. Li CX, Poznansky MJ (1990) Characterization of the ZO-1 protein in endothelial and other cell lines. J Cell Sci 97: 231-237
73. Gardner TW, Lesher T, Khin S et al. (1996) Histamine reduces ZO-1 tight-junction protein expression in cultured retinal microvascular endothelial cells. Biochem J 320: 717-721
74. Staddon JM, Herrenknecht K, Smales C, Rubin LL (1995) Evidence that tyrosine phosphorylation may increase tight junction permeability. J Cell Sci 108: 609-619
75. Satoh H, Zhong Y, Isomura H et al. (1996) Localization of 7H6 tight junction-associated antigen along the cell border of vascular endothelial cells correlates with paracellular barrier function against ions, large molecules, and cancer cells. Exp Cell Res 222: 269-274
76. Gottardi CJ, Arpin M, Fanning AS, Louvard D (1996) The junction-associated protein, zonula occludens-1, localizes to the nucleus before the maturation and during the remodeling of cell-cell contacts. Proc Natl Acad Sci USA 93: 10779-10784
77. Valiron O, Chevrier V, Usson Y et al. (1996) Desmoplakin expression and organization at human umbilical vein endothelial cell-to-cell junctions. J Cell Sci 109: 2141-2149
78. Schmelz M, Franke WW (1993) Complexus adhaerentes, a new group of desmoplakin-containing junctions in endothelial cells: the syndesmos connecting retothelial cells of lymph nodes. Eur J Cell Biol 61: 274-289

Chapter 7

Chemokines and Chemokine Receptors

A. Mantovani[1,2], P. Allavena[1], C. Garlanda[1], S. Ramponi[1], C. Paganini[1], A. Vecchi[1], S. Sozzani[1]

Introduction

Classic chemoattractants include complement components, formyl peptides and leukotriene B4. In addition, various cytokines are able to elicit directional migration of leukocytes. While molecules such as monocyte-colony stimulating factor or tumor necrosis factor also exert chemotactic activity, the main chemotactic cytokines are a superfamily of molecules known as chemokines (for chemotactic cytokines) [1].

Chemokines play a central role in the multistep process of leukocyte recruitment and are involved in a variety of disease processes including inflammation, allergy and neoplasia. As such they now represent a prime target for the development of novel therapeutic strategies. Chemical antagonists remain the holy grail for the future.

Leukocyte recruitment has been recognized as an early event in inflammatory processes since the late nineteenth century. Accumulation and trafficking of leukocytes in tissues under physiological and pathological conditions are orderly (typically neutrophils preceed mononuclear cells) and selective in that in certain states one or more leukocyte subset is recruited preferentially (e.g. eosinophils in allergy). The current paradigm of recruitment is that of a multistep process involving the action of chemotactic signals [2, 3].

Here we will concisely summarize the main features of chemokines and their receptors and focus on emerging evidence of how regulation of receptor expression is a crucial determinant of the function of the chemokine system. Emphasis will be on myelomonocytic and dendritic cells as well as on emerging implications for polarized T helper 1 (Th1) versus Th2 responses.

Chemokines are a superfamily of small proteins which play a crucial role in immune and inflammatory reactions and in viral infection [4-6]. Most chemokines cause chemotactic migration of leukocytes, but these molecules also affect angiogenesis, collagen production and the proliferation of hematopoietic precursors. Based on the cysteine motifs CXC, CC, C and CX3C, four chemokine families have been identified (Fig. 1). The chemokine scaffold consists of an N-

[1] Mario Negri Institute of Pharmacological Research, Via Eritrea 62 - 20157 Milan, Italy. e-mail: mantovani@irfmn.mnegri.it
[2] Department of Biotechnology, Section of General Pathology, University of Brescia, Italy

Chemokine family (Chromosome)	Proteic structure	Prototypic molecules	Cellular targets	Receptor
CXC (4, 10)	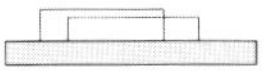	IL-8 IP-10, SDF-1 BCA-1	Neu T, NK B	CXCR1,2 CXCR3,4 CXCR5
CC (17, 16, 9, 2)	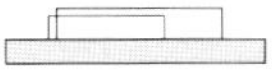	MCPs, MIPs	Mo, DC, T, B, NK, Eo, Ba	CCR1-8/9
C (1)		Lymphotactin	T, NK	Unknown
CX3C (16)	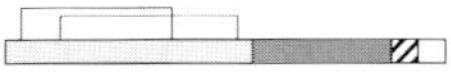	Fractalkine	T, NK, Mo	CX3CR1

Fig. 1. Four chemokine families, according to cysteine motifs. All chemokine are composed of a homologous, disulfide-bonded structure (*light gray bar* with one or two cross-links indicating approximate cysteine position). In addition, lymphotactin and fractalkine contain a carboxyl-terminal mucin-like domain (*dark gray bar*). Fractalkine, a membrane-associated molecule, also possesses a transmembrane domain (*striped bar*) and a carboxyl-terminal cytoplasmic domain (*open bar*). Gene locations refer to human chromosomes. *Mo*, Monocytes; *Neu*, Neutrophils, *Eo*, Eosinophils; *Ba*, Basophils; *DC*, Dentdritic cells

terminal loop connected via disulfide bonds to the more structured core of the molecule (3 β sheets) and a C-terminal α helix. About 50 human chemokines have been identified.

The major eponymous function of chemokines is chemotaxis for leukocytes. Schematically (Fig. 1), CXC (or α) chemokines are active on neutrophils (PMN) and lymphocytes. CC (or β) chemokines (e.g. monocyte chemotactic proteins (MCPs) and macrophage inflammatory proteins (MIPs)) exert their action on multiple leukocyte subtypes, including monocytes, basophils, eosinophils, T lymphocytes, dendritic cells and natural killer (NK) cells, but they are generally inactive on PMNs. Eotaxins (CC), the chemokines with the most restricted spectrum of action, are selectively active on eosinophilic and basophilic granulocytes [4, 7-9]. Lymphotactin and fractalkine are the only proteins so far described with the C or CX3C motif, respectively [10-12]. These molecules both act on lymphoid cells (T lymphocytes and NK cells), while fractalkine is also active on monocytes [10-13] (S. Sozzani, P. Allavena, T.N. Wells, unpublished results).

Chemokines are redundant in their action on target cells. No chemokine is active only on one leukocyte population, and usually a given leukocyte population has receptors for and responds to different molecules (Table 1). Interestingly, mononuclear phagocytes, the most evolutionary ancient cell type of innate immunity, respond to the widest range of chemokines. These include most CC chemokines, fractalkine (CX3C) and certain CXC molecules (e.g. SDF-1 and, under certain conditions, interleukin (IL)-8).

Table 1. Expression of chemokine receptors in human leukocyte populations

Receptor	Main ligands	Main cells
CCR1	MCP-3, RANTES, MIP-1α	Mo, T, NK, DC, PMN
CCR2 B/A	MCPs (1-4)	Mo, T(act), NK (act)
CCR3	Eotaxin, MCP-3, RANTES	Eo, Ba, T (Th2)
CCR4	TARC, MDC	T (Th2,Tc2), NK, DC
CCR5	MIP-1β, MIP-1α, RANTES	Mo, T (Th1, Tc1), DC
CCR6	MIP-3α/LARC/Exodus	T, CD34-DC
CCR7	ELC	T, Mo
CCR8	I309, TARC	T (Th2), Mo
CCR9	MCPs, RANTES, MIP-1α, MIP-1β	?
CXCR1	IL-8	PMN
CXCR2	IL-8, gro, NAP-2	PMN
CXCR3	IP-10, MIG	T (Th1)
CXCR4	SDF-1	Widely expressed
CXCR5	BCA-1	B
CX3CR1	Fractalkine	Mo, NK, T

Mo, Monocytes; *DC*, Dendritic cells; *CD34-DC*, DC-derived from CD34 cells in vitro; *PMN*, Neutrophils; *Eo*, Eosinophils; *Ba*, Basophils; *Th*, T helper; *Tc*, T cytotoxic; (*act*), activated; *TARC*, Thymus and activation-regulated chemokine; *ELC*, EBI1-ligand chemokine; *NAP-2*, Neutrophil activating protein-2; *BCA-1*, β cell attractant-1.

Receptors and Signal Transduction

All chemokine receptors identified to date are seven transmembrane domain proteins coupled to GTP-binding proteins with homology to the family of chemotactic receptors. Five receptors for CXC chemokines (CXCR1-5) and nine for CC chemokines (CCR1-9) have recently been cloned (Table 1). These receptors show a promiscuous pattern of ligand recognition and are differentially expressed and regulated in leukocytes [4, 6, 14-16]. The receptor for fractalkine has been recently characterized [17], while that for lymphotactin is still unknown.

Activation of chemotactic receptors results in an increase of intracellular free calcium concentration and in the activation of phospholipases C, D and A2. Calcium fluxes were observed after receptor binding by most of the CC chemokines [8,18-21]. This response is rapid, transient and sensitive to *Bordetella pertussis* toxin (PTox) [22-26], suggesting that chemokine receptors are associated with PTox-sensitive GTP-binding proteins. In support of this observation, it was reported that monocyte chemotaxis in response to MCP-1, MCP-3, RANTES, and MIP-1α is inhibited in a concentration-dependent manner by PTox [22, 27, 28], while in the same experimental conditions cholera toxin was ineffective [22,28]. Studies performed in transfected cells have shown that both CXCR1 and CXCR2, as well as CCR1 and CCR2, can efficiently couple with PTox-insensitive G proteins, such as Gα14 and Gα16 [29,30]. The biological relevance of this finding for circulating leukocytes is still unclear.

In human monocytes activated by MCP-1, the influx of Ca^{2+} across the plasma membrane rather than the release from intracellular stores appeared to be the main mechanism responsible for intracellular Ca^{2+} elevation [27, 28]. Calcium influx was required for arachidonate accumulation by MCP-1 in human monocytes [31]. Bioactive products of arachidonic acid metabolism (platelet-activating factors (PAF) and 5-oxo-ETE) increased in a synergistic fashion both arachidonic acid release and chemotactic response induced by CC chemokines but not by formyl-Meth-Leu-Phe (fMLP) [31, 32]. Alternatively, phospholipase A2 (PLA2) inhibitors blocked both monocyte polarization and chemotaxis [31]. An antisense oligonucleotide for cellular PLA2 (cPLA2) induced a concentration-dependent inhibition of monocyte chemotaxis to all the CC chemokines tested (MCP-1, MCP-2, RANTES and MIP-1α). On the contrary, chemotactic response to two "classic" chemotactic factors, fMLP and C5a, was not affected by this treatment [33]. This finding shows that cPLA2 is indeed an important effector enzyme for chemokine-induced monocyte migration. Recent findings have shown that both CC and CXC chemokines activate Pyk2/RAFTK tyrosine kinase and the Ras/Raf/MAP kinase pathways [34-36]. It is important to note that MAP kinases (ERKs, p38 and JNK) are upstream of cPLA2 activation, and inhibitors of MAP kinases reduced the chemotactic response to MCP-1 [35]. There is also evidence for a role of phosphatidylinositol 3-kinase (PI3K) in chemokine receptor signal transduction [37]. PI3K is activated by interaction with βγ subunits of heterotrimeric G proteins or by low molecular weight G proteins. Using CXCR1-transfected pre-B cells, IL-8 was shown to activate Rho, a low molecular weight G protein, and this effect was implicated in IL-8-induced adhesion to fibrinogen [38].

Regulation of Receptor Expression During Activation and Deactivation of Mononuclear Phagocytes

Leukocyte infiltration into tissues is regulated by local production of chemotactic signals. Chemokine receptors are expressed on different types of leukocytes. Some receptors are restricted to certain cells (e.g. CXCR1 on PMN, and CCR3 on eosinophils and basophils), while others, such as CCR1 and CCR2, are expressed on different types of leukocytes. In addition, chemokine receptors are constitutively expressed on some cells, whereas they are inducible in others. Regulation of chemokine receptors is emerging as an alternative mechanism to control the level and specificity of leukocyte migration. IL-2-activated, but not resting, T lymphocytes and NK cells migrate in response to MCPs. CXCR3 is expressed only in IL-2-activated T lymphocytes, and IL-2 upregulates CCR6 expression [4]. In PMNs, IL-8 receptors can be upregulated or downregulated by granulocyte-colony stimulating factor (G-CSF) and lipopolysaccharide (LPS), respectively [39].

It was recently described that inflammatory and anti-inflammatory agonists regulated in opposite ways CC chemokine receptor expression in human monocytes. LPS and other microbial agents caused a rapid and drastic reduction of CCR2 mRNA levels. The rate of nuclear transcription of CCR2 was not affected by

LPS, whereas the mRNA half-life was reduced. The effect was drastic and rapid with an ED_{50} of approximately 1 ng/ml, and the half-maximal effect was reached with an optimal dose in approximately 45 min. Inhibition of MCP-1 receptor expression was functionally relevant since LPS-treated monocytes showed a reduced capacity to bind and respond to MCP-1 chemotactically. The action of LPS on CC chemokine receptors was specific in that CXCR2 was unaffected [40]. Interferon gamma (IFN-γ), a potent activator of mononuclear phagocytes, also inhibited CCR2 expression in a rapid (1 h) and selective manner by reducing the half-life the CCR2 mRNA. This effect was synergistic with the action of other proinflammatory molecules, such as IL-1, tumor necrosis factor (TNF) and LPS [41].

Conversely, incubation of human monocytes with IL-10 increased the expression of CCR1, 2 and 5 as evaluated by northern blot analysis. No major variations in the expression of CXCR2 were detectable while CXCR4 mRNA levels were reduced [42]. The effect of IL-10 was concentration-dependent (EC_{50}=0.3 ng/ml) and fast, already detectable after 30 min and reaching a plateau after 2 h of stimulation. The estimated half-life of CCR5 mRNA was doubled after exposure to IL-10. In contrast, the rate of nuclear transcription of the gene, as investigated by nuclear run-off analysis, was not affected. Accordingly, IL-10-treated monocytes responded better to CC chemokines in terms of chemotactic migration and intracellular calcium transients and the effect was best observed when suboptimal agonist concentrations were used. IL-10-treated monocytes were also more easily infected by the macrophage-tropic HIV strain BAL [42]. This result is consistent with the use of CCR5 as major fusion coreceptor by BAL.

Regulation of receptor expression, in addition to agonist production, is likely a crucial point in modulating of the chemokine system. An emerging paradigm indicates that at least some proinflammatory and anti-inflammatory molecules exert reciprocal and opposing influences on chemokine agonist production and receptor expression. We speculate that the divergent effect of certain proinflammatory signals on agonist versus receptor expression may serve to retain mononuclear phagocytes at sites of inflammation, to prevent their reverse transmigration, and possibly, to limit excessive recruitment.

Receptor Expression in Immature and Mature Dendritic Cells

Dendritic cells (DC) have the unique capacity to initiate primary and secondary immune responses. DC take up antigens in peripheral tissues and migrate to lymphoid organs where they present processed peptides to T cells. During migration, DC undergo maturation from a "processing" to a "presenting" functional phenotype, characterized by the expression of costimulatory molecules, cytokine production, and the ability to stimulate T cell proliferation. For their central role in the regulation of immunity, DC are considered interesting tools and targets for immunotherapeutic interventions.

Response of DC to chemokines has been extensively characterized in vitro. Classic chemotactic agonists (e.g. formylated peptides), C5a, the CC chemokines

MCP-3, MCP-4, RANTES, MIP-1α, MIP-1β, and MIP-5, and the CXC chemokine SDF-1 induce directional migration and calcium fluxes of monocyte-derived DC (mono-DC) and CD34$^+$ cell-derived DC (CD34-DC) [43, 44]. On the contrary, three CXC chemokines – IL-8, Gro and interferon-γ-induced protein-10 (IP-10) – eotaxin (CC), and lymphotactin (C) were inactive. PAF, a weak chemotactic factor for neutrophils and monocytes, efficiently induced chemotaxis and calcium transients in both mono-DC and CD34-DC [45]. MDC (macrophage-derived chemokine), a new CC chemokine, preferentially activated DC, being two orders of magnitude more potent on DC than on mononuclear cells [46]. Mono-DC express high levels of mRNA for CCR1, CCR2 and CCR5 receptors. Among the CXC chemokine receptors investigated, CXCR1, CXCR2, and CXCR4 were also found to be expressed by DC [44]. A similar pattern of chemokine receptor expression was observed in CD34-DC (S. Sozzani, P. Allavena, unpublished results), with CCR6 being the only exception. This receptor is selectively expressed in CD34-DC that migrate in response to MIP-3α/LARC/Exodus, the CCR6 ligand [47]. Active chemokines increase the adhesion of mono-DC to endothelial cells and promote their transmigration across the endothelial cell monolayer. The latter response is inhibited by blocking antibodies to β1-integrins and by anti-CD31 [48]. On the contrary, brief exposure of mono-DC to chemokines did not affect their ability to take up fluorescein-labelled dextran or to promote allogenic mixed lymphocyte reaction [44]. In vivo intradermal administration of GM-CSF led to an increased number of DC within the human dermis. TNF, and possibly other LPS-induced cytokines, quickly recruited DC in the airway epithelia in a model of respiratory infection, and systemic administration of LPS induced a profound loss of major histocompatibility complex (MHC) class II$^+$ cells from heart and kidney in the mouse [49]. Since these proinflammatory agonists are inactive in promoting cell migration in vitro, it is likely that the observed DC mobilization observed in vivo is the result of secondary mediators, such as chemokines [44]. Recent results indicate that maturation differentially affects expression of chemokine receptors in DC [50].

Finally, DC in the mucosal epithelium are likely the initial target for HIV-1. Following migration to the nearest lymphoid station, DC likely contribute to viral spread. DC generated in vitro, as well as Langherans cells, express and can be infected through CCR5 and CXCR4, the two major fusion coreceptors for HIV macrophage and T-cell tropic strains, respectively [51, 52].

Chemokines in Polarized Th1 and Th2 Responses

T cells can be subdivided into polarized type I and type II cells, depending on the spectrum of cytokines which they produce. T helper 1 cells (Th1 cells) are characterized by production of TNF and IFN-γ and promote immunity based on macrophage activation and effector functions. At the other extreme of the spectrum, Th2 cells are characterized by IL-4 and IL-5 production and elicit immune responses based on the effector function of mastocytes and eosinophils. The latter cell types are typically involved in allergic inflammation.

Recent results indicate that chemokines are part of the Th1 and Th2 paradigm. It was found that polarized Th1 and Th2 populations differentially express chemokine receptors. In particular, Th1 cells characteristically express high levels of CCR5 and CXCR3 whereas Th2 cells express CCR4, CCR8 and to a lesser degree, CCR3. In accordance with receptor expression, polarized Th1/Th2 cells (as well as $CD8^+$ T cells with a similar cytokine profile, unpublished observations), differentially respond to appropriate agonists for these receptors, including MIP-1β and interferon-γ-induced protein-10 (IP-10) for Th1 cells and MDC, I 309 and eotaxin for Th2 cells [53, 54]. Production of IP-10 and similar CXCR3 agonists such as ITAC is induced by IFN-γ. Conversely, production of eotaxin and MDC is induced by IL-4 and IL-13, typical Th2 cytokines. Thus, chemokines are probably an essential part of an amplification circuit of polarized Th1 and Th2 responses. Consistent with this view, the CCR4 agonist MDC was recently shown to be induced by IL-4 and IL-3 and to be inhibited by IFN-γ [55, 56].

References

1. Mantovani A, Allavena P, Vecchi A, Sozzani S (1998) Chemokines and chemokine receptors during activation and deactivation of monocytes and dendritic cells and in amplification of Th1 versus Th2 responses. Int J Clin Lab Res 28: 77-82
2. Butcher EC (1991)Leukocyte-endothelial cell recognition: Three (or more) steps to specificity and diversity. Cell 67: 1033-1036
3. Mantovani A, Bussolino F, Dejana E (1992) Cytokine regulation of endothelial cell function. FASEB J 6: 2591-2599
4. Baggiolini M, Dewald B, Moser B (1997) Human chemokines: An update. Annu Rev Immunol 15: 675-705
5. Hedrick JA, Zlotnik A (1996) Chemokines and lymphocyte biology. Curr Opin Immunol 8: 343-347
6. Rollins BJ (1997) Chemokines. Blood 90: 909-928
7. Ben-Baruch A, Michiel DF, Oppenheim JJ (1995) Signals and receptors involved in recruitment of inflammatory cells. J Biol Chem 270: 11703-11706
8. Schall TJ (1994) The chemokines. In: Thomson A (ed) The cytokine handbook. Academic Press, London, pp 419-460
9. Sozzani S, Locati M, Allavena P et al. (1996) Chemokines: A superfamily of chemotactic cytokines. Int J Clin Lab Res 26: 69-82
10. Kelner GS, Kennedy J, Bacon KB et al. (1994) Lymphotactin: A cytokine that represents a new class of chemokine. Science 266: 1395-1399
11. Bazan JF, Bacon KB, Hardiman G et al. (1997) A new class of membrane-bound chemokine with a CX3C motif. Nature 385: 640-644
12. Pan Y, Lloyd C, Zhou H et al. (1997) Neurotactin, a membrane-anchored chemokine upregulated in brain inflammation. Nature 387: 611-617
13. Bianchi G, Sozzani S, Zlotnik A et al. (1996) Migratory response of human NK cells to lymphotactin. Eur J Immunol 26: 3238-3241
14. Izumi S, Hirai K, Miyamasu M et al. (1997) Expression and regulation of monocyte chemoattractant protein-1 by human eosinophils. Eur J Immunol 27: 816-824
15. Legler DF, Loetscher M, Roos RS et al. (1998) B cell-attracting chemokine 1, a human CXC chemokine expressed in lymphoid tissues, selectively attracts B lymphocytes via BLR1/CXCR5. J Exp Med 187: 655-660

16. Nibbs RJB, Wylie SM, Yang JY et al. (1997) Cloning and characterization of a novel promiscuous human beta-chemokine receptor D6. J Biol Chem 272: 32078-32083
17. Imai T, Hieshima K, Haskell C et al. (1997) Identification and molecular characterization of fractalkine receptor CX3CR1, which mediates both leukocyte migration and adhesion. Cell 91: 521-530
18. Oppenheim JJ, Zachariae CO, Mukaida N, Matsushima K (1991) Properties of the novel proinflammatory supergene "intercrine" cytokine family. Annu Rev Immunol 9: 617-648
19. Baggiolini M, Dewald B, Moser B (1994) Interleukin-8 and related chemotactic cytokines - CXC and CC chemokines. Adv Immunol 55: 99-179
20. Miller MD, Krangel MS (1992) Biology and biochemistry of the chemokines: a family of chemotactic and inflammatory cytokines. Crit Rev Immunol 12: 17-46
21. Sozzani S, Locati M, Zhou D et al. (1995) Receptors, signal transduction and spectrum of action of monocyte chemotactic protein-1 and related chemokines. J Leukoc Biol 57: 788-794
22. Sozzani S, Luini W, Molino M et al. (1991) The signal transduction pathway involved in the migration induced by a monocyte chemotactic cytokine. J Immunol 147: 2215-2221
23. McColl SR, Hachicha M, Levasseur S et al. (1993) Uncoupling of early signal transduction events from effector function in human peripheral blood neutrophils in response to recombinant macrophage inflammatory proteins-1 alpha and -1 beta. J Immunol 150: 4550-4560
24. Heinrich JN, Ryseck RP, Macdonald-Bravo H, Bravo R (1993) The product of a novel growth factor-activated gene, fic, is a biologically active C-C-type cytokine. Mol Cell Biol 13: 2020-2030
25. Bischoff SC, Krieger M, Brunner T, Dahinden CA (1992) Monocyte chemotactic protein 1 is a potent activator of human basophils. J Exp Med 175: 1271-1275
26. Myers SJ, Wong LM, Charo IF (1995) Signal transduction and ligand specificity of the human monocyte chemoattractant protein-1 receptor in transfected embryonic kidney cells. J Biol Chem 270: 5786-5792
27. Sozzani S, Molino M, Locati M et al. (1993) Receptor-activated calcium influx in human monocytes exposed to monocyte chemotactic protein-1 and related cytokines. J Immunol 150: 1544-1553
28. Sozzani S, Zhou D, Locati M et al. (1994) Receptors and transduction pathways for monocyte chemotactic protein-2 and monocyte chemotactic protein-3 - Similarities and differences with MCP-1. J Immunol 152: 3615-3622
29. Wu D, LaRosa GJ, Simon MI (1993) G protein-coupled signal transduction pathways for interleukin-8. Science 261: 101-103
30. Kuang YN, Wu YP, Jiang HP, Wu DQ (1996) Selective G protein coupling by C-C chemokine receptors. J Biol Chem 271: 3975-3978
31. Locati M, Zhou D, Luini W et al. (1994) Rapid induction of arachidonic acid release by monocyte chemotactic protein-1 and related chemokines - Role of Ca^{2+} influx, synergism with platelet-activating factor and significance for chemotaxis. J Biol Chem 269: 4746-4753
32. Sozzani S, Rieppi M, Locati M et al. (1994) Synergism between platelet activating factor and C-C chemokines for arachidonate release in human monocyte. Biochem Biophys Res Commun 199: 761-766
33. Locati M, Lamorte G, Luini W et al. (1996) Inhibition of monocyte chemotaxis to C-C chemokines by antisense oligonucleotide for cytosolic phospholipase A2. J Biol Chem 271: 6010-6016

34. Knall C, Young S, Nick JA et al. (1996) Interleukin-8 regulation of the Ras/Raf/mitogen-activated protein kinase pathway in human neutrophils. J Biol Chem 271: 2832-2838
35. Yen HH, Zhang YJ, Penfold S, Rollins BJ (1997) MCP-1-mediated chemotaxis requires activation of non-overlapping signal transduction pathways. J Leukocyte Biol 61: 529-532
36. Davis CB, Dikic I, Unutmaz D et al. (1997) Signal transduction due to HIV-1 evelope interactions with chemokine receptors CXCR4 or CCR5. J Exp Med 186: 1793-1798
37. Turner L, Ward SG, Westwick J (1995) RANTES-activated human T lymphocytes - A role for phosphoinositide 3-kinase. J Immunol 155: 2437-2444
38. Laudanna C, Campbell JJ, Butcher EC (1996) Role of rho in chemoattractant-activated leukocyte adhesion through integrins. Science 271: 981-983
39. Lloyd AR, Biragyn A, Johnston JA et al. (1995) Granulocyte-colony stimulating factor and lipopolysaccharide regulate the expression of interleukin 8 receptors on polymorphonuclear leukocytes. J Biol Chem 270: 28188-28192
40. Sica A, Saccani A, Borsatti A et al. (1997) Bacterial lipopolysaccharide rapidly inhibits expression of C-C chemokine receptors in human monocytes. J Exp Med 185: 969-974
41. Penton-Rol G, Polentarutti N, Luini W et al. (1998) Selective inhibition of expression of the chemokine receptor CCR2 in human monocytes by IFN-γ. J Immunol 160: 3869-3873
42. Sozzani S, Ghezzi S, Iannolo G et al. (1998) Interleukin-10 increases CCR5 expression and HIV infection in human monocytes. J Exp Med 187: 439-444
43. Sozzani S, Sallusto F, Luini W et al. (1995) Migration of dendritic cells in response to formyl peptides, C5a and a distinct set of chemokines. J Immunol 155: 3292-3295
44. Sozzani S, Luini W, Borsatti A et al. (1997) Receptor expression and responsiveness of human dendritic cells to a defined set of CC and CXC chemokines. J Immunol 159: 1993-2000
45. Sozzani S, Longoni D, Bonecchi R et al. (1997) Human monocyte-derived and CD34+ cell-derived dendritic cells express functional receptors for platelet activating factor. FEBS Lett 418: 98-100
46. Godiska R, Chantry D, Raport CJ et al. (1997) Human macrophage derived chemokine (MDC) a novel chemoattractant for monocytes, monocyte derived dendritic cells, and natural killer cells. J Exp Med 185: 1595-1604
47. Power CA, Church DJ, Meyer A et al. (1997) TNC Cloning and characterization of a specific receptor for the novel CC chemokine MIP-3 alpha from lung dendritic cells. J Exp Med 186: 825-835
48. D'Amico G, Bianchi G, Bernasconi S et al. (1998) Adhesion, transendothelial migration and reverse transmigration of in vitro cultured dendritic cells. Blood 92: 207-214
49. Austyn JM (1996) New insights into the mobilization and phagocytic activity of dendritic cells. J Exp Med 183: 1287-1292
50. Sozzani S, Allavena P, D'Amico G et al. (1998) Differential regulation of chemokine receptors during dendritic cell maturation: a model for their trafficking properties. J Immunol 161: 1083-1086
51. Granelli-Piperno A, Moser B, Pope M et al. (1996) Efficient interaction of HIV-1 with purified dendritic cells via multiple chemokine coreceptors. J Exp Med 184: 2433-2438
52. Zaitseva M, Blauvelt A, Lee S et al. (1997) Expression and function of CCR5 and CXCR4 on human Langerhans cells and macrophages: implication for HIV primary infection. Nature Med 3: 1369-1375
53. Bonecchi R, Bianchi G, Bordignon PP et al. (1998) Differential expression of chemokine receptors and chemotactic responsiveness of type 1 T helper cells (Th1s) and Th2s. J Exp Med 187: 129-134

54. Sallusto F, Mackay CR, Lanzavecchia A (1997) Selective expression of the eotaxin receptor CCR3 by human T helper 2 cells. Science 277: 2005-2007
55. Sozzani S, Luini W, Bianchi G et al. (1998) The viral chemokine macrophage inflammatory protein-II is a selective Th2 chemoattractant. Blood 92: 4036-4039
56. Bonecchi R, Sozzani S, Stine J et al. (1998) Divergent effects of IL-4 and interferon gamma on macrophage-derived chemokine (MDC) production: an amplification circuit of polarized T helper 2 responses. Blood 92: 2668-2671

Chapter 8

Th1/Th2 Cytokine Network

M.M. D'Elios, G. Del Prete

Introduction

The immune system has evolved different defensive mechanisms against pathogens. The first defensive line is provided by "natural" immunity, including phagocytes, T cell receptor (TCR) $\gamma\delta^+$ T cells, natural killer (NK) cells, mast cells, neutrophils and eosinophils, as well as complement components and proinflammatory cytokines, such as interferons (IFNs), interleukin (IL)-1, IL-6, IL-12, IL-18 and tumor necrosis factor (TNF)-α. The more specialized TCR $\alpha\beta^+$ T lymphocytes provide the second defense wall. These cells account for the "specific immunity", which results in specialized types of immune responses which allow vertebrates to recognize and eliminate, or at least control, infectious agents in different body compartments. Viruses growing within infected cells are faced through the killing of their host cells by $CD8^+$ class I MHC-restricted cytotoxic T lymphocytes. Most microbial components are endocytosed by antigen-presenting cells (APC), processed and presented preferentially to $CD4^+$ class II MHC-restricted T helper (Th) cells. $CD4^+$ T cells co-operate with B cells for the production of antibodies which opsonize extracellular microbes and neutralize their exotoxins. This branch of the specific Th cell-mediated immune response is known as "humoral immunity". Other microbes, however, survive within macrophages despite the unfavorable microenvironment. Antigen-activated $CD4^+$ Th cells are required to activate macrophages, whose reactive metabolites and TNF-α finally lead to the destruction of pathogens. This latter branch of the specific Th cell-mediated response is known as "cell-mediated immunity" (CMI).

Most successful immune responses involve both humoral and cell-mediated immunity, but in some conditions the two prototypes of effector reactions tend to be mutually exclusive. The mechanisms by which $CD4^+$ Th cells are responsible for this dichotomy remained unclear until Mosmann et al. [1] provided evidence that repeated antigen stimulation of murine $CD4^+$ Th cells in vitro resulted in the development of polarized patterns of cytokine production (type 1 or Th1 and type 2 or Th2) that could account for effector reactions characterized by prevalent CMI or antibody response, respectively.

Institute of Internal Medicine and Immunoallergology, University of Florence, Viale Morgagni 85 - 50134 Florence, Italy. e-mail: delios@cesit1.unifi.it

Th1/Th2 Cytokine Network

Th1 cells produce IFN-γ, IL-2 and TNF-β, and elicit macrophage activation and delayed-type hypersensitivity (DTH) reactions. Th2 cells produce IL-4, IL-5, IL-6, IL-10 and IL-13, which act as growth/differentiation factors for B cells, eosinophils and mast cells and inhibit several macrophage functions [2, 3] (Table 1). A similar heterogeneity in the cytokine profile was also observed in $CD8^+$ cytotoxic T cells (Tc1, Tc2), γδ T cells and NK cells [4, 5]. However, most T cells do not express a polarized type 1 or type 2 cytokine profile; these T cells (coded as Th0) represent a heterogenous population of partially differentiated effector cells consisting of multiple subsets which secrete different combinations of both Th1 and Th2 cytokines [6-8]. The cytokine response at effector level can remain mixed or further differentiate into the Th1 or Th2 pathway under the influence of polarizing signals from the microenvironment. Human Th1 and Th2 cells also differ for their responsiveness to cytokines [9]. Both Th1 and Th2 cells proliferate in response to IL-2, but Th2 are more responsive to IL-4 than are Th1; on the other hand IFN-γ tends to inhibit the proliferative response of Th2 cells [10].

Table 1. Functional properties of human Th1 and Th2 cells

Function	Th1	Th2
Cytokine production		
IFN-γ	+++	–
TNF-β	+++	–
IL-2	+++	+
TNF-α	+++	+
GM-CSF	++	++
IL-3	++	+++
IL-6	+	+++
IL-13	+	+++
IL-10	+	+++
IL-4	–	+++
IL-5	–	+++
Cytolytic activity	+++	–
Help to B cells for Ig synthesis		
IgM, IgG, IgA (low T/B ratio)	+++	++
IgM, IgG, IgA (high T/B ratio)	–	+++
IgE	–	+++
Interaction with monocytes		
Procoagulant activity (PCA)	+++	–
Tissue factor production	+++	–
Delayed-type hypersensitivity (DTH)	+++	–
Inhibition of PCA and DTH	–	+++

Table 2. Signals involved in Th1 or Th2 development

Signals	Th1	Th2
Cytokines		
IL-12	++++	–
IFN-γ, IFN-α	+++	
IL-18	++	–
IL-4	–	++++
IL-11, PGE$_2$	–	+++
IL-1, IL-10	–	++
Physical form of the antigen		
Corpuscular	+	–
Soluble	+	+
Dose of the antigen		
High	–	+
Medium/low	+	–
Very low	–	+
Adjuvants		
Complete Freund's	+	–
Alum	–	+
B. pertussis or *V. cholera* toxin	–	+
Antigen-presenting cells		
Professional (macrophages, dendritic, B cells)	+	+
Keratinocytes	–	+
Costimulatory molecules		
B7.1-CD28	+	+
B7.2-CD28	–	+
CD30-CD30L	–	+

Th1 and Th2 cells substantially differ for their cytolytic potential and mode of help for B-cell antibody synthesis (Table 2). Th2 clones, usually devoid of cytolytic activity, induce IgM, IgG, IgA, and IgE synthesis by autologous B cells in the presence of the specific antigen, with a response which is proportional to the number of Th2 cells added to B cells [11]. In contrast, Th1 clones, most of which are cytolytic, provide B-cell help for IgM, IgG, IgA (but not IgE) synthesis at low T-cell/B-cell ratios. At T-cell/B-cell ratios higher than 1/1, there is a decline in B-cell help related to the Th1-mediated lytic activity against antigen-presenting autologous B cells [11]. This may represent an important mechanism for downregulating antibody responses in vivo as well. Th1 and Th2 cells exhibit different abilities to activate monocytic cells [12]. Th1, but not Th2, help monocytes to express tissue factor (TF) production and procoagulant activity (Table 1). In this type of Th cell-monocyte cooperation, both cell-to-cell contact with activated T cells and Th1 cytokines (namely IFN-γ), are required for optimal TF synthesis, whereas Th2-derived cytokines (IL-4, IL-10 and IL-13) are strongly inhibitory [12].

Polarized human Th1 and Th2 cells also exhibit preferential expression of some activation molecules or chemokine/cytokine receptors. CD30 (a member of the TNF-receptor family) is mainly expressed by IL-4-producing Th0/Th2 cells both in vitro [13, 14] and in vivo [15], whereas lymphocyte activation gene (LAG)-3, a member of the immunoglobulin superfamily, is preferentially expressed by IFN-γ producing Th0/Th1-like cells [16]. LAG-3 expression and release is upregulated by IFN-γ and downregulated by IL-4, whereas CD30 expression/release, which is strictly dependent on IL-4, is lost by fully differentiated, IL-4-receptor-lacking Th1 cells [17]. The expression of CD26 associates with Th1-induced reaction in granulomatous diseases [18], whereas the expression of the L-selectin CD62L would preferentially associate with the production of Th2 cytokines [19]. Since IFN-γ is transiently expressed on the cell surface during secretion, its detection may be a marker of an ongoing Th1 function [20]. CCR3 (the receptor for eotaxin and MCP-3) and CCR4 (the receptor for TARC and MDC) preferentially associate with Th2 cells [21, 22], whereas CXCR3 (the receptor for IP-10, I-TAC and Mig) and CCR5 (the MIP-1β receptor) are preferentially expressed by Th1 cells [22, 23]. It has recently been shown that the β-chain of the IL-12 receptor is selectively expressed by activated Th1 cells since this protein is induced during the differentiation of naive T cells into the Th1, but not the Th2, pathway [24].

The factors responsible for the Th cell polarization into a predominant Th1 or Th2 profile have been extensively investigated. Current evidence suggests that Th1 and Th2 cells develop from the same Th-cell precursor under the influence of mechanisms associated with antigen presentation [25]. Both environmental and genetic factors influence Th1 or Th2 differentiation mainly by determining the "leader cytokine" in the microenvironment of the responding Th cell (Table 2). IL-4 is the most powerful stimulus for Th2 differentiation, whereas IL-12, IL-18 and IFNs favor Th1 development [26–30]. A role has been demonstrated for the site of antigen presentation, the physical form of the immunogen, the type of adjuvant, and the dose of antigen [31]. Microbial products (particularly from intracellular bacteria) induce Th1-dominated responses because they stimulate IL-12 production. IFN-γ and IFN-α favor Th1 development by enhancing IL-12 secretion by macrophages and maintaining the expression of functional IL-12 receptors on Th cells [32]. On other hand IL-11 and prostaglandin E_2 (PGE_2) would promote Th2 cell polarization [33, 34]. The genetic mechanisms that concur with environmental factors in controlling the type of Th cell differentiation still remain elusive. Recently, a role has been suggested for the T cell phenotype switch-1 (Tps-1) locus on murine chromosome 11, which controls the maintenance of IL-12 responsiveness and therefore the subsequent Th1/Th2 development [35]. Binding of IL-12 to its receptor on naive Th cells results in the triggering of intracellular signals (rapidly transduced into the nucleus), such as the Jak-signal transducer and activator of transcription (STAT)-4 [36]. Other transcription factors specific for Th1 differentiation are Hlx and ERM [37, 38].

On the other hand, signaling by IL-4 occurs through activation of STAT-6; disruption of STAT-6 gene results in deficient Th2 responses [39]. Two other genes involved in the differentiation of Th2 cells have been discovered. These genes code for GATA-3 and c-*maf* transcription factors, which probably regulate the production of Th2-type cytokines (GATA-3) or more selectively (c-*maf*) the production of IL-4 [40, 41].

Th1/Th2 Network in Health and Disease

Th1/Th2 cytokine production naturally occurs during immune responses and can be detected in a variety of infectious or immunopathological human disorders [42] (Table 3). In most human infections, specific immunity is of crucial importance, but an inappropriate response may not only result in lack of protection, but even contribute to the induction of immunopathology. In human leishmaniasis, lack of IFN-γ and high IL-4 production predict progression into fulminant visceral disease, whereas individuals whose cells produce large amounts of IFN-γ usually remain asymptomatic [43, 44]. Strong Th1 response was found in uncomplicated cutaneous leishmaniasis or subclinical infection, whereas individuals cured from visceral leishmaniasis recognized *Leishmania* antigens with both Th1 and Th2 cells [45]. Th1 cytokine mRNA was found in the skin of patients with localized and mucocutaneous leishmaniasis, whereas Th2 cytokine

Table 3. Preferential Th1 or Th2 responses in humans

Th1	Th2
Multiple sclerosis	Systemic sclerosis
Type 1 diabetes mellitus	Helminth infections
Hashimoto's thyroiditis	Allergic diseases
Crohn's disease	Systemic lupus erythematosus
Acute graft rejection	Transplantation tolerance
Unexplained recurrent abortions	Successful pregnancy
Lyme arthritis	Omenn's syndrome
Yersinia-induced arthritis	Primary hypereosinophilic syndrome
Primary sclerosing colangitis	Sézary syndrome
H. pylori-induced peptic ulcer	Measles virus infection & vaccination
HCV-induced chronic hepatitis	Progression to AIDS in HIV infection

HCV, Hepatitis C virus; *AIDS*, Acquired immunodeficiency syndrome; *HIV*, Human immunodeficiency syndrome.

mRNA was highly expressed in the skin of patients with destructive forms of cutaneous or active visceral disease [46]. Interestingly, IFN-γ in combination with pentavalent antimony was effective in treating severe or refractory visceral leishmaniasis [47].

Parasitic infections, characterized by eosinophilia and elevated IgE levels, usually elicit Th2 cytokines [48]. Th2 responses, which downregulate host protective Th1 functions, are less detrimental to parasites; on other hand, the host would avoid immunopathological reactions related to strong, but harmful, Th1 responses. The pathology resulting from *S. mansoni* infection is indeed predominantly caused by the host Th2 response leading to chronic granulomatous reaction and consequent damage of intestine and liver [49].

In the immune response to bacterial infections, Th2 cells seem to be appropriate opponents against toxin-producing bacteria, since Th2 cytokines favor B-cell maturation and production of neutralizing antibodies. In contrast, intracellular bacteria (e.g. *Listeria monocytogenes*, *Mycobacteria*, *Salmonellae*) are more properly faced by Th1 cells, which produce cytokines able to activate macrophages and cytotoxic T cells. Mice with disrupted IFN-γ or IFN-γ receptor (R) gene and producing high levels of IL-4 succumb to mycobacterial infections [50], whereas mice resistant to *M. bovis* produce high levels of IFN-γ and IL-2, and low amounts of IL-4 [51]. Likewise, patients with IFN-γR or IL-12R deficiency are extremely sensitive to mycobacterial infections and develop severe and often fatal disease [52, 53].

Th0 cells, which secrete a combination of both Th2- and Th1-type cytokines, should be the best effector cells in the immune response to extracellular bacteria since antibodies (which neutralize adhesion/invasion and opsonize bacteria) and phagocytosis are both required. The predominance of the Th1 or Th2 response in any infectious disease is probably modulated by both the pathogen and the genetic background of the host, whose innate immunity plays a key role. Since bacteria possess several components which can trigger IL-12 production by macrophages, it is not surprising that most of them favor Th1 development (Table 2). These "Th1 inducers"' include the lipoarabinomannan of mycobacteria, teichoic acids of gram-positive bacteria, lipopolysaccharides of gram-negative bacteria, and viral polynucleotides [54]. In genetically predisposed individuals, some strong and persistent Th1 responses against bacteria may often result in immunopathological reactions, such as Lyme disease induced by *Borrelia burgdorferi* [55], or reactive arthritis following infection with *Yersinia enterocolitica* [56]. A further example of Th1 polarization of the specific immune response is provided by *Helicobacter pylori* infection. Mice infected with *H. felis* show a Th1 response with local and systemic production of IFN-γ and undetectable levels of IL-4 and IL-5. In vivo neutralization of IFN-γ resulted in a significant reduction of gastric inflammation [57]. *H. pylori*-infected patients with duodenal ulcer showed IFN-γ, TNF-α and IL-12, but not IL-4, expression in vivo [58].

Strong evidence supports the concept that allergic disorders result from a Th2-type response to certain groups of ubiquitous antigens that can activate the immune system after inhalation, ingestion, or penetration through the skin (allergens). In atopic subjects, in vitro T-cell response to allergens is preferentially due to Th2-like cells [59, 60]. T cells that accumulate in the target organs of allergic patients preferentially express the Th0/Th2 profile [61-64]. In vivo challenge with the clinically relevant allergen results in the local activation and recruitment of allergen-specific Th2-like cells [65, 66]. Successful specific immunotherapy associates with upregulation of Th1 responses [67] and downregulation of allergen-reactive Th2 cells [67-69]. Either alterations in IL-4 gene regulation, or in cytokine genes responsible for inhibition of Th2-cell development (e.g. IFN-α or γ, IL-12 and IL-18) may account for the the preferential Th2-type response to environmental allergens in atopic subjects. Overexpression of the genes for IL-3, IL-5, or granulocyte-macrophage colony-stimulating factor (GM-CSF), located together with IL-4 and IL-13 genes within the same cluster on chromosome 5, can also explain the Th2-dominated response of allergic subjects [70].

Th1-type effector mechanisms, such as DTH and Cytolytic T Lymphocyte (CTL) activities, are supposed to play a central role in acute allograft rejection [71]. Proteins and transcripts for intragraft IL-2, IFN-γ and the CTL-associated marker, granzyme B, were found in allografts undergoing acute rejection [71]. In a recent study, T-cell clones derived from kidney grafts of patients during acute phase of rejection showed a clear-cut Th1 cytokine profile and effector functions [72]. Interestingly, the amounts of IFN-γ produced by graft-derived T-cell clones significantly correlatated with the degree of interstitial graft infiltration, suggesting that IFN-γ contributes, at least in part, to the degree of graft infiltration and to the severity of the rejection episode [72]. The production of Th2-type cytokines may be critical for the induction and maintenance of allograft tolerance. Several in vivo studies on the cytokine network during tolerance induction showed decreased IL-2/IFN-γ and increased IL-4/IL-10 transcription [73].

The embryo, because of expression of paternal major histocompatibility complex (MHC) antigens, resembles an allograft which is not rejected by the maternal immune system until the time of delivery. This apparent paradox has recently been ascribed to a Th2 switch occurring at the fetomaternal interface, allowing fetal survival through inhibition of Th1 response [74]. Recent data indicate that progesterone promotes the preferential development of a local Th2-type atmosphere that favors fetal survival by inhibiting Th1-mediated rejection [75]. Leukemia inhibitory factor (LIF) and IL-11, both Th2-related cytokines, are essential at maternal decidual level for female reproduction, the former being important at the time of embryo implantation [76] and the latter soon after implantation [77]. Interestingly, concentrations of progesterone or LIF comparable to those present at the fetomaternal interface were found to favor the preferential development of Th2 cells in vitro, whereas relaxin (another corpus luteum-derived hormone) favors the development of a predominant Th1 response, which is required at time of delivery. Thus, defective production of Th2 cytokines and LIF at the fetomaternal interface may contribute, at least in part, to premature rejection (unexplained abortion) of the conceptus [78].

Th1/Th2 Network in Autoimmune Diseases

Evidence from animal models and human disorders suggests that Th1 cytokines are involved in the genesis of organ-specific autoimmune diseases, whereas a less polarized cytokine pattern has been found in systemic autoimmune diseases. A prevalent Th1 cytokine profile was found in target organs of patients with Hashimoto's thyroiditis, multiple sclerosis, or type 1 diabetes mellitus, whereas a less restricted pattern of Th response was found in rheumatoid arthritis and Sjogren's syndrome. Patients with systemic lupus erythematosus or systemic sclerosis rather show a prevalent Th2 pattern (reviewed in [79]) (Table 3).

Active lesions of multiple sclerosis (MS) patients are characterized by lymphocyte (mainly CD4$^+$ T cells) and macrophage infiltration in the brain [80]. The possible role of Th1/Th2 imbalance has been extensively investigated in experimental autoimmune encephalomyelitis (EAE). Most T-cell clones derived from mice immunized with myelin basic protein peptides exhibited a Th1 pattern (reviewed in [79]). At the peak of disease severity, perivascular infiltrates in the brain stained positive for IL-2 and IFN-γ, whereas recovery was associated with the appearance of TGF-β1 and IL-4 [81]. The presence of IL-4 in the brain of animals recovering from EAE led to the hypothesis that disease remission might be related to the presence of antigen-specific Th2 cells. Accordingly, Th2 clones specific for myelin basic protein or proteolipid protein (PLP) and producing high amounts of IL-4 and IL-10 did not induce EAE, but rather suppressed the induction of EAE when mixed in adoptive transfer with Th1 clones [82]. It has been shown that mice with disrupted IFN-γ gene are still susceptible to the induction of EAE [83]. Several studies have suggested a pathogenic role for TNF-α and IFN-γ in human MS. High levels of TNF-α were found in the cerebrospinal fluid (CSF) of patients with chronic progressive MS [84]. Most T-cell clones derived from the CSF of patients with MS showed a Th1 profile [85], and increased numbers of IFN-γ-producing T cells were found in CSF of MS patients [86]. T-cell clones specific for PLP peptides generated from MS during an acute attack showed the Th1 pattern, whereas a more heterogenous cytokine profile was found in the same patients during remission. High LAG-3 expression and soluble LAG-3 (sLAG-3) production by T-cell clones generated from the CSF of MS patients were found and high levels of sLAG-3 were detected in the serum of patients with relapsing-remitting MS [16]. Interestingly, IL-12p40, but not other cytokines, was upregulated in acute MS plaques in early disease [80], suggesting a preeminent role for this cytokine in the induction of a Th1-type autoimmune response.

Several studies showed that T cells from lymphocytic thyroid infiltrates of patients with Hashimoto's thyroiditis or Graves' disease had a clear-cut Th1 profile, with high production of TNF-α, IFN-γ, and strong cytolytic potential [87-89]. Likewise, a homogenous Th1 profile was also observed in CD4$^+$ T-cell clones derived from retroorbital infiltrates of patients with Graves' ophthalmopathy [90]. Recent studies on the cytokine profile of thyroid antigen-specific T-cell clones showed that the majority of thyroid peroxidase (TPO)-specific clones had a Th1 profile, whereas the cytokine pattern of clones specific for thyroid-stimulating hormone receptor resembled that of Th0 or Th2 cells [91]. Even though the initiating events are still unclear, it is reasonable to suggest that the thyroid microenvironment allows the expansion of autoreactive clones with either a Th1 pattern or a less polarized profile, which promote the synthesis of pathogenic autoantibodies.

Crohn's disease, a disorder characterized by chronic inflammation of the gastrointestinal tract unrelated to a specific pathogen [92], provides another example of Th1 polarization. Increased expression of mRNA for IL-2 and IFN-γ was found in the gut mucosa of patients with active Crohn's disease, whereas IL-4 mRNA expression was frequently undetectable [93]. T-cell clones generated from the colonic mucosa of patients with Crohn's disease were found to produce higher levels of IFN-γ, but significantly reduced IL-4 and IL-5, in comparison with T-cell clones derived from patients with noninflammatory bowel disorders. High numbers of both activated $CD4^+$ T cells showing LAG-3 and IFN-γ reactivity, and IL-12-containing macrophages were found in inflammatory gut infiltrates from patients with Crohn's disease, but not in controls. Addition of anti-IL-12 antibody to cultures of T cells recovered from the gut mucosa of Crohn's disease patients inhibited the development of IFN-γ-producing T cells, suggesting that in situ IL-12 production plays a role in the development of Th1 cells at intestinal level [94].

An imbalance between Th1 and Th2 cytokines has been suggested to occur in patients with systemic lupus erythematosus (SLE). A balance in favor of Th2 cytokines may contribute to the increased B cell activation characteristic of SLE. This hypothesis was raised by the notion that Th2 cells are essential in the induction of the autoimmune alterations observed in animals following exposure to some chemicals, such as gold salts and mercurials [95]. The mechanism by which these chemicals induce Th2 cells may be related to their ability to trigger IL-4 production in genetically susceptible animals [96]. Analysis of cytokine production at single cell level showed that SLE patients have lower numbers of cells producing Th1 cytokines, but higher proportions of cells secreting Th2 cytokines than normal controls [97, 98]. Further support to the hypothesis of a preferential activation of cells producing Th2-type cytokines was recently provided by the observation in SLE patients of high serum levels of sCD30, which appeared to correlate with disease activity [99]. However, the predominant activation and pathogenic role of Th2 cells in human SLE still remains to be established. More solid evidence for a predominant activation of Th2 cells has been obtained in progressive systemic sclerosis (SSc), a disorder characterized by inflammatory and fibrotic changes in the skin (scleroderma) and other organs. Different cytokines, including IL-1α and IL-1β, IL-6, TNF-β, and TGF-β, may modulate fibrosis or promote vascular damage due to their ability to affect fibroblast activities, such as growth or production of extracellular matrix, collagenase and prostaglandins [100]. IL-4 as well induces human fibroblasts to synthesize extracellular matrix proteins and stimulates the growth of subconfluent fibroblasts [101]. Interestingly, peripheral blood mononuclear cells from patients with SSc tend to produce higher amounts of IL-2 and IL-4 than do controls [102]. Recent studies showed spontaneous IL-4, IL-5 and CD30 mRNA expression in peripheral blood T cells from the majority of

SSc patients, whereas no spontaneous mRNA expression for IFN-γ or IL-10 was found. Accordingly, high proportions of $CD4^+$ T cells in the perivascular infiltrates of the skin from SSc patients were $CD30^+$, and high levels of sCD30 were detected in the serum of the majority of SSc patients, particularly in those with active disease [103]. These data suggest predominant activation of Th2 cells in SSc and support the view that abnormal and persistent IL-4 production may play an important role in the induction of fibrosis.

Conclusions

The Th1/Th2 cytokine network provides a useful model for explaining both the different types of immune protection and the pathogenetic mechanisms of several immunopathological disorders. The development of polarized Th1 or Th2 responses depends on both individual genetic background and environmental factors, especially cytokines of the "natural immunity" at time of antigen presentation. Th1-dominated responses are potentially effective in eradicating infectious agents, particularly those hidden within host cells. When Th1 response is poorly effective or exhaustively prolonged, it may result in host pathology. However, polarized Th2 responses provide incomplete protection against the majority of infectious agents, and thus, they may be regarded as part of down-regulatory or suppressor mechanisms for exaggerated and inappropriate Th1 responses.

The Th1/Th2 concept applied to the study of immunopathology revealed that a number of diseases are mediated by Th1 cells, the clearest examples being multiple sclerosis, type 1 diabetes mellitus and thyroid autoimmunity. In other disorders, Th1/Th2 polarization is less prominent, whereas Th2 responses tend to predominate in progressive systemic sclerosis or allergic diseases. In animal models, Th1/Th2 switching resulted in prevention of different diseases. Therefore, modulation of the Th1/Th2 cytokine network might represent a potentially useful therapeutic tool also for human diseases.

References

1. Mosmann TR, Cherwinski H, Bond MW et al. (1986) Two types of murine T cell clone. I. Definition according to profiles of lymphokine activities and secreted proteins. J Immunol 136: 2348-2357
2. Mosmann TR, Coffman RL (1989) TH1 and TH2 cells: Different patterns of lymphokine secretion lead to different functional properties. Ann Rev Immunol 7: 145-173
3. Del Prete G, De Carli M, Mastromauro C et al. (1991) Purified protein derivative of *Mycobacterium tuberculosis* and excretory-secretory antigen(s) of *Toxocara canis* expand in vitro human T cells with stable and opposite (type 1 T helper or type 2 T helper) profile of cytokine production. J Clin Invest 88: 346-351
4. Mosmann TR, Sad S (1996) The expanding universe of T-cell subsets: Th1, Th2 and more. Immunol Today 17: 138-146

5. Ferrick DA, Schrenzel MD, Mulvania T (1995) Differential production of IFN-γ and IL-4 in response to Th1- and Th2-stimulating pathogens by gamma delta T cells in vivo. Nature 373: 255-258
6. Paliard X, de Waal Malefijt R, Yssel H, et al. (1988) Simultaneous production of IL-2, IL-4, and IFN-γ by activated human CD4+ and CD8+ T cell clones. J Immunol 141: 849-858
7. Openshaw P, Murphy EE, Hosken NA et al. (1995) Heterogeneity of intracellular cytokine synthesis at the single-cell level in polarized T helper 1 and T helper 2 populations. J Exp Med 182: 1357-1367
8. Salgame P, Abrams JS, Clayberger C et al. (1995) Differing lymphokine profiles and functional subsets of human CD4+ and CD8+ T cell clones. Science 254: 279-281
9. Romagnani S (1995) Biology of human Th1 and Th2 cells. J Clin Immunol 15: 121-129
10. Del Prete G, De Carli M, Almerigogna F et al. (1993) Human IL-10 is produced by both type 1 helper (Th1) and type 2 helper (Th2) T cell clones and inhibits their antigen-specific proliferation and cytokine production. J Immunol 150: 353-360
11. Del Prete G, De Carli M, Ricci M, Romagnani S (1991) Helper activity for immunoglobulin synthesis by Th1 and Th2 human T-cell clones: The help of Th1 clones is limited by their cytolytic capacity. J Exp Med 174: 809-813
12. Del Prete G, De Carli M, Lammel RM et al. (1995) Th1 and Th2 T-helper cells exert opposite regulatory effects on procoagulant activity and tissue factor production by human monocytes. Blood 86: 250-257
13. Del Prete G, De Carli M, Almerigogna F et al. (1995) Preferential expression of CD30 by human CD4+ T cells producing Th2-type cytokines. FASEB J 9: 81-86
14. Del Prete G, De Carli M, D'Elios MM et al. (1995) CD30-mediated signalling promotes the development of human Th2-like T cells. J Exp Med 182: 1655-1661
15. D'Elios MM, Romagnani P, Scaletti C et al. (1997) In vivo CD30 expression in human diseases with predominant activation of Th2-like T cells. J Leukoc Biol 61: 539-544
16. Annunziato F, Manetti R, Tomasevic L et al. (1996) Expression and release of LAG-3-associated protein by human CD4+ T cells are associated with IFN-γ. FASEB J 10: 767-776
17. Nakamura T, Lee RK, Nam SY et al. (1997) Reciprocal regulation of CD30 expression on CD4+ T cells by IL-4 and IFN-γ. J Immunol 158: 2090-2098
18. Scheel D, Richter E, Toellner K-M et al. (1995) Correlation of CD26 expression with Th1-like reactions in granulomatous diseases. In: Schlossmann SF (ed) Leukocyte typing V. White cell differentiation antigens. Oxford University, Oxford, pp 1101-1111
19. Kanegane H, Kasahara Y, Niida Y et al. (1996) Expression of L-selectin (CD62L) discriminates Th1- and Th2-like cytokine-producing memory CD4+ T cells. Immunology 87: 186-190
20. Assenmacher M, Scheffold A, Schmitz J et al. (1996) Specific expression of surface interferon-γ on interferon-γ-producing T cells from mouse and man. Eur J Immunol 26: 263-267
21. Sallusto F, Mackay CR, Lanzavecchia A (1997) Selective expression of the eotaxin receptor CCR3 by human T helper 2 cells. Science 277: 2005-2007
22. Bonecchi R, Bianchi R, Panina-Bordignon P et al. (1998) Differential expression of chemokine receptors and chemotactic responsiveness of Th1 and Th2 cells. J Exp Med 187: 129-134
23. Loetscher P, Uguccioni MG, Bordoli L et al. (1998) CCR5 is characteristic of Th1 lymphocytes. Nature 391: 344-345
24. Rogge L, Barberis-Maino L, Biffi M et al. (1997) Selective expression of an interleukin-12 receptor component by human T helper 1 cells. J Exp Med 185: 825-831
25. Kamogawa Y, Minasi LE, Carding SR et al. (1993) The relationship of IL-4- and IFN-γ-producing T cells studied by lineage ablation of IL-4-producing cells. Cell 75: 985-995

26. Swain SL, Weinberg AD, English M (1990) CD4⁺ T cell subsets. Lymphokine secretion of memory cells and of effector cells that develop from precursors in vitro. J Immunol 144: 1788-1799
27. Parronchi P, De Carli M, Manetti R et al. (1992) IL-4 and IFNs (alpha and gamma) exert opposite regulatory effects on the development of cytolytic potential by Th1 and Th2 human T cell clones. J Immunol 149: 2977-2982
28. Hsieh CS, Macatonia SE, Tripp CS et al. (1993) Development of Th1 CD4⁺ T cells through IL-12 produced by *Listeria*-induced macrophages. Science 260: 547-549
29. Manetti R, Parronchi P, Giudizi MG et al. (1993) Natural killer cell stimulatory factor (IL-12) induces T helper type 1 (Th1)-specific immune responses and inhibits the development of IL-4-producing Th cells. J Exp Med 177: 1199-1204
31. Okamura H, Kashiwamura S, Tsutsui H (1998) Regulation of IFN-γ production by IL-12 and IL-18. Curr Opin Immunol 10: 259-264
31. Constant SL, Bottomly K (1997) Induction of Th1 and Th2 CD4⁺ T cell responses: The alternative approaches. Annu Rev Immunol 15: 297-322
32. Szabo S, Jacobson NG, Dighe AS et al. (1995) Developmental committment to the Th2 lineage by extinction of IL-12 signaling. Immunity 2: 665-675
33. Hill GR, Cooke KR, Teshima T, Ferrara JL (1998) IL-11 promotes T cell polarization and prevents acute graft-versus-host disease after allogeneic bone marrow transplantation. J Clin Invest 102: 115-123
34. Hilkens CM, Snijders A, Kapsenberg M (1996) Accessory cell-derived IL-12 and PGE$_2$ determine the IFN-γ level of activated human CD4⁺ T cells. J Immunol 156: 1722-1727
35. Gorham JD, Guler ML, Steen RG et al. (1996) Genetic mapping of a locus controlling development of Th1/Th2 type responses. Proc Natl Acad Sci USA 93: 12467-12472
36. Bacon CM, Petricoin EF, Ortaldo JE et al. (1995) Interleukin 12 induces tyrosine phosphorylation and activation of STAT4 in human lymphocytes. Proc Natl Acad Sci USA 92: 7307-7311
37. Rincon M, Flavell RA (1997) T cell subsets: Transcriptional control in the Th1/Th2 decision. Curr Biol 7: 729-732
38. Murphy KM (1998) T lymhpocyte differentiation in the periphery. Curr Opin Immunol 10: 226-232
39. Takeda K, Tanaka T, Shi W et al. (1996) Essential role of STAT6 in IL-4 signalling. Nature 380: 627-633
40. Szabo SJ, Glimcher LH, Ho IC (1997) Genes that regulate interleukin-4 expression in T cells. Curr Opin Immunol 9: 775-781
41. Zheng W, Flavell RA (1997) The transcription factor GATA-3 is necessary and sufficient for Th2 cytokine gene expression in CD4 T cells Cell 89: 587-596
42. Romagnani S (1994) Lymphokine production by human T cells in disease states. Annu Rev Immunol 12:227-257
43. Reiner SL, Locksley RM (1995) The regulation of immunity to *Leishmania major*. Annu Rev Immunol 13: 151-177
44. Carvalho EM, Bacellar O, Brownell C et al. (1994) Restoration of IFN-γ production and lymphocyte proliferation in visceral leishmaniasis. J Immunol 152: 5949-5956
45. Kemp M, Theander TG, Kharazmi A (1996) The contrasting roles of CD4⁺ T cells in intracellular infections in humans: Leishmaniasis as an example. Immunol Today 17: 13-16
46. Pirmez S, Yamamura M, Uyemura K et al. (1993) Cytokine patterns in the pathogenesis of human leishmaniasis. J Clin Invest 91: 1390-1395
47. Badaró R, Johnson WD (1993) The role of interferon-γ in the treatment of visceral and diffuse cutaneous leishmaniasis. J Infect Dis 167(S1): S13-17

48. Finkelman FD, Pearce EJ, Urban JF Jr, Sher A (1991) Regulation and biological function of helminth-induced cytokine responses. Immunoparasitol Today 12: A62-66
49. Sher A, Gazzinelli RT, Oswald IP et al. (1992) Role of T cell derived cytokines in the downregulation of immune responses in parasitic and retroviral infection. Immunol Rev 127: 183-204
50. Cooper MA, Dalton DK, Stewart TA, Orme JM (1993) Disseminated tuberculosis in IFN-γ gene-disrupted mice. J Exp Med 178: 2249-2254
51. Flynn JL, Chan J, Trieblod KJ et al. (1993) An essential role for IFN-γ in resistance to *Mycobacterium tuberculosis*. J Exp Med 178: 2249-2254
52. Newport MJ, Huxley CM, Huston S, Levin M (1996) A mutation in IFN-γ receptor gene and sensitivity to mycobacterial infection. N Engl J Med 335: 1941-1949
53. de Jong R, Altare F, Elferink DG, Ottenhoff TH (1998) Severe mycobacterial and salmonella infections in IL-12 receptor-deficient patients. Science 280: 1435-1438
54. Daugelat S, Kauffman SH (1995) Role of Th1 and Th2 cells in bacterial infections. Chem Immunol 63: 66-97
55. Yssel H, Shanafelt MC, Soderberg C et al. (1992) *Borrelia burgdorferi* activates a T helper type 1-like T cell subset in Lyme arthritis. J Exp Med 174: 593-601
56. Lahesmaa R, Yssel H, Batsford S et al. (1992) *Yersinia enterocolitica* activates a T helper type 1-like T cell subset in reactive arthritis. J Immunol 148: 3079-3085
57. Mohammadi M, Czinn S, Nedrud J (1996) *Helicobacter*-specific cell-mediated immune responses display a predominant Th1 phenotype and promote a delayed-type hypersensitivity response in the stomachs of mice. J Immunol 156: 4729-4736
58. D'Elios MM, Manghetti M, De Carli M et al. (1997) Th1 effector cells specific for *Helicobacter pylori* in the gastric antrum of patients with peptic ulcer disease. J Immunol 158: 962-967
59. Wierenga EA, Snoek M, de Groot C et al. (1990) Evidence for compartmentalization of functional subsets of CD4$^+$ T lymphocytes in atopic patients. J Immunol 144: 4651-4656
60. Parronchi P, Macchia D, Piccinni M-P et al. (1991) Allergen- and bacterial antigen-specific T-cell clones established from atopic donors show a different profile of cytokine production. Proc Natl Acad Sci USA 88: 4538-4542
61. Kay AB, Ying S, Varney V et al. (1991) Messenger RNA expression of the cytokine gene cluster interleukin-3 (IL-3), IL-4, IL-5, and granulocyte/macrophage colony-stimulating factor, in allergen-induced late-phase reactions in atopic subjects. J Exp Med 173: 775-778
62. Maggi E, Biswas P, Del Prete G et al. (1991) Accumulation of Th2-like helper T cells in the conjunctiva of patients with vernal conjunctivitis. J Immunol 146: 1169-1174
63. van der Heijden FL, Wierenga EA, Bos JD, Kapsenberg ML (1991) High frequency of IL-4-producing CD4$^+$ allergen-specific T lymphocytes in atopic dermatitis lesional skin. J Invest Dermatol 97: 389-394
64. Robinson DS, Hamid Q, Ying S et al. (1992) Predominant Th2-like bronchoalveolar T-lymphocyte population in atopic asthma. N Engl J Med 326: 295-304
65. Del Prete G, De Carli M, D'Elios MM et al. (1993) Allergen exposure induces the activation of allergen-specific Th2 cells in the airway mucosa of patients with allergic respiratory disorders. Eur J Immunol 23: 1445-1449
66. Robinson D, Hamid Q, Bentley A et al. (1993) Activation of CD4$^+$ T cells, increased Th2-type cytokine mRNA expression, and eosinophil recruitment in bronchoalveolar lavage after allergen inhalation challenge in patients with atopic asthma. J Allergy Clin Immunol 92: 313-324
67. Varney VA, Hamid Q, Gaga M et al. (1993) Influence of grass pollen immunotherapy on cellular infiltration and cytokine mRNA expression during allergen-induced late-phase cutaneous responses. J Clin Invest 92: 644-651

68. Secrist H, Chelen CJ, Wen Y et al. (1993) Allergen immunotherapy decreases interleukin 4 production in CD4⁺ T cells from allergic individuals. J Exp Med 178: 2123-2130
69. Jutel M, Pichler WJ, Skrbic D et al. (1995) Bee venom immunotherapy results in decrease of IL-4 and IL-5 and increase of IFN-γ secretion in specific allergen-stimulated T cell cultures. J Immunol 154: 4187-4194
70. Romagnani S (1995) Atopic allergy and other hypersensitivities. Curr Opin Immunol 7: 745-750
71. Suthantiran M, Strom TB (1995) Immunobiology and immunopharmacology of organ allograft rejection. J Clin Immunol 15: 161-171
72. D'Elios MM, Josien R, Manghetti M et al. (1997) Predominant Th1 infiltration in acute rejection episodes of human kidney grafts. Kidney Int 51: 1876-1884
73. Dallman MJ (1995) Cytokines and transplantation: Th1/Th2 regulation of the immune response to solid organ transplants in the adult. Curr Opin Immunol 7: 632-638
74. Wegmann TG, Lin H, Gulbert L, Mossman TR (1993) Bidirectional cytokine interactions in the maternal-fetal relationship: is successful pregnancy a Th2 phenomenon? Immunol Today 14: 353-356
75. Piccinni M-P, Beloni L, Giannarini L et al. (1995) Progesterone favors the development of human T helper cells producing Th2-type cytokines and promotes both IL-4 production and membrane CD30 expression in established Th1 cell clones. J Immunol 155: 128-133
76. Piccinni M-P, Beloni L, Livi C et al. (1998) Role of type 2 T helper (Th2) cytokines and leukemia inhibitory factor (LIF) produced by decidual T cells in unexplained recurrent abortions. Nat Med (in press)
77. Robb L, Li R, Hartely L, Begley (1998) Infertility in female mice lacking the receptor for IL-11 is due to a defective uterine response to implantation. Nat Med 4: 303-308
78. Piccinni M-P, Romagnani S (1996) Regulation of fetal allograft survival by hormone-controlled Th1- and Th2-type cytokines. Immunol Res 15: 141-150
79. Romagnani S (1997) The Th1/Th2 paradigm in disease. Springer-Verlag, Berlin Heidelberg New York
80. Windhagen A, Nicholson LB, Weiner HL et al. (1996) Role of Th1 and Th2 cells in neurologic disorders. Chem Immunol 63: 171-186
81. Khoury SJ, Hancock WW, Weiner HL (1992) Oral tolerance to myelin basic protein and natural recovery from experimental autoimmune encephalomyelitis are associated with downregulation of inflammatory cytokines and differential upregulation of transforming growth factor β, interleukin 4 and prostaglandin E expression in the brain. J Exp Med 176: 1335-1364
82. Cua DJ, Hinton DR, Stohlman SA (1995) Self-antigen-induced Th2 responses in experimental allergic encephalomyelitis (EAE)-resistant mice. J Immunol 155: 4052-4059
83. Ferber IA, Brocke S, Taylor-Edwards C et al. (1996) Mice with a disrupted IFN-γ gene are susceptible to the induction of experimental autoimmune encephalomyelitis (EAE). J Immunol 156: 5-7
84. Sharief MK, Hentges A (1991) Association between tumor necrosis factor alpha and disease progression in patients with multiple sclerosis. N Engl J Med 325: 467-472
85. Brod SA, Benjamin D, Hafler DA (1991) Restricted T cell expression of IL2/IFN-γ mRNA in human inflammatory disease. J Immunol 147: 810-815
86. Olsson T (1993) Cytokines in neuroinflammatory disease: Role of myelin-autoreactive T cell production of interferon-gamma. J Neuroimmunol 40: 211-218
87. Del Prete G, Tiri A, Mariotti S et al. (1987) Enhanced production of gamma-interferon by thyroid-derived T cell clones from patients with Hashimoto's thyroiditis. Clin Exp Immunol 69: 323-331

88. Margolick JB, Weetman AP, Burman KD (1988) Immunohistochemical analysis of intrathyroidal lymphocytes in Graves' disease: evidence of activated T cells and production of interferon-gamma. Clin Immunol Immunopathol 47: 208-218
89. Zheng RQH, Abney ER, Chu CQ et al. (1992) Detection of in vivo production of tumor necrosis factor-alpha by human thyroid epithelial cells. Immunology 75: 456-462
90. De Carli M, D'Elios MM, Mariotti S et al. (1993) Cytolytic T cells with Th1-like cytokine profile predominate in retroorbital lymphocytic infiltrates of Graves' ophthalmopathy. J Clin Endocrinol Metab 77: 1120-1124
91. Mullins RJ, Cohen SBA, Webb LMC et al. (1995) Identification of thyroid stimulating hormone receptor-specific T cells in Graves' disease thyroid using autoantigen-transfected Epstein Barr virus-transformed B cell lines. J Clin Invest 96: 30-37
92. Kirsner JB, Shorter RG (1982) Recent developments in "nonspecific" inflammatory bowel disease. N Engl J Med 306: 775-837
93. Niessner M, Volk BA (1995) Altered Th1/Th2 cytokine profiles in the intestinal mucosa of patients with inflammatory bowel disease as assessed by quantitative transcribed polymerase chain reaction (RT-PCR). Clin Exp Immunol 101: 428-435
94. Parronchi P, Romagnani P, Annunziato F et al. (1997) Type 1 T-helper cells predominance and interleukin-12 expression in the gut of patients with Crohn's disease. Am J Pathol 150: 823-831
95. Goldman M, Druet P, Gleichmann E (1991) Th2 cells in systemic autoimmunity: Insights from allogenic diseases and chemically-induced autoimmunity. Immunol Today 12: 223-227
96. Prigent P, Saoudi A, Pannetier C et al. (1995) Mercuric chloride, a chemical responsible for T helper cell (Th)2-mediated autoimmunity in Brown Norway rats, directly triggers T cells to produce interleukin-4. J Clin Invest 96: 1484-1489
97. Dueymes M, Barrier J, Bescancenot JF et al. (1993) Relationship of IL-4 to isotypic distribution of anti-DS DNA antibodies in SLE. Int Arch Allergy Appl Immunol 101: 408-415
98. Hagiwara E, Gourley MF, Lee M, Klinman DM (1996) Disease severity in patients with systemic lupus erythematosus correlates with an increased ratio of IL-10/IFN-γ secreting cells in the peripheral blood. Arthritis Rheum 39: 379-385
99. Caligaris-Cappio F, Bertero MT, Converso M et al. (1995) Circulating levels of soluble CD30, a marker of cells producing Th2-type cytokines, are increased in patients with systemic lupus erythematosus and correlate with disease activity. Clin Exp Rheum 13: 339-343
100. Kovacs EJ (1991) Fibrogenic cytokines: The role of immune mediators in the development of scar tissue. Immunol Today 12: 17-23
101. Postletwhite AE, Holness MA, Katai H, Raghow R (1992) Human fibroblasts synthesize elevated levels of extracellular matrix proteins in response to interleukin 4. J Clin Invest 90: 1479-1485
102. Famularo G, Procopio A, Giacomelli R et al. (1990) Soluble interleukin-2 receptor, interleukin-2 and interleukin-4 in sera and supernatants from patients with progressive systemic sclerosis. Clin Exp Immunol 81: 368-372
103. Mavilia C, Scaletti C, Romagnani P et al. (1997) Type 2 helper T-cell predominance and high CD30 expression in systemic sclerosis. Am J Pathol 151: 1751-1758

Chapter 9

Lymphocyte Trafficking in the Central Nervous System

H. LASSMANN

Introduction

The central nervous system (CNS) is separated from the peripheral immune system by the blood-brain barrier. This tight barrier system impedes the exchange of cells and inflammatory mediators between the nervous tissue and the blood. Thus, for our understanding of the pathogenesis of brain inflammation, essential questions deal with the mechanisms of normal immune surveillance, the routes and mechanisms by which immune cells are recruited into the brain, how immune cells recognize their specific brain targets, and finally how they are cleared from established lesions during recovery. Basic principles regarding these questions have been recognized during recent years, and these will be discussed in this short review.

Immune Surveillance

Although the intact blood-brain barrier is impermeable for most cellular components of the immune system, activated T lymphocytes can pass quite readily [1, 2]. Thus T cell activation, which occurs in conditions of peripheral immune activation, allows the cells to enter the CNS and search for their specific antigen. In the absence of antigen they are quickly cleared from the nervous system [2]. It is, however, not clear whether they actively leave the CNS or are locally destroyed. In contrast, in situations in which antigen is recognized, a cascade of events leads to the secondary recruitment of other immune cells (e.g. macrophages, B cells and in some instances granulocytes), the entry of antibodies and inflammatory mediators, and thus the initiation of an inflammatory process.

Migration of Immune Cells Through the Blood-Brain Barrier

At present, little is known about the molecular mechanisms that guide the primary activated T cells, which start the inflammatory cascade, into the brain. It is

---

Institute of Neurology, University of Vienna, Schwarzspanierstrasse 17 - 1090 Vienna, Austria. e-mail: hans.lassmann@univie.ac.at

likely that adhesion molecules on brain endothelial cells are involved, as they are in secondary recruitment of inflammatory cells into established lesions. However, under normal conditions the expression of adhesion molecules is very low [3–5] and the specific binding partners for the subset of activated T cells have not been identified.

In contrast, when an inflammatory focus is established, there is massive endothelial expression of different adhesion molecules, the most important being selectins, ICAM-1 and VCAM [4–7]. In addition, the local expression of chemokines by hematogenous and resident cells [8–10] as well as the expression of chemokine receptors on the respective infiltrating immune cells apparently potentiate the inflammatory response. In addition to the classic chemokines, certain neuropeptides can exert chemotactic functions [11, 12] and modify the quality and quantity of the inflammatory response in the brain [13].

The routes of leukocyte egress in brain vessels may be different for different cell populations. Whereas lymphocytes seem to travel through channels formed by the endothelial cells themselves [14, 15], granulocytes and monocytes may preferentially pass through disrupted interendothelial tight junctions [16].

Local Antigen Recognition in the CNS

When T cells pass through the blood-brain barrier, they are confronted with local antigen. In the CNS there is a hierarchy of major histocompatibility complex (MHC) expression. In normal conditions MHC antigens are restricted to perivascular and meningeal macrophages. In a mild proinflammatory environment, it is primarily the microglia cell population that is activated and expresses class I and II MHC antigens. Only when extremely severe proinflammatory stimuli are present can MHC antigens additionally be found on neuroectodermal cells, in particular astrocytes and ependymal cells [17].

Interestingly, MHC expression on meningeal and perivascular macrophages is sufficient to trigger the complete spectrum of T cell-mediated brain inflammation, even in situations when the respective T cells are directed against an antigen of myelin or oligodendrocytes [18, 19]. This implies that CNS antigen has to be released from its local stores into the brain extracellular space, diffuse to the perivascular or meningeal compartment, and be recognized there by the specific T cells. To what extent local MHC expression on microglia and neuroectodermal cells may modify the pattern of brain inflammation is so far undetermined in vivo.

An important question relates to the fact that under normal conditions MHC expression is practically absent on cells residing in the nervous parenchyma. It has recently been shown that MHC expression on neurons as well as on neighboring glia cells is downregulated as long as electrical activity of neurons is maintained [20, 21]. In contrast, when electrical activity is blocked – for instance by tetrodotoxin – a prominent upregulation of MHC expression occurs not only in glia cells but even in the neurons themselves. This may be an important aspect of immune privilege of the nervous system by protecting functionally active nervous elements against a cytotoxic assault of immune cells in both autoimmune and infectious conditions.

Clearance of T Cells from Established Inflammatory CNS Lesions

In the model of passive transfer of autoreactive T cell lines, the paradigm of acute monophasic T-cell-mediated brain inflammation, a peak of inflammatory reaction is reached 4-6 days after intravenous transfer of the cells. Thereafter, inflammation subsides rapidly, leaving only few scattered T cells within the brain lesions at day nine. Thus, within a short time span of 3 days, huge numbers of T lymphocytes are cleared from the lesions. How do these cells disappear? Do they migrate back into the circulation and the lymphatics or are they locally destroyed?

Pender et al. [22] first drew attention to the abundance of cells undergoing apoptosis in acute experimental allergic encephalomyelitis (EAE) lesions, suggesting that most of the dying cells are either T cells or oligodendrocytes. In a more detailed quantitative study it became clear that most of the apoptotic cells in fact are T lymphocytes, and that at the peak of disease 30%-40% of the local T cell infiltrate dies simultaneously [23]. Considering the short time lapse between earliest morphological signs of apoptosis and complete dissolution of apoptotic cells in vitro, it is likely that over a 24-hour time period more than three times the total infiltrating T cell population in these acute inflammatory lesions is cleared by programmed cell death. This suggests that the inflammatory reaction can only be maintained as long as there is a continuous influx of T cells from the peripheral circulation. A similar process of T cell clearance has been observed at other so-called immunoprivileged sites, such as the peripheral nervous system and the eye [24, 25], but not in inflammatory conditions in skin, muscle or other organs [23, 26, 27].

The question of which T cell populations are affected by apoptosis in the CNS has recently been addressed by Bauer et al. [28]. Transfer experiments with pre-labeled T cell populations showed that the autoantigen-specific as well as the secondarily recruited T cell populations are equally destroyed. Furthermore, T cells with specificities which cannot be found at all in the CNS, such as ovalbumin, undergo apoptosis when cotransferred together with myelin basic protein (MBP)-reactive autoimmune T cells. The relative proportion of dying T cells was also independent from their state of activation at the time of transfer.

Similar patterns of T cell elimination have recently been described in the eye, where the expression of Fas ligand on local resident cells may provide the death signal for infiltrating T cells expressing Fas receptor [25, 29]. Indeed, defective T cell elimination from eye lesions has been reported in animals with genetic deletion of Fas ligand or Fas receptor. In contrast, the situation appears to be different in the brain. T cell apoptosis is qualitatively and quantitatively similar in Fas ligand-deficient animals with autoimmune encephalomyelitis compared to wild-type animals (Bachmann et al., unpublished). Similarly, deletion of the genes for perforin, inducible nitric oxide (i-NOS) or interleukin-1 converting enzyme (ICE) showed no effect on apoptotic elimination of T cells. These data suggest that the CNS harbors a specific basic mechanism of T cell clearance which follows different mechanisms as compared to those in the peripheral immune system.

In addition to this basic mechanism of T cell clearance, there appears to be another way of T cell elimination affecting the antigen-specific population. Treatment of EAE animals with high doses of soluble antigenic peptide not only ameliorates disease but also dramatically increases the rate of T cell apoptosis within the lesions [30]. This effect appears to depend upon functional signaling through the tumor necrosis factor (TNF) receptor-1 (TNFR1) pathway, since anti-TNF treatment abolishes the effect of soluble antigen treatment on T cell apoptosis (A. Weishaupt et al., unpublished). In mice with chronic demyelinating EAE, the rate of T cell apoptosis in the lesions is significantly increased in animals with active demyelination in comparison to those with inactive disease. In contrast, EAE animals with impaired TNFR1 signaling pathway show a significant 50% reduction of T cell apoptosis in the lesion, and its increase during demyelinating activity is abolished (Bachmann et al., unpublished).

These data suggest that within the nervous system, two independent pathways of T cell elimination exist. A basic mechanism involves all incoming T cells and possibly occurs by intrinsic signals provided by resident cells in the CNS. This basic mechanism can be augmented by a second pathway which affects the antigen-specific population selectively, depends upon the local liberation of antigen, and requires intact TNFR1 signaling. It has, however, to be emphasized that clearance of inflammation by local destruction of inflammatory cells has so far only been shown unequivocally for T lymphocytes. The pathways by which other immune cells, such as macrophages or B lymphocytes, are cleared from inflammatory CNS lesions are unknown.

Conclusions

The data discussed above suggest that immune privilege of the CNS is accomplished by three major mechanisms. First, passage of hematogenous cells through the normal blood-brain barrier is restricted to a small population of activated T cells. This is apparently due to the low expression of adhesion molecules and chemokines under normal conditions and guarantees selective immune surveillance. However, once inflammation starts, an immunological cascade similar to that in other vessels of the body occurs. Second, there is active downregulation of MHC expression and antigen presentation associated with electrical activity of intact CNS tissue. Finally, immune control is further maintained by efficient and selective destruction of T lymphocytes within the CNS parenchyma. The latter mechanism may be instrumental to avoid generalized autoimmunity following inflammation of the CNS.

References

1. Wekerle H, Linington C, Lassmann H, Meyermann R (1986) Cellular immune reactivity within the CNS. Trends Neurosci 9: 271-277
2. Hickey WF, Hsu BL, Kimura H (1991) T lymphocyte entry into the central nervous system in experimental allergic encephalomyelitis. J Neurosci Res 28: 254-260
3. Male D, Pryce G, Rahman J (1990) Comparison of immunological properties of rat cerebral and aortic endothelium. J Neuroimmunol 30: 161-168
4. Sobel RA, Mitchell ME, Fondren G (1990) Intercellular adhesion molecule-1 (ICAM-1) in cellular immune reactions in the human central nervous system. Am J Pathol 136: 1309-1316
5. Yednock TA, Cannon C, Fritz LC et al. (1992) Prevention of experimental autoimmune encephalomyelitis by antibodies against alpha4-beta1 integrin. Nature 356: 63-66
6. Cannella B, Cross AH, Raine CS (1991) Adhesion-related molecules in the central nervous system. Upregulation correlates with inflammatory cell influx during relapsing autoimmune encephalomyelitis. Lab Invest 65: 23-31
7. Lassmann H, Rössler K, Zimprich F, Vass K (1991) Expression of adhesion molecules and histocompatibility antigens at the blood-brain barrier. Brain Pathol 1: 115-123
8. Schlüsener HJ, Meyermann R (1993) Intercrines in brain pathology. Expression of intercrines in a multiple sclerosis and a Creutzfeld-Jacob lesion. Acta Neuropathol (Berl) 86: 393-396
9. Tani M, Ransohoff RM (1994) Do chemokines mediate inflammatory cell invasion of the central nervous system parenchyma? Brain Pathol 4: 135-143
10. Ransohoff RM (1997) Chemokines in neurological disease models: Correlation between chemokine expression patterns and inflammatory pathology. J Leukoc Biol 62: 645-652
11. Ruff M, Wahl SM, Pert CB (1985) Substance P receptor mediated chemotaxis of human monocytes. Peptide 6: 107-111
12. Smith CH, Barker JN, Morris RW et al. (1993) Neuropeptides induce rapid expression of endothelial adhesion molecules and elicit granulocytic infiltration in human skin. J Immunol 151: 3274-3282
13. Storch MK, Fischer-Colbrie R, Smith T et al. (1996) Co-localization of secretoneurin immunoreactivity and macrophage infiltration in the lesions of experimental autoimmune encephalomyelitis. Neuroscience 71: 885-893
14. Raine CS, Cannella B, Duijvestijn AM, Cross AH (1990) Homing to central nervous system vasculature by antigen-specific lymphocytes. II. Lymphocyte/endothelial cell adhesion during initial stages of autoimmune demyelination. Lab Invest 63: 476-489
15. Wisniewski HM, Lossinsky AS (1991) Structural and functional aspects of the interaction of inflammatory cells with the blood-brain barrier in experimental brain inflammation. Brain Pathol 1: 89-96
16. Cross AH, Raine CS (1991) Central nervous system endothelial cell-polymorphonuclear cell interactions during autoimmune demyelination. Am J Pathol 139: 1401-1409
17. Vass K, Lassmann H (1990) Intrathecal application of interferon gamma: progressive appearance of MHC antigens within the rat nervous system. Am J Pathol 137: 789-800
18. Hickey WF, Kimura H (1988) Perivascular microglial cells of the CNS are bone marrow derived and present antigen in vivo. Science 239: 290-292
19. Lassmann H, Schmied M, Vass K, Hickey WF (1993) Bone marrow derived elements and resident microglia in brain inflammation. Glia 7: 19-24

20. Neumann H, Boucraut J, Hahnel C et al. (1996) Neuronal control of MHC class II inducibility in rat astrocytes and microglia. Eur J Neurosci 8: 2582-2590
21. Neumann H, Cavalle A, Jenne DE, Wekerle H (1995) Induction of MHC I genes in neurons. Science 269: 549-551
22. Pender MP, Nguyen KB, McCombe PA, Kerr JFR (1991) Apoptosis in the nervous system in experimental allergic encephalomyelitis. J Neurol Sci 104: 81-87
23. Schmied M, Breitschopf H, Gold R et al. (1993) Apoptosis of T lymphocytes in experimental autoimmune encephalomyelitis: Evidence for programmed cell death as a mechanism to control inflammation in the brain. Am J Pathol 143: 446-452
24. Zettl UW, Gold R, Hartung HP, Toyka KV (1994) Apoptotic cell death of T-lymphocytes in experimental autoimmune neuritis of the Lewis rat. Neurosci Lett 176: 75-79
25. Griffith TS, Brunner T, Fletcher SM et al. (1995) Fas ligand-induced apoptosis as a mechanism of immune privilege. Science 270: 1189-1192
26. Schneider C, Gold R, Dalakas MC et al. (1996) MHC class-I-mediated cytotoxicity does not induce apoptosis in muscle fibers nor in inflammatory T cells: Studies in patients with polymyositis, dermatomyositis and inclusion body myositis. J Neuropathol Exp Neurol 55: 1205-1209
27. Gold R, Hartung HP, Lassmann H (1997) T-cell apoptosis in autoimmune diseases: Termination of inflammation in the nervous system and other sites with specialized immune-defense mechanisms. Trends Neurosci 20: 399-404
28. Bauer J, Bradl M, Hickey WF et al. (1998) T cell apoptosis in inflammatory brain lesions. Destruction of T-cells does not depend on antigen recognition. Am J Pathol 153: 715-724
29. Griffith TS, Ferguson TA (1997) The role of Fas L-induced apoptosis in immune privilege. Immunol Today 18: 240-244
30. Weishaupt A, Gold R, Gaupp S et al. (1997) Antigen therapy eliminates T cell inflammation by apoptosis: effective treatment of experimental autoimmune neuritis with recombinant myelin protein P2. Proc Natl Acad Sci USA 94: 1338-1343

Chapter 10

Antigen Presentation in the Central Nervous System

F. ALOISI

Relative Immune Privilege of the Central Nervous System

Immune-privileged sites such as the central nervous system (CNS), eye and testes, are physiologically adapted to protect their delicate structures and functions from damaging inflammatory responses. Major features contributing to the immune privilege of the CNS include: the blood-brain barrier (BBB) – the tight endothelial junctions of the brain vasculature that limit access of plasma proteins and blood-derived cells to the CNS; the lack of a conventional lymphatic system; the absence within the CNS parenchyma of dendritic cells, the most potent antigen-presenting cells (APC) for initiation of T cell responses; and the paucity of class I and class II major histocompatibility complex (MHC) molecules on resident CNS cells.

These features are indicative of a deficient immune environment in the CNS. However, immune responses directed against neurotropic pathogens and CNS autoimmune diseases can develop naturally or be induced in laboratory animals, indicating that the immune system can detect and destroy antigens within the CNS. Experimental evidence supports the view that the CNS is continuously and effectively patrolled by the immune system. The demonstration that monocytes and activated T cells traffic slowly through the normal CNS [1-3] and that the CNS microenvironment is able to regulate immune responses [4-6] has shed light on the mechanisms underlying immune surveillance and active maintenance of immune privilege in the CNS.

Antigen Recognition in the CNS

Because humoral and cell-mediated immune responses are initiated in lymphoid organs and because the CNS parenchyma lacks both lymphatic drainage and dendritic cells (that in most other tissues take-up potential antigens and transport them to lymphoid organs to initiate immune responses [7]), the issue of whether antigens leave the CNS to reach the lymphoid organs has been the focus of sever-

---

Neurophysiology Unit, Laboratory of Organ and System Pathophysiology, Istituto Superiore di Sanità, V. le Regina Elena, 299 - 00161 Rome, Italy. e-mail: fos4@iss.it

al studies. Cserr and Knopf [4] have demonstrated that soluble antigen injected into the CNS parenchyma can escape the CNS via non-lymphatic routes (mainly the lamina cribrosa of the olfactory nerve and the arachnoid villi), reach the cervical lymph nodes and elicit a humoral immune response. Soluble antigen delivered to the normal CNS also elicits an intrathecal antibody response, suggesting that B cells can traffic through the intact BBB and be retained at the site of antigen deposition [8]. B cell trafficking to the CNS and intrathecal antibody responses are well documented in infectious and autoimmune CNS diseases with a compromised BBB [9, 10], indicating that the local humoral immune response may serve to neutralize or opsonize antigen.

In contrast to soluble antigens, tissue grafts or pathogens delivered to the CNS parenchyma are inefficiently or not at all recognized by the immune system. Allogeneic transplants placed in the CNS are rapidly rejected only when the same tissues are concomitantly grafted to a peripheral site [11]. Matyszak and Perry showed that heat-killed bacillus Calmette-Guèrin (BCG) does not elicit a cell-mediated immune response when injected into the CNS parenchyma [12]. However, subsequent peripheral immunization with BCG elicits a strong delayed-type hypersensitivity (DTH) response, and leads to T cell and macrophage recruitment into the CNS, disruption of the BBB, myelin breakdown, and axonal damage [12]. These experiments demonstrate that the immune system is able to recognize particulate antigens in the CNS only after peripheral sensitization to the same antigens.

In most naturally occurring infections of the CNS, pathogens first infect other tissues. Therefore, it is likely that dendritic cells in the periphery take-up neurotropic infectious agents for presentation to naive T cells and only subsequently are peripherally primed T cells recruited into the infected CNS where they recognize antigen on local APC. Although dendritic cells are not present in the CNS parenchyma proper, small numbers of dendritic-like cells have been detected in some CNS-associated compartments, such as the meninges and the choroid plexus, a vascular formation ensheathed by epithelial cells which projects into the cerebral ventricles and is responsible for the secretion of cerebrospinal fluid (CSF) [13, 14]. In the stroma of the choroid plexus from patients with acquired immunodeficiency syndrome (AIDS), a population of MHC class II-positive dendritic cells infected by HIV-1 has been described [15]. These observations suggest not only that dendritic cells of the choroid plexus represent a potential reservoir of HIV-1 in the CNS but also that these cells may initiate virus-specific immune responses.

Some dendritic cells have also been detected in the perivascular inflammatory infiltrates in experimental autoimmune encephalomyelitis (EAE) and in DTH lesions induced in response to BCG sequestrated in the CNS parenchyma [13]. Dendritic cells recruited into the CNS may be involved in the presentation of pathogenic antigen or of auto-antigen generated during tissue damage, and thus may contribute to the perpetuation of the inflammatory process. Detailed studies on the expression of adhesion/costimulatory molecules or other markers associated with dendritic cell maturation and on the intracerebral synthesis of chemokines active on dendritic cells are still lacking. This could provide important information on the functional status and mechanisms of recruitment of dendritic cells in the inflamed CNS.

The recruitment and fate of CD4+ T cells targeted to CNS auto-antigens have been extensively investigated in EAE, an animal model of CNS autoimmune disease. EAE is induced either by immunization with CNS antigens or by adoptive transfer of T cells sensitized in vitro to myelin basic protein (MBP) or to other myelin components [16]. It is mediated by T helper (Th) 1 cells which, after migration across the BBB, recognize the target antigen on local CNS APC and secrete the proinflammatory cytokines interferon (IFN)-γ, interleukin (IL)-2, and tumor necrosis factor (TNF)-β, leading to neurological deficits and myelin destruction [17]. Recognition of CNS antigens by autoreactive CD4+ T lymphocytes is thought to be involved in the pathogenesis of multiple sclerosis (MS), a chronic disease characterized by the presence of inflammatory infiltrates in the white matter, breakdown of the BBB, macrophage/microglia activation, myelin destruction, astrocytosis and eventually axonal loss [18]. Although the etiology of MS is still unknown, viral infections are believed to play a role in the initiation of CNS autoimmunity, either because a virus shares amino acid sequences or conformations with CNS proteins (molecular mimicry) or because a viral infection may induce an immune response leading to the aberrant activation of autoreactive T cells [19, 20]. Identification of the cell types that function as APC in the CNS has long been considered a crucial issue for understanding the immunopathogenesis of autoimmune demyelinating diseases.

Antigen Presentation by CNS Macrophages

Perivascular Cells and Other CNS-Associated Macrophages

Perivascular macrophages located within the basal lamina of microvessels in the CNS parenchyma, and macrophages present in the stroma of the choroid plexus and meninges are thought to play an important role in CNS immunosurveillance. Studies in radiation bone marrow-chimeric animals have shown that perivascular and meningeal macrophages are continuously replaced by hematogenous cells, indicating that the CNS is ensheathed by a renewable, small pool of phagocytic cells which act as first-line scavengers [2]. Perivascular macrophages are also the first cells interacting with leukocytes entering the CNS during infection or autoimmunity. These phagocytes show constitutive expression of class I and class II MHC molecules, which is enhanced during CNS inflammation [5, 21]. Perivascular cells also express T-cell costimulatory molecules in autoimmune CNS diseases [22]. Studies performed in bone marrow-chimeric animals after induction of EAE have shown that perivascular cells and other CNS-associated macrophages play an important role in presenting antigen to autoreactive T cells [23]. Moreover, when these cells were isolated from the CNS of normal rats, they acted as APC capable of stimulating T cell proliferation and cytokine secretion ex vivo [24]. Whether these phagocytes migrate out of the CNS and prime T cells in peripheral organs has not been demonstrated and is indeed difficult to prove experimentally.

Microglia

In addition to the above-mentioned mononuclear phagocytes, which are associated with the CNS but do not reside in the brain parenchyma proper, the CNS harbors a network of resident, silenced tissue macrophages called microglia which can transform into potent immune effectors cells [5]. Unlike all other cell types in the CNS, microglia derive from bone marrow progenitors that enter the CNS during embryogenesis. In normal conditions, microglia are ramified non-phagocytic cells that lack or express low levels of typical macrophage markers (e.g. CD14, CD45, Fc receptors). In contrast to perivascular cells, microglia represent a stable cellular pool and show a slow turnover with hematogenous cells [2]. The main function of microglia in the adult CNS is to respond to even minor alterations in brain homeostasis and to undergo progressive activation leading to the acquisition of a macrophage phenotype [5, 25]. Activated microglia secrete cytokines (e.g. IL-1, TNF-α, transforming growth factor beta (TGF-β), and chemokines), display cytostatic and cytotoxic potential, and become phagocytic when neuronal or myelin damage occurs.

Microglia are the only intraparenchymal cell type expressing class II MHC molecules in the normal adult CNS [26, 27]. In the human CNS, class II MHC expression has been detected mainly on microglia in the white matter, whereas in different rodent strains class II MHC levels are either undetectable or detectable only on a small percentage of microglia. Upregulation of MHC molecules is one of the first signs of microglia activation in response to a wide range of damaging stimuli. In humans, high levels of class II MHC molecules have been detected on activated microglia in the CNS of individuals with MS and neurodegenerative or virus-induced diseases [26-30].

Activated microglia display several features that allow them to function as efficient APC. They take-up particles and microbes by phagocytosis, express receptors such as Fcγ that mediate active endocytosis, and process protein antigens efficiently [5, 31-33]. Microglia express several adhesion/costimulatory molecules that interact with receptors on T cells to promote APC-T cell adhesion and activation, such as CD11a, CD40, CD54, CD80 and CD86 [22, 34-36]. Moreover, microglia can secrete IL-12 [37, 38], a cytokine which enhances both innate and acquired immunity and has a pivotal role in the generation of Th1 responses [39]. These properties are readily upregulated in vitro by inflammatory stimuli (e.g. IFN-γ or bacterial components) or can be demonstrated in activated microglia in MS lesions or in EAE. Cultured microglia from rat or human brain act as efficient APC for the activation of naive, memory and differentiated T cells [32, 33, 40-42]. Microglia from mouse brain restimulate very efficiently both Th1 and Th2 cells, inducing T cell proliferation and cytokine secretion [43]. This suggests that antigen-presenting microglia enhance pro-inflammatory Th1 responses – involved in host defense, virus-induced tissue damage and autoimmunity – as well as Th2 responses with anti-inflammatory and Th1-downregulatory effects [44]. In a hierarchy of APC, microglia appear to be as efficient as dendritic cells and much more efficient than B cells in the restimulation of effector T helper cells, but are less potent than dendritic cells in T cell priming [45]. Interactions between microglia and T cells represent an important trigger for microglia activation and upregulation of APC function [46, 47]. In the Th1-mediated disease EAE [38], and during antigen presentation to Th1 cells in vitro [48], microglia secrete IL-12 that may favor the development of Th1 responses during CNS infection and autoimmunity.

Collectively, the available in situ and in vitro evidence indicates that microglia may contribute to the reactivation of T cells infiltrating the CNS. This function is particularly relevant in MS where phagocytosis of myelin by class II MHC-positive microglia could play an important role in the presentation of myelin antigens to $CD4^+$ T cells [19, 49]. According to Ford et al. [46] who studied the APC function of microglia in a rat model of graft-versus-host disease, activated microglia only partially induce T cells, leading to: increased T cell volume but no proliferation; secretion of IFN-γ and TNF-α but not IL-2; and T cell death by apoptosis. T cell apoptosis following antigen-specific microglia-T cell interactions may be involved in the termination of effector T cell responses within the CNS.

MHC Molecule Expression on Neuroectodermally-Derived Cells

Astrocytes

Astrocytes are the most abundant CNS glial cell type. The main function of astrocytes in the adult brain is to support neuronal activity and metabolism [50]. By extending their processes around brain microvessels, astrocytes together with microglia, constitute the perivascular glia limitans. Expression of class I and class II MHC molecules on astrocytes is absent in the normal brain and generally low or undetectable in neuroinflammatory diseases. Few studies, however, have reported the presence of some class II MHC-positive, reactive astrocytes in MS lesions and during viral infection in mice [51, 52]. Intracerebral injection of IFN-γ induces class II MHC expression mainly on microglia, but also on a few astrocytes [53]. From these in situ studies it appears that astrocytes have a marginal role as APC in the inflamed CNS. In the mid 1980s, astrocytes were the first CNS cell type found to act as APC and to stimulate T cell proliferation [54]. At least in vitro, class I and class II MHC and adhesion molecules (CD54, CD106) are induced on astrocytes by IFN-γ, IL-1 or TNF-α [55]. Despite conflicting results, the actual view is that cultured astrocytes are inefficient in processing and presenting protein antigen for the restimulation of T cells [41, 43], are unable to prime T cells [45, 56], do not express costimulatory molecules [22, 43], and fail to secrete IL-12 [37]. Astrocytes are however able to present synthetic antigenic peptide and to activate Th2 cells as efficiently as dendritic cells and microglia, in contrast to their poor capacity to restimulate Th1 cells [43]. This suggests that, in the presence of appropriate peptides (possibly generated during immune-mediated CNS damage), astrocytes could contribute to the restimulation of Th2 cells secreting anti-inflammatory cytokines (IL-4, IL-10). Cultured astrocytes can induce T-cell apoptosis [57], suppress activation of T cells triggered by other APC [58, 59], and inhibit class II MHC, CD54 [60], and IL-12 [37] expression in macrophages and microglia, possibly via secretion of anti-inflammatory mediators such as TGF-β and prostaglandin E_2. Collectively, these in vitro findings suggest that astrocytes may play a role in preventing or limiting CNS inflammation.

Oligodendrocytes

Oligodendrocytes, the myelin-forming cells of the CNS, can be infected by several neurotropic viruses. Myelin is thought to be the primary target of autoimmune attack in MS [61]. At present there is no convincing evidence that oligodendrocytes express class I or class II MHC molecules in situ, suggesting that oligodendrocytes may escape immune recognition by $CD8^+$ and $CD4^+$ T lymphocytes. However, cultured oligodendrocytes express class I MHC molecules after treatment with IFN-γ and TNF-α or virus infection ([62] and references therein), and are lysed by MBP-specific $CD8^+$ T cell lines [63]. According to some studies [64, 65], oligodendrocyte damage in MS could be mediated by γδ T cells which may recognize antigen (possibly heat shock proteins) in a non-MHC-restricted manner.

Work by Linington and collaborators in the EAE model indicates that antibodies against proteins expressed on the myelin surface (particularly myelin oligodendrocyte glycoprotein) may play a major role in CNS demyelination during T-cell-mediated inflammation [66]. Antigen-antibody complexes on the myelin surface may bind to complement or to Fc receptors expressed on activated macrophages and microglia, leading to myelin phagocytosis and destruction.

Neurons

Lack of class I MHC expression on neurons in situ and in primary cultures has been implicated as a possible mechanism for viral persistence in the CNS [67]. Neumann et al. [68, 69] have shown that exposure of hippocampal neurons to IFN-γ induces class I MHC heavy chain, β2-microglobulin and TAP1/TAP2 peptide transporter gene transcription and cell-surface expression of class I MHC molecules. This occurs in a minority of cultured hippocampal neurons after stimulation with IFN-γ and is strongly enhanced when the electrical activity of neurons is suppressed with the sodium channel blocker tetrodotoxin. These findings suggest that negative signals provided during normal bioelectric activity suppress the inducible expression of class I MHC molecules in neurons; this in turn would allow neurons to escape immune surveillance. In contrast, infection of neurons leading to functional damage and loss of bioelectric activity would render neurons susceptible to recognition and elimination by cytotoxic $CD8^+$ T lymphocytes.

The limited expression of MHC molecules throughout the CNS parenchyma led Wekerle and coworkers to hypothesize that neurons might play a role in inhibiting MHC expression on surrounding glial cells [70, 71]. Using hippocampal slices, Neumann et al. [71] have shown that electrically active neurons almost completely suppress the IFN-γ-inducible expression of class II MHC molecules on microglia and astrocytes. Neurotrophins, particularly nerve growth factor, mediated this inhibitory effect [72], indicating that molecules released during normal neuronal activity can limit the potential for antigen presentation in the CNS. This implies that neuronal loss or dysfunction results in upregulation of MHC molecules. Futhermore, these observations are consistent with increased class II MHC expression on activated microglia in human neurodegenerative diseases (Alzheimer's disease, Parkinson's disease, amyotropic lateral sclerosis) [28, 29] and with the preferential accumulation of encephalitogenic T cells in areas with neuronal lesions [73].

Based on the available evidence, possible interactions between different CNS cell types and T cell subpopulations are schematically illustrated in Fig. 1.

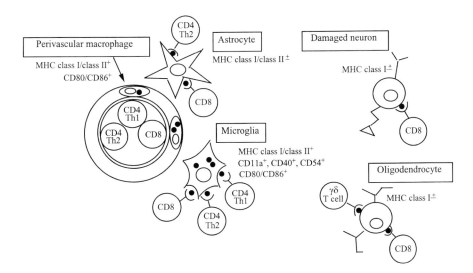

Fig. 1. Possible interactions between T cells extravasating from a CNS capillary and different CNS cell populations. Due to their strategic location close to the BBB and constitutive expression of MHC molecules, perivascular macrophages are likely to be the first APC interacting with CD8+ or CD4+ T cells crossing the endothelium of the cerebral microvessels. Astrocytes microglia contribute to the formation of the perivascular glia limitans and may readily interact with T cells infiltrating the CNS parenchyma. Activated microglia upregulate expression of MHC and adhesion/costimulatory molecules in the inflamed CNS and function as efficient APC for the activation of CD8+ and CD4+ (both Th1 and Th2) T cells in vitro. MHC expression is scarcely induced on astrocytes in vivo but at least in vitro astrocytes are able to present antigenic peptides and to stimulate Th2 cells efficiently. Although there is no clear evidence for MHC expression on oligodendrocytes and neurons in situ, in vitro studies suggest that both oligodendrocytes and damaged, inactive neurons could express MHC class I after exposure to IFN-γ and present viral or auto-antigens to CD8+ cytototoxic T cells. Oligodendrocyte-associated antigens could also be recognized by γδ T cells in a non-MHC restricted manner

Influence of the CNS Microenvironment on T Cells

Increasing experimental evidence supports the view that the CNS provides an unfavorable environment for the propagation of T cell responses. T cells infiltrating the CNS during EAE do not proliferate [74], whereas T cells recovered from the CNS of virus-infected mice secrete IFN-γ but not IL-2 and do not proliferate when stimulated in vitro [75]. In EAE, autoreactive encephalitogenic CD4[+] T cells undergo apoptosis after being recruited into the CNS parenchyma [76]. T-cell apoptosis could be the result of activation-induced cell death following antigen recognition on CNS APC, although it appears that antigen-independent mechanisms are also involved in T cell elimination [77]. Immunoregulatory molecules present in the CNS parenchyma or CSF, such as TGF-β, α-melanocyte stimulating hormone, vasoactive intestinal peptide and gangliosides, are implicated in suppressing T cell responses [4, 75, 78]. Deletion of peripherally activated T cells recognizing CNS antigen could also occur via Fas-Fas ligand interactions [79], since Fas ligand appears to be expressed on glial cells in the inflamed CNS parenchyma, on brain tumors, and possibly also on neurons [80-82]. All the above-mentioned mechanisms may contribute to terminate T cell responses once they have been initiated in the CNS and to actively maintain immune privilege in the normal CNS. Which of the T-cell inhibitory mechanisms operating in the CNS might fail and contribute to the development of autoimmune CNS diseases is an important issue which remains to be explored.

References

1. Wekerle H, Linington C, Lassmann H, Meyermann R (1986) Cellular immune reactivity within the CNS. Trends Neurosci 9: 271-277
2. Lassmann H, Schmied M, Vass K, Hickey WF (1993) Bone marrow derived elements and resident microglia in brain inflammation. Glia 7: 19-24
3. Hickey WF, Hsu BL, Kimura H (1991) T-cell entry into the rat central nervous system. J Neurosci Res 28: 254-260
4. Cserr HF, Knopf PN (1992) Cervical lymphatics, the blood-brain barrier and the immunoreactivity of the brain: A new view. Immunol Today 13: 507-512
5. Kreutzberg GW (1996) Microglia: A sensor for pathological events in the CNS. Trends Neurosci 19: 312-318
6. Ransohoff RM (1997) Chemokines in neurological disease models: Correlation between chemokine expression patterns and inflammatory pathology. J Leukoc Biol 62: 645-652
7. Banchereau J, Steinman RM (1998) Dendritic cells and the control of immunity. Nature 392: 245-252
8. Knopf PN, Harling-Berg CJ, Cserr HF et al. (1998) Antigen-dependent intrathecal antibody synthesis in the normal rat brain: Tissue entry and local retention of antigen-specific B cells. J Immunol 161: 692-701
9. Tourtellotte WW, Walsh MJ, Baumhefner RW et al. (1984) The current status of multiple sclerosis intra-blood-brain-barrier IgG synthesis. Ann N Y Acad Sci 436: 52-67
10. Tyor WR, Griffin DE (1993) Virus specificity and isotype expression of intra-

parenchymal antibody-secreting cells during Sindbis virus encephalitis in mice. J Neuroimmunol 48: 37-44
11. Poltorak M, Freed WJ (1995) Transplantation into the central nervous system. In: Keane RW, Hickey WF (eds) Immunology of the nervous system. Oxford University, Oxford, pp 611-641
12. Matyszak MK, Perry VH (1995) Demyelination in the central nervous system following a delayed-type hypersensitivity response to bacillus Calmette-Guèrin. Neuroscience 64: 967-977
13. Matyszak MK, Perry VH (1996) The potential role of dendritic cells in immune-mediated inflammatory responses in the central nervous system. Neuroscience 74: 599-608
14. Serot JM, Foliquet B, Bene MC, Faure GC (1997) Ultrastructural and immunohistological evidence of dendritic-like cells within human choroid plexus epithelium. Neuroreport 8: 1995-1998
15. Hanly A, Petito CK (1998) HLA-DR positive dendritic cells of the human choroid plexus: A potential reservoir of HIV in the central nervous system. Hum Pathol 29: 88-93
16. Raine CS (1985) Experimental allergic encephalomyelitis and experimental allergic neuritis. In: Koetsier JC (ed) Demyelinating diseases. Elsevier Science, Amsterdam, pp 429-503 (Handbook of clinical neurology, vol 47)
17. Owens T, Renno T, Taupin V, Krakowski M (1995) Inflammatory cytokines in the brain: does the CNS shape immune responses? Immunol Today 15: 566-571
18. Hafler DA, Weiner HL (1995) Immunologic mechanisms and therapy in multiple sclerosis. Immunol Rev 144: 75-107
19. Wucherpfennig KW (1995) Autoimmunity in the central nervous system: mechanisms of antigen presentation and recognition. Clin Immunol Immunopathol 72: 293-306
20. Oldstone MBA (1990) Molecular mimicry and autoimmune disesase. Cell 50: 819-820
21. Graeber MB, Streit WJ, Buringer D et al. (1992) Ultrastructural location of major histocompatibility complex (MHC) class II perivascular cells in histologically normal human brain. J Neuropathol Exp Neurol 51: 303-311
22. De Simone R, Giampaolo A, Giometto B et al. (1995) The costimulatory molecule B7 is expressed on human microglia in culture and in multiple sclerosis acute lesions. J Neuropathol Exp Neurol 54: 175-187
23. Hickey WF, Kimura H (1988) Perivascular microglial cells of the CNS are bone-marrow derived and present antigen in vivo. Science 239: 290-292
24. Ford AL, Goodsall AL, Hickey WF, Sedgwick JD (1995) Normal adult ramified microglia separated from other central nervous system macrophages by flow cytometric sorting. Phenotypic differences defined and direct ex vivo antigen presentation to myelin basic protein-reactive CD4$^+$ T cells compared. J Immunol 154: 4309-4321
25. Perry VH, Gordon S (1988) Macrophages and microglia in the nervous system. Trends Neurosci 11: 273-279
26. Hayes GM, Woodroofe MN, Cuzner LM (1987) Microglia are the major cell type expressing MHC class II in human white matter. J Neurol Sci 80: 25-37
27. Ulvestad E, Williams K, Bö L et al. (1994) HLA class II molecules (HLA-DR, -DP, -DQ) on cells in the human CNS in situ and in vitro. Immunology 82: 535-541
28. McGeer P, Itagaki S, Boyes BE, McGeer EG (1988) Reactive microglia are positive for HLA-DR in the substantia nigra of Parkinson's and Alzheimer's disease. Neurobiology 38: 1285-1291
29. McGeer P, Kawamato T, Walker DG et al. (1993) Microglia in degenerative neurological disease. Glia 7: 84-92
30. An SF, Ciardi A, Giometto B et al. (1996) Investigation on the expression of major histocompatibility complex class II and cytokines and detection of HIV-1 DNA within

brains of asymptomatic and symptomatic HIV-1-positive patients. Acta Neuropathol 91: 494-503
31. Ulvestad E, Williams K, Vedeler C et al. (1994) Reactive microglia in multiple sclerosis lesions have an increased expression of receptors for the Fc part of IgG. J Neurol Sci 121: 125-131
32. Cash E, Rott O (1994) Microglial cells qualify as the stimulators of unprimed $CD4^+$ and $CD8^+$ T lymphocytes in the central nervous system. Clin Exp Immunol 98: 313-318
33. Frei K, Siepl C, Groscurth P et al. (1987) Antigen presentation and tumor cytotoxicity by interferon-γ treated microglial cells. Eur J Immunol 17: 1271-1278
34. Cannella B, Raine CS (1995) The adhesion molecule and cytokine profile of multiple sclerosis lesions. Ann Neurol 37: 424-435
35. Gerritse K, Laman JD, Noelle RJ et al. (1996) CD40-CD40 ligand interactions in experimental allergic encephalomyelitis and multiple sclerosis. Proc Natl Acad Sci USA 93: 2499-2504
36. Issazadeh S, Navikas V, Schaub M et al. (1998) Kinetics of expression of costimulatory molecules and their ligands in murine relapsing autoimmune encephalomyelitis in vivo. J Immunol 161: 1104-1112
37. Aloisi F, Penna G, Cerase J et al. (1997) IL-12 production by central nervous system microglia is inhibited by astrocytes. J Immunol 159: 1604-1612
38. Krakowski ML, Owens T (1997) The central nervous system environment controls effector $CD4^+$ T cell cytokine profile in experimental allergic encephalomyelitis. Eur J Immunol 27: 2840-2847
39. Trinchieri G (1995) Interleukin-12: A proinflammatory cytokine with immunoregulatory functions that bridge innate resistance and antigen-specific adaptive immunity. Annu Rev Immunol 13: 251-276
40. Williams K, Ulvestad E, Cragg L et al. (1993) Induction of primary T cell responses by human glial cells. J Neurosci Res 36: 382-390
41. Matsumoto Y, Ohmori K, Fujiwara M (1992) Immune regulation by brain cells in the central nervous system: Microglia but not astrocytes present myelin basic protein to encephalitogenic T cells under in vivo-mimicking conditions. Immunology 76: 209-216
42. Dhib-Jalbut S, Gogate N, Jiang H et al. (1996) Human microglia activate lymphoproliferative responses to recall viral antigens. J Neuroimmunol 65: 67-73
43. Aloisi F, Ria F, Penna G, Adorini L (1998) Microglia are more efficient than astrocytes in antigen processing and in Th1 but not Th2 cell activation. J Immunol 160: 4671-4680
44. Abbas AK, Murphy KM, Sher A (1996) Functional diversity of helper T lymphocytes. Nature 383: 787-793
45. Aloisi F, Ria F, De Simone R et al. (1998) Relative efficiency of microglia, astrocytes, dendritic cells and B cells in naive $CD4^+$ T cell priming and Th1/Th2 cell restimulation. (*submitted*)
46. Ford AL, Foulcher E, Lemckert FA, Sedgwick JD (1996) Microglia induce CD4 T lymphocyte final effector function and death. J Exp Med 184: 1737-1745
47. Renno T, Krakowski M, Piccirillo C et al. (1995) TNF-alpha expression by resident microglia and infiltrating leukocytes in the central nervous system of mice with experimental allergic encephalomyelitis. Regulation by Th1 cytokines. J Immunol 154: 944-953
48. Aloisi F, Penna G, Polazzi E et al. (1999) CD40-CD154 interaction and IFN-γ are required for IL-12 but not prostaglandin E_2 secretion by microglia during antigen presentation to Th1 cells. J Immunol (*in press*)

49. Li H, Newcombe J, Groome P, Cuzner ML (1993) Characterization and distribution of phagocytic macrophages in multiple sclerosis plaques. Neuropathol Appl Neurobiol 19: 214-223
50. Kimelberg HK, Norenberg MD (1989) Astrocytes. Sci Am 260: 66-72
51. Lee SC, Moore GRW, Golenwsky G, Raine CS (1990) Multiple sclerosis: A role for astroglia in active demyelination suggested by class II MHC expression and ultrastructural study. J Neuropathol Exp Neurol 49: 122-136
52. Morris MM, Dyson H, Baker D et al. (1997) Characterization of the cellular and cytokine response in the central nervous system following Semliki Forest virus infection. J Neuroimmunol 74: 185-197
53. Vass K, Lassmann H (1990) Intrathecal application of interferon gamma: Progressive appearance of MHC antigens within the rat nervous system. Am J Pathol 137: 789-800
54. Fontana A, Fierz W, Wekerle H (1984) Astrocytes present myelin basic protein to encephalitogenic T-cell lines. Nature 307: 273-275
55. Merrill, JE, Benveniste EN (1996) Cytokines in inflammatory brain lesions: Helpful and harmful. Trends Neurosci 19: 331-338
56. Sedgwick JD, Mössner R, Schwender S, ter Meulen V (1991) Major histocompatibility complex-expressing nonhematopoietic astroglial cells prime only $CD8^+$ T lymphocytes: Astroglial cells as perpetuators but not initiators of $CD4^+$ T cell responses in the central nervous system. J Exp Med 173: 1235-1246
57. Gold R, Schmied M, Tontsch U et al. (1996) Antigen presentation by astrocytes primes rat T lymphocytes for apoptotic cell death: A model for T cell apoptosis in vivo. Brain 119: 651-659
58. Matsumoto Y, Hanawa H, Tsuchida M, Abo T (1993) In situ inactivation of infiltrating T cells in the central nervous system with autoimmune encephalomyelitis. The role of astrocytes. Immunology 79: 381-388
59. Meinl E, Aloisi F, Ertl B et al. (1994) Multiple sclerosis. Immunomodulatory effects of human astrocytes on T cells. Brain 117: 1323-1330
60. Hailer NP, Heppner FL, Haas D, Nitsch R (1998) Astrocytic factors deactivate antigen presenting cells that invade the central nervous system. Brain Pathol 8: 459-474
61. Steinman L (1996) Multiple sclerosis: a coordinated attack against myelin in the central nervous system. Cell 85: 299-302
62. Agresti C, Bernardo A, Del Russo N et al. (1998) Synergistic stimulation of MHC class I and IRF-1 gene expression by IFN-γ and TNF-α in oligodendrocytes. Eur J Neurosci 10: 2975-2983
63. Jurewicz A, Biddison WE, Antel JP (1998) MHC class I-restricted lysis of human oligodendrocytes by myelin basic protein peptide-specific CD8 T lymphocytes. J Immunol 160: 3056-3059
64. Selmaj K, Brosnan CF, Raine CS (1991) Colocalization of lymphocytes bearing γδ T-cell receptor and heat shock protein hsp65-positive oligodendrocytes in multiple sclerosis. Proc Natl Acad Sci USA 88: 6452-6456
65. Freedman MS, Bitar R, Antel JP (1993) γδ T-cell-human glial cell interactions. II. Relationship between heat shock protein expression and susceptibility to cytolysis. J Neuroimmunol 74: 143-148
66. Linington C, Bradl M, Lassmann H et al. (1988) Augmentation of demyelination in rat acute allergic encephalomyelitis by circulating mouse monoclonal antibodies directed against a myelin/oligodendrocyte glycoprotein. Am J Pathol 130: 443-454
67. Joly E, Mucke L, Oldstone MBA (1991) Viral persistence in neurons explained by lack of major histocompatibility class I expression. Science 253: 1283-1285

68. Neumann H, Cavaliè A, Jenne DE, Wekerle H (1995) Induction of MHC class I genes in neurons. Science 269: 549-552
69. Neumann H, Schmidt H, Cavaliè A et al. (1997) Major histocompatibility complex (MHC) class I gene expression in single neurons of the central nervous system: Differential regulation by IFN-γ and tumor necrosis factor-α. J Exp Med 185: 305-316
70. Aloisi F, Wekerle H (1990) Immune reactivity in the central nervous system: Intercellular control of the expression of major histocompatibility antigens. In: Levi G (ed) Differentiation and functions of glial cells. Wiley Liss, New York, pp 371-378
71. Neumann H, Boucraut J, Hahnel C et al. (1996) Neuronal control of MHC class II inducibility in rat astrocytes and microglia. Eur J Neurosci 8: 2582-2590
72. Neumann H, Misgeld T, Matsumuro K, Wekerle H (1998) Neurotrophins inhibit major histocompatibility class II inducibility of microglia: Involvement of the p75 neurotrophin receptor. Proc Natl Acad Sci USA 95: 5779-5784
73. Maehlen J, Olsson T, Zachau A et al. (1989) Local enhancement of major histocompatibility complex (MHC) class I and class II expression and cell infiltration in experimental allergic encephalomyelitis around axotomized motor neurons. J Neuroimmunol 23: 125-132
74. Ohmori, K, Hong Y, Fujiwara M, Matsumoto Y (1992) In situ demonstration of proliferating cells in the rat central nervous system during experimental autoimmune encephalomyelitis. Evidence suggesting that most infiltrating T cells do not proliferate in the target organ. Lab Invest 66: 54-62
75. Irani DN, Lin K-I, Griffin DE (1997) Regulation of brain-derived T cells during acute central nervous system inflammation. J Immunol 158: 2318-2326
76. Bauer J, Wekerle H, Lassmann H (1995) Apoptosis in brain-specific autoimmune diseases. Curr Opin Immunol 7: 839-843
77. Bauer J, Bradl M, Hickey WF et al. (1998) T-cell apoptosis in inflammatory brain lesions. Destruction of T cells does not depend on antigen recognition. Am J Pathol 153: 715-724
78. Taylor AW, Streilein JW (1996) Inhibition of antigen-stimulated effector T cells by human cerebrospinal fluid. Neuroimmunomodul 3: 112-118
79. Sakata K, Sakata A, Kong L et al. (1998) Role of Fas/FasL interaction in physiology and pathology: the good and the bad. Clin Immunol Immunopathol 87: 1-7
80. D'Souza SD, Bonetti B, Balasingam V et al. (1996) Multiple sclerosis: Fas signaling in oligodendrocyte cell death. J Exp Med 184: 2361-2370
81. Dowling P, Shang G, Raval S et al. (1996) Involvement of the CD95 (APO-1/Fas) receptor/ligand system in multiple sclerosis brain. J Exp Med 184: 1513-1518
82. Saas P, Walker PR, Hahne R et al. (1997) Fas ligand expression by astrocytoma in vivo: maintaining immune privilege in the brain? J Clin Invest 99: 1173-1178

Chapter 11

Myelination of the Central Nervous System

G.G. Consalez[1], V. Avellana-Adalid[2], C. Alli[1], A. Baron Van Evercooren[2]

Developmental CNS Myelination: How Relevant Is It to Multiple Sclerosis?

Multiple sclerosis (MS) results from a combination of genetically determined susceptibility, environmental factors, viral or bacterial agents, soluble cytokines released during the inflammatory and autoimmune responses, and probably other, as yet undetermined etiologic agents. The disease can cause variable degrees of tissue destruction in the central nervous system (CNS), ranging from marginal demyelination to complete oligodendrocyte loss, severe glial scarring [1], and axonal transection. In some instances oligodendrocytes are morphologically preserved in demyelination plaques and remain capable of differentiating and remyelinating, as shown in humans and various model systems [1, 3]. In other cases, however, oligodendrocytes vanish and progenitors have to migrate into the plaque and proliferate. Adult oligodendrocyte progenitors are different from their post-natal counterparts, but seem to retain the ability to migrate and proliferate under the influence of specific growth factors [4]. However, the proliferating pool is limited, as is its ability to migrate [5, 6].

Besides the intrinsic, immediate effects of demyelination on nervous conduction, the early downregulation of myelin gene products in response to various pathogenetic mechanisms may exert its harmful effects by making distal extensions of the glial cell vulnerable, and by causing the presentation of novel antigens to the immune system. Furthermore, in response to a local inflammatory reaction, a switch may occur in the differentiation of common glial cell progenitors, leading to preferential differentiation into astrocytes. Astrocyte hyperplasia and scarring is a prominent feature of MS lesions, and represents a potential obstacle to remyelination. Oligodendrocyte depletion and astrocyte proliferation may therefore compromise the clinical picture, ultimately leading to a complete and permanent inhibition of axonal conduction.

Thus, one important factor in the clinical evolution of the disease lies in the ability of oligodendrocytes or their precursors to proliferate, differentiate and

---

[1] Department of Neuroscience, San Raffaele Scientific Institute, Via Olgettina 58 - Milan, Italy. e-mail: g.consalez@hsr.it
[2] INSERM 134, Cellular, Molecular and Clinical Neurobiology, Hôpital de la Salpetrière, Paris, France

produce remyelination. The physiologic modulation of oligodendrocyte gene expression in the course of MS can play a major role in setting the outcome of demyelinating diseases. It is a commonly accepted notion that remyelination in MS lesions recapitulates some of the molecular mechanisms active during late embryonic and post-natal CNS development. Therefore, an improved knowledge of the mechanisms that regulate developmental CNS myelination, including cell type specification and terminal differentiation of CNS glial cells, is a prerequisite for the elaboration of rationally designed drugs or gene therapy vectors promoting a remyelinating response.

What Is Glia?

The term neuroglia means "nerve glue", and reflects Rudolf Virchow's original belief that glia is the connective tissue element of the central nervous system [7]. Virchow believed glia to be of mesodermal origin, a conviction later corrected by Wilhelm His who, in 1889, established the notion that glial cells, just like neurons, originate from the ectodermal germ layer.

A possible exception to this rule may be represented by microglia, one of the three main glial cell types present in the CNS. According to the original view [8], the term microglia defines cells of mesenchymal (not epithelial) origin, that arise in the pia mater during the prenatal period and remain quiescent until activated and transformed into macrophages by trauma or infection. More recent work, however, suggests that these cells can also arise from monocytes leaving the bloodstream during embryogenesis and going through a series of intermediate phenotypes, to eventually differentiate into microglia [9]. Thus, microglia might in fact derive from cells of mesodermal origin. The recognized functions of microglia in the CNS (Table 1) include phagocytosis of cellular debris, induction of the chromatolytic response to axotomy (i.e. disappearance of the Nissl substance coupled with swelling of the neuronal cell body), and a contribution to the immune response within neural structures.

The term macroglia, on the contrary, designates cells – astrocytes and oligodendrocytes – that are uniquely of neuroectodermal origin, arising in the neuroepithelium, or ventricular zone of the neural tube. Although a fundamental role is played by microglia in the acute phase of demyelinating disorders, here we will mainly deal with the mechanisms regulating the development of macroglia.

Table 1. Functions of microglial cells (modified from [87] with permission)

Initiation of Schwann cell mitosis
Immune response
Chromatolytic response to axotomy
Phagocytosis of cellular debris

Recent payments may not be reflected on this statement.

When inquiring about this account or when informing us of changes in personal information, (insurance coverage, address, etc.) please indicate the Medical Record Number and Date of Service.

INTERNET ADDRESS:
www.henryford.com

EMAIL:
bills@hfhs.org

DIRECT LETTERS TO:

CUSTOMER SERVICES
BOX 339
TROY, MI 48099-0339

DIRECT PAYMENTS TO:

DEPT. 55115
HENRY FORD HEALTH SYSTEM
P.O. BOX 55000
DETROIT, MI 48255-0115

Please pay amount due before the due date. Any account balance that extends beyond that date may be subject to the collection process.

$ 15.00 will be charged for returned checks

▼ **PLEASE DETACH AT PERFORATION** ▼

Table 2. Functions of astrocytes (modified from [87] with permission)

Transport from blood to neurons
Induction of vascular endothelial blood-brain barrier
Axonal guidance
Stimulation of neurite outgrowth
Regulation of neuron morphogenesis
Compartmentalization of neurons
Phagocytosis of cellular debris
Uptake of neuroactive peptides and extracellular glutamate
Production of PDGF and CNTF
Production of IGF
Production of a and b FGF
Production of GMF
Production of NGF
Regulation of extracellular K^+
Regulation of blood supply

PDGF, plateled-derived growth factor; *CNTF*, ciliary neurotrophic factor; *IGF*, insulin-related growth factor; *bFGF*, basic fibroblast growth factor; *GMF*, glial maturation factor; *NGF*, nerve growth factor.

Astrocytes were named by Lenhossék [10], 18 years after their original identification by Camillo Golgi in 1873. Parallel studies by Koelliker and Andriezen, in 1893, led to their subdivision into protoplasmic and fibrous types. The former inhabit the white matter, while the latter, characterized by the presence of fibers in the cytoplasm, are seen mostly in the gray matter. Functions performed by astrocytes are listed in Table 2.

Lenhossék and Cajal, and in recent times Rakic [11, 12] and Choi [13], demonstrated that astrocytes are the cellular constituents of radial glia, and provided the earliest example in development of differentiating glial cells. Radial glia have an early embryonic role in guiding the migration of young neurons to their destination in the mammalian brain [11, 14] or cerebellum [15, 16], where these cells are referred to as Bergmann glia. The neural-glial interaction involves reciprocal recognition between glia and newborn neurons. Interactions between neurons and glia occur through a number of surface adhesion molecules, mainly astrotactin, a protein that provides a receptor system for glia-guided neuronal migration [17, 18]. At the end of development, astrocytes are active in a wide variety of functions, including the production of trophic factors and of factors affecting the growth and guidance of axons, as well as the uptake of neuroactive peptides.

Oligodendrocytes were first discovered by W. Robertson [19, 20], who initially missed their role in myelination and postulated their mesodermal origin. Later, they were rediscovered by Rio-Hortega [21] who first distinguished them from astrocytes by virtue of their shorter and sparser processes. He was the first to conjecture that oligodendrocytes might be implicated in myelination of the central nervous system. He also postulated correctly that astrocytes and oligodendro-

Table 3. Functions of oligodendrocytes (modified from [87] with permission)

Myelination in CNS
Inhibition of neurite outgrowth
Uptake of neuroactive peptides

CNS, central nervous system.

cytes might stem from a common progenitor migrated into the white matter to proliferate and differentiate.

The functions of oligodendrocytes are listed in Table 3. Fundamentally, oligodendrocytes are in charge of ensheathing nerve fibers with myelin [22], a key event in the establishment of efficient nerve conduction. Myelination is the end point of oligodendrocyte differentiation, and is tightly regulated in post-natal development. In CNS development, this process is inhibited until the end of neuronal differentiation and axonogenesis, as premature myelination would lead to inhibition of neurite outgrowth [23].

How Do Progenitors Decide to Become Glia and Not Neurons?

In the CNS, neuronal and glial progenitors are both generated in the neuroepithelium, a pseudostratified epithelium in which cells, tethered at their basal and apical ends, proliferate by means of a cell-cycle-tied motion of the nucleus and cell body, that shuttles from the ventricular to the apical end of the layer, and back (Fig. 1). This process is named interkinetic nuclear migration [24]. The neuroepithelium occupies the ventricular zone of the neural tube, i.e. the layer of columnar cells adjacent to the ventricular cavities and, in the spinal cord, to the central canal, or prospective ependymal canal. Neuroepithelial cells are referred to as neuroblasts or neural progenitors, where the term neural designates cells fated to become glia or neurons.

In reality, different cell-types are present in the ventricular zone, including multipotent neuron-glia progenitors, as well as cells restricted to form either cell type (reviewed in [25, 26]). While clonal analysis has demonstrated the existence of neuronal progenitors in the mouse neural tube very early in development, the earliest precursors of spinal cord oligodendrocytes are found in the ventral portion of the mouse neural tube at embryonic days 12-14 [27, 28].

Several factors have been implicated in the choice between neuronal and glial cell fate in mammalians as well as in other model systems more or less distant in phylogeny (e.g. the fruifly *Drosophila melanogaster*, the grasshopper, the chick). In one study by the group of Sally Temple at Albany Medical College [29], murine neuronal/glial progenitors in cell culture were exposed to increasing levels of the 18 kDa isoform of basic fibroblast growth factor (bFGF, also known as FGF2), a soluble protein released into the extracellular space either as a result of plasma membrane damage [30] or through an energy-dependent pathway [31]. In this

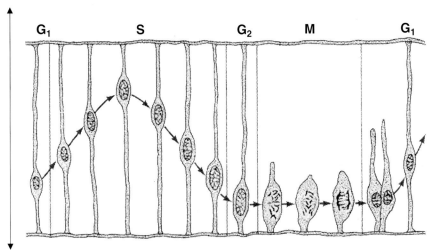

Fig. 1. The proliferation of neural (neuronal *and* glial) precursors takes place initially in the ventricular zone, a pseudostratified epithelium immediately adjacent to ventricular cavities or to the prospective ependymal canal. In this epithelial layer, neural precursors shuttle from the ventricular to the apical end as they cycle mitotically. $G1, S, G2$, and M represent phases of the cell cycle. Cells divide (M phase) as they reach the ventricular extremity of their shuttling range. Proliferating neural progenitors receive important cell type specification signals in this cell layer. Cells that exit the ventricular zone after mitosis are found in the thin subventricular zone, where they still proliferate and undergo further differentiation. Eventually, they migrate to the mantle region of the neural tube, or to the telencephalic cortex. (Modified from [88] with permission)

study, neuroepithelial cell clones exposed to low concentrations of bFGF developed mostly, if not uniquely, as clusters of neuronal progenitors; on the contrary, neuroblast clones exposed to higher levels of bFGF contained a remarkably greater percentage of glial cells. Thus, in mammalian systems, bFGF may be part of a cell-type specification switch controlling the fate of bipotential (neural/glial) progenitors. A similar role is proposed for thyroid hormone (T3), that orients embryonic neural stem cells to the oligodendrocyte phenotype [32, 33].

Additional factors affecting the cell fate choice between glia and neurons have been identified and studied in other model organisms. Perhaps the best characterized one was identified in *Drosophila melanogaster*, and involves the *glial cell missing* (*gcm*) gene. In parallel, the groups of Corey Goodman at UCB [34], Yoshiki Hotta in Tokyo [35], and Angela Giangrande in Strasbourg [36] isolated a factor, eventually classified as *gcm*, necessary to direct glial fate commitment and the cell fate switch between neurons and glia. This gene, which encodes a novel type of nuclear protein without obvious homologies in protein sequence databases, is active both in the CNS and peripheral nervous system (PNS). An analysis carried out in *Drosophila* demonstrated that all progeny of cells expressing the *gcm* gene become glial cells in the PNS and CNS, alike. In *gcm* mutants, i.e. in flies carrying molecular alterations of the gene, progenitors fated to become glial cells became neurons instead; conversely, when the *gcm* gene was overexpressed in vivo by transgenic techniques, progenitor cells fated to develop as neurons switched to become glia. This is true both for the PNS and CNS. Thus, in *Drosophila*, *gcm* represents a master gene in gliogenesis, and commits neuroepithelial progenitors to a nonneuronal fate, in other words it tells them bluntly that their job is to become glia. This binary switch lays a foundation for the chain of cell-type specification and differentiation events leading to the production of mature glial cells in the adult fly.

How does the *gcm* gene function? Work by the group of Christian Klämbt in Köln demonstrated that *gcm* acts by positively regulating two genes, namely *pointed* (*pnt*) [37] and *tramtrack* (*ttk*) [38]. The former, encoding two ETS-type transcription factors, has a role in activating the glial pathway; the latter, encoding two zinc finger transcription factors that function as transcriptional repressors [39], operates in a parallel pathway to suppress neuronal cell fate.

To what extent can *Drosophila melanogaster* give us clues as to the mechanisms regulating gliogenesis in mammalians, especially humans? In very general terms, developmental studies in *Drosophila* can teach us lessons on the genetic pathways used by virtually all multicellular organisms in evolution, including humans, to lay out the basic plan of their bodies and to dictate the cell fate of multipotential progenitors. Of course, profound differences in the end structures originating from these developmental processes – in organisms as divergent as humans and flies – prevent the establishment of precise correlations at the anatomical level; however, and not surprisingly, homologs of the *gcm* and *pointed* genes are present in the human genome. Further studies of these genes' expression in mammalian CNS development will be required to determine whether one can postulate a role for these factors in mammalian gliogenesis.

How Does a Glial Progenitor Decide Whether to Be an Oligodendrocyte or an Astrocyte?

So far, we have dealt with some of the mechanisms in glial development up to the cell fate choice between becoming a neuron and becoming glia. Now, what are the factors affecting subsequent cell fate choices that are just as dramatic, such as the choice between turning into cells that make myelin to ensheathe nerve fibers (oligodendrocytes) or becoming structural and trophic constituents of the white matter (astrocytes)? The cell fate switch between the two is regulated by influxes mediated by secreted proteins, and in general by pathways that control the cell fate specification of neuronal and glial progenitors in vivo during early development.

The neural tube, from which the CNS originates from head to tail, is subdivided into major domains along the anteroposterior axis (forebrain, midbrain, hindbrain and spinal cord) which, in turn, are further partitioned into numerous segments, each displaying specific identities and performing specific tasks. An equally complex, albeit somewhat less obvious subdivision also takes place to specify the dorsoventral polarity of the nervous system. Diffusible signals originate from the axial mesoderm as well as the lateral mesoderm and nonneural ectoderm to create concentration gradients of different morphogens and affect the final destiny of pluripotent neural progenitors according to their physical positions along the dorsoventral axis of the neural tube (Fig. 2).

This developmentally regulated combinatorial code affects the future of neuronal progenitors. Those located on the ventral side of the neural tube receive diffusible signals from the notochord and switch to a motoneuron or ventral interneuron fate, whereas those positioned close to the dorsal roof plate of the neural tube acquire fates of commissural and association neurons, as well as neural crest [40]. Similarly, the specification of dopaminergic and serotoninergic neurons in the ventral midbrain requires diffusible signals from the ventral mesoderm and from the midbrain-hindbrain boundary [41, 42]. Do similar mechanisms regulate the fate of committed glial progenitor cells?

In 1991, the pioneering study performed in Robert Miller's laboratory, at Case Western University, provided the first evidence that oligodendrocytes originate only in ventral regions of the rat spinal cord, to subsequently migrate to other sites, including the developing dorsal columns [28]. In a later paper [43], the same group pinpointed the specific place of birth of oligodendrocyte progenitors. Spinal cord explants from E16.5 rat embryos marked with a thymidine analog (BrdU) were analyzed. A monoclonal antibody specific for BrdU and a battery of markers specific for oligodendrocyte progenitors, mature oligodendrocytes and astrocytes, allowed the authors to demonstrate that oligodendrocytes stem from a distinct cluster of embryonic spinal cord cells located astride the midline, in the neuroepithelium ventral to the central canal (the prospective ependymal canal). The stage at which these cells originate corresponds to late gestation embryogenesis (embryonic days 16-17), with a peak 4 days before birth in the rat embryo. This stage is clearly posterior to the bulk of neuronal proliferation that peaks before embryonic day 13 in the same organism. At later stages, this clustering becomes less apparent, and proliferating cells in the CNS are more uniformly distributed in the ventral and dorsal spinal cord. These cells are capable of migrating into different areas of the ventricular zone and developing gray matter, increasing in number and diffusing both ventrally and dorsally. Once cultured at low density [44], a large majority the labeled cells originating from the ventral midline differentiate into oligodendrocytes [43]. Further studies conducted by the same group in the chick confirmed the nature of these progenitors by employing oligodendrocyte lineage-specific markers [45]. In 1996, the group of Monique Dubois-Dalcq obtained comparable results by analyzing spinal cords from human fetuses at 45-83 days post-conception [46].

Two papers published in 1996 [47, 48] uncovered the nature of the cell-type specification signals required to generate oligodendrocytes in the spinal cord.

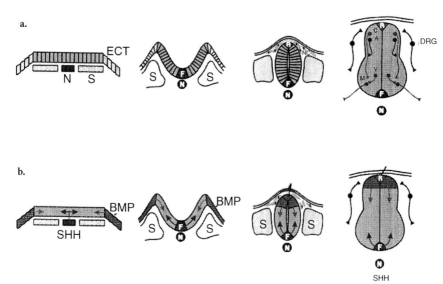

Fig. 2a, b. *Spinal cord development.* **a** Developmental stages. Transverse sections are shown, dorsal side up. At the end of gastrulation, the neural plate forms. It is comprised of columnar epithelial cells flanked by nonneural ectoderm (*ECT*) and underlain by axial (notochord, *N*) and paraxial (somite, *S*) mesoderm. During neurulation, the neural plate buckles at its midline, forming the neural folds and a floor plate (*F*). The neural tube forms by fusion of the dorsal tips of the neural folds, generating roof plate cells (*R*) at its dorsal midline, and neural crest (*NC*) cells that migrate to form a wide variety of neural and mesenchymal structures. Dorsal root ganglia (*DRG*) derive from postmigratory neural crest, and contain the cell bodies of sensory neurons. *C*, commissural neuron; *A*, association neuron; *V*, ventral interneuron; *M*, motoneuron. **b** The cell-type specification processes that attribute different identities to neuronal and glial cells along the dorsoventral axis of the neural tube are largely governed by gradients of diffusible morphogens, primarily sonic hedgehog (*SHH*) ventrally, and bone morphogenetic proteins (*BMPs*) dorsally. At the neural plate stage, SHH is produced in the notochord and diffuses dorsally, whereas BMPs are produced in the epidermal ectoderm flanking the plate and diffuse ventrally. At the neural fold stage, SHH is produced in the notochord and in the newly formed floor plate, and BMPs at the tip of the neural folds. After closure of the neural tube, SHH is produced in the floor plate, while BMP expression is lost from the epidermal ectoderm and maintained in the roof plate and dorsal neural tube. Oligodendrocyte progenitors originate near the floor plate in the presence of high concentrations of SHH. (Modified from [40] with permission)

These papers, through classic neurobiology techniques combined with molecular approaches, revealed that the appearance of oligodendrocyte progenitors on the ventral side of the ependymal canal requires diffusible signals originating in the notochord, a transient mesodermal structure located alongside the ventral aspect of the spinal cord. In particular, the diffusible protein sonic hedgehog, secreted by the notochord and floor plate of the neural tube, is sufficient to induce the appearance of oligodendrocyte progenitors within its range of diffusion. Conversely, astrocytes can originate both ventrally and dorsally, as demonstrated by classic neurobiology techniques (quail-chick chimeras), that further confirmed the exclusive ventral origin of oligodendrocytes [49]. Although controversy exists on these points [50, 51], we expect that further evidence and plenty of open, unprejudiced debate will clarify these issues of considerable value in the understanding of CNS developmental myelination.

How Do Glia Differentiate?

In the brain, oligodendrocytes arise postnatally from pre-progenitors located in the subventricular zone (SVZ) from where they migrate into the white matter. Subsequently, they proliferate and differentiate into mature myelinating oligodendrocytes. The various stages of the oligodendroglial lineage have been defined in vivo and especially in vitro (Fig. 3).

The O-2A glial progenitor was first isolated from the newborn rat optic nerve [52] and then from other regions of the rodent CNS (reviewed in [53]). The O2-A progenitor arises from a pre-progenitor cell and, depending on culture conditions, it generates oligodendrocytes or type-2 astrocytes in vitro, hence its designation. The oligodendroglial lineage is characterized by the expression of stage-specific morphological and antigenic phenotypes [52, 54]. The pre-progenitor cells are small, round, proliferative cells expressing the embryonic polysialylated form of the neural adhesion molecule (PSA-NCAM) [55].

The bipolar O-2A cells express membrane gangliosides such as GD3 [56] and those recognized by the monoclonal antibody A2B5 [52, 57]. The type-2 astrocytes that express specific gangliosides as well as glial fibrillary protein (GFAP) in vitro are rarely encountered during CNS development in vivo. As O-2A progenitors mature, they become multipolar, express sulfatide and other membrane-associated antigens, and become recognized by the monoclonal antibody O4 [58, 59]. They differentiate into GalC-expressing pro-oligodendrocytes before entering their later stage of maturation by expressing structural myelin proteins such as MBP, PLP, and MOG (for review see [60]).

The development of the oligodendroglial lineage is controlled by growth factors and hormones acting via membrane-associated and nuclear receptors to promote or modulate proliferation and differentiation. In vivo, these soluble factors probably arise from the neuronal and astroglial microenvironment of oligodendrocyte progenitors (reviewed in [60]). In vitro, it is possible to dissect the different effects produced by growth factors and hormones at different stages of the oligodendroglial lineage. In this way, it was recently shown that pre-progenitors responsive to platelet-derived growth factor (PDGF) [55] expressed both FGF [61] and thyroid hormone receptor genes (THRs) and proliferated in response to epidermal growth factor (EGF) and FGF [33], a behavior similar to that of self-renewing stem cells in cultured neurospheres [62].

In these experiments, addition of active thyroid hormone T3 further enhances the effect of FGF2 and favors the oligodendrocyte fate. Thyroid hormone limits the number of O-2A cell divisions and allows them to enter the differentiation

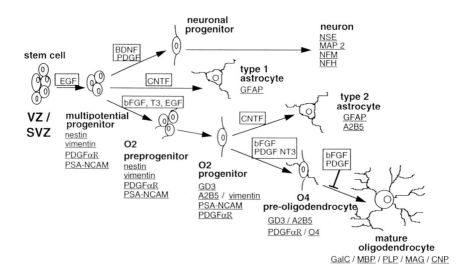

Fig. 3. Differentiation of neural progenitors in vitro. *In bold*: cell types; *underlined*: stage or type-specific cell markers; *boxed*: growth and differentiation factors promoting various stages of progenitor maturation (see text). bFGF and PDGF inhibit terminal differentiation of oligodendrocytes. *VZ*, ventricular zone of the neural tube; *SVZ*, subventricular zone of the neural tube; *EGF*, epidermal growth factor; *BDNF*, brain-derived neurotrophic factor; *PDGF*, platelet-derived growth factor; *CNTF*, ciliary neurotrophic factor; *bFGF*, basic fibroblast growth factor; *T3*, triiodothyronine; *NT3*, neurotrophin 3; *PDGFαR*, PDGF receptor alpha; *PSA-NCAM*, polysialylated neural cell adhesion molecule; *GD3*, a membrane ganglioside; *GFAP*, glial fibrillary acidic protein; *A2B5*, a membrane ganglioside recognized by the homonimous monoclonal antibody; *NSE*, neuron-specific enolase; *MAP2*, microtubule associated protein 2; *NFH*, high molecular weight neurofilament protein subunit; *NFM*, middle molecular weight neurofilament protein subunit; *GalC*, galactocerebroside; *MBP*, myelin basic protein; *PLP*, proteolipid protein; *MAG*, myelin-associated glycoprotein; *CNP*, 2',3'-cyclic nucleotide 3'-phosphodiesterase

stage [63]. This hormone also promotes myelination and synthesis of the myelin proteolipid protein (PLP) and myelin basic protein (MBP) in oligodendrocytes [64].

It is established that O-2A survival, proliferation and migration are mediated by PDGF [65, 67]. FGF2 prevents O-2A maturation [61] and promotes their proliferation by itself or in cooperation with PDGF while blocking differentiation [68]. Other factors that promote oligodendrocyte progenitors proliferation are insulin-like growth factor (IGF-1) [69] and the neurotrophin NT-3 [70], as well as the more recently identified oligodendroglial mitogen, neuregulin/glial growth factor-2 (GGF2) [71]. Neuregulin mutants exhibit embryonic lethality [72], while IGF-1 knockout mutants display severe hypomyelination in the brain [73], suggesting a role for this factor in the development of the oligodendroglial lineage. PDGFα receptor mutants generated by homologous recombination die in utero, failing to clarify in vivo the role of PDFGα in CNS myelination [74]. However, in transgenic mice overexpressing PDGF, oligodendrocyte progenitors hyperproliferate [75].

While most of the effects on O-2A proliferation and differentiation can be mediated by astrocytes, neurons also profoundly influence the biology of the oligodendroglial lineage [71, 76]. A mitogenic contact-mediated effect is related to membrane-bound FGF and to some as yet uncharacterized mitogen [77]. Neuronal-conditioned medium from the B103 [78] or B104 [79] neuroblastoma cell lines, or from cerebellar granular neurons [80] are also mitogenic for oligodendrocyte progenitors in vitro; some of these effects are mediated by GGF2 [71]. The B104-conditioned medium is a powerful mitogen for oligodendrocyte progenitors, allowing these cell populations to be expanded in the form of oligospheres [81]. It also promotes differentiation of oligodendrocyte pre-progenitors into progenitors (V. Avellana-Adalid, inpublished data) and of stem caells into progenitors [82]. It contains the PDGF-AA homodimer as well as the β1 and β2 forms of the transforming growth factor (TGF) [83], which may explain partially its regulatory/mitogenic effect.

Identifying growth factor requirements to expand the glial progenitor population ex vivo could be instrumental in developing approaches for therapeutic transplantation to repair demyelinated lesions as those present in MS. While this approach may be hindered by several limitations (reviewed in [84]), promoting re-activation or proliferation of the endogenous pool of oligodendrocyte progenitors may provide replacement to those lost in demyelination. Recently, evidence has been provided showing that cells derived from rodent adult subventricular zone (SVZ) can differentiate into myelin-producing oligodendrocytes, and that this capacity may be enhanced by treating them with FGF [85, 86]. These results suggest that cellular therapies implying growth factor-mediated modulation of cell fate specification and differentiation of oligodendrocyte progenitors could represent new strategies to repair the demyelinated adult CNS.

Acknowledgements: We thank Associazione Italiana Sclerosi Multipla (AISM), the Armenise Harvard Foundation, and Istituto Superiore di Sanità (ISS) for their support of our work on CNS myelination.

References

1. Lucchinetti CF, Brück W, Rodriguez M, Lassmann H (1996) Distinct patterns of multiple sclerosis pathology indicated heterogeneity in pathogenesis. Brain Pathol 6: 259-274
2. Trapp BD, Peterson J, Ransohoff RM et al. (1998) Axonal transection in the lesions of multiple sclerosis. New England Journal of Medicine 338(5): 278-285
3. Miller DJ, Asakura K, Rodriguez M (1996) Central nervous system remyelination - Clinical application of basic neuroscience principles. Brain Pathol 6: 331-344
4. Shi J, Marinovich A, Barres BA (1998) Purification and characterization of adult oligodendrocyte precursor cells from the rat optic nerve. J Neurosci 18: 4627-4636
5. Keirstead HS, Blakemore WF (1998) Response of the oligodendrocyte progenitor cell population (defined by NG2) to demyelination of the adult spinal cord. Glia 22: 161-170

6. Gensert JM, Goldman JE (1997) Endogenous progenitors remyelinate demyelinated axons in the adult CNS. Neuron 19: 197-203
7. Virchow R (1854) Ueber die ausgebreitete Vorkommen einer dem Nervenmark analogen Substanz in den tierischen Geweben. Arch Pat Anat Physiol Klin Med 6: 562-572
8. Rio-Hortega P (1919) El tercer elemento de los centros nerviosos. I. La microglia normal. II. Intervención de la microglia en los procesos patológicos (Celulas en bastocito y cuerposgranulo-adiposos). III. Naturaleza probable de la microglia. Bol Soc Esp Biol 9: 69-129
9. Perry VH, Gordon S (1988) Macrophages and microglia in the nervous system. Trends Neurosci 11: 273-277
10. von Lenhossék M (1891) Zur Kenntnis der Neuroglia des menschlihen Rückenmarkes. Verh Anat Ges 5: 193-221
11. Rakic P (1972) Mode of cell migration to superficial layers of fetal monkey neocortex. J Comp Neurol 145: 61-84
12. Rakic P (1981) Neuronal-glial interaction during brain development. Trends Neurosci 4: 184-187
13. Choi BH (1981) Radial glia of developing human fetal spinal cord: Golgi, immunohistochemical and electron microscopic study. Brain Res 227(2): 249-267
14. Hatten ME (1990) Riding the glial monorail: a common mechanism for glial-guided neuronal migration in different regions of the mammalian brain. Trends Neurosci 13: 179-184
15. Rakic P, Sidman R (1973) Organization of cerebellar cortex secondary to deficit of granule cells in *weaver* mutant mice. J Comp Neurol 152: 133-162
16. Rakic P (1975) Synaptic specificity in the cerebellar cortex: study of anomalous circuits induced by single gene mutations in mice. Cold Spring Harbor Symp Quant Biol 40: 333-346
17. Fishell G, Hatten ME (1991) Astrotactin provides a receptor system for glia-guided neuronal migration. Development 113: 755-765
18. Edmunson JC, Liem RKH, Kuster JC, Hatten MC (1988) Astrotactin: a novel neuronal cell surface antigen that mediates neuronal-astroglial interactions in cerebellar microcultures. J Cell Biol 106: 505-517
19. Robertson W (1899) On a new method of obtaining a black reaction in certain tissue elements of the central nervous system (platinum method). Scottish Med Surg J 4: 23
20. Robertson W (1900) A microscopic demonstration of the normal and pathological-histology of mesoglia cells. J Ment Sci 46: 733-752
21. Rio-Hortega P (1921) Estudios sobre la neuroglia. La glia de escarsas radiaciones (oligodendroglia). Bol Soc Esp Biol 21: 64-92
22. Wood P, Bunge RP (1984) The biology of the oligodendrocyte. In: Norton WT (ed): Oligodendroglia, advances in Neurochemistry. Plenum, New York, pp 1-46
23. Caroni P, Schwab ME (1989) Codistribution of neurite growth inhibitors and oligodendrocytes in rat CNS: appearance follows nerve fiber growth and precedes myelination. Dev Biol 136: 287-295
24. Sauer FC (1936) The interkinetic migration of embryonic epithelial nuclei. J Morphol 60: 1-11
25. Kilpatrick TJ, Richards LJ, Bartlett PF (1995) The regulation of neural precursor cells within the mammalian brain. Mol Cell Neurosci 6: 2-5
26. Temple S, Quian X (1996) Vertebrate neural progenitor cells: subtypes and regulation. Curr Opin Neurobiol 6: 11-17

27. Yu W-P, Collarini EJ, Pringle NP, Richardson WD (1994) Embryonic expression of myelin genes: evidence for a focal source of oligodendrocyte precursors in the ventricular zone of the neural tube. Neuron 12: 1353-1362
28. Warf BC, Fok-Seang J, Miller RH (1991) Evidence for the ventral origin of oligodendrocyte precursors in the rat spinal cord. J Neurosci 11: 2477-2488
29. Qian X, Davis AA, Goderie SK, Temple S (1997) FGF2 concentration regulates the generation of neurons and glia from multipotent cortical stem cells. Neuron 18(1): 81-93
30. Mason I (1994) The ins and outs of fibroblast growth factors. Cell 78: 547-552
31. Mignatti P, Morimoto T, Rifkin DB (1992) Basic fibroblast growth factr, a protein devoid of secretory signal sequence, is released by cells via a pathwayindependent of the endoplasmic reticulum-golgi complex. J Cell Physiol 151: 181-193
32. Johe KK, Hazel TG, Muller T, Dugich-Djordjevic MM, McKay RDG (1996) Single factors direct the differentiation of stem cells from the fetal and adult central nervous system. Gene Dev 10: 3129-3140
33. Ben-Hur T, Rogister B, Murray K, Rougon G, Dubois-Dalcq M (1998) Growth and fate of PSA-NCAM+ precursors of the postnatal brain. J Neurosci 18(15): 5777-5788
34. Jones BW, Fetter RD, Tear G, Goodman CS (1995) Glial cells missing: a genetic switch that controls glial versus neuronal fate. Cell 82(6): 1013-1023
35. Hosoya T, Takizawa K, Nitta K, Hotta Y (1995) Glial cells missing: a binary switch between neuronal and glial determination in Drosophila. Cell 82(6): 1025-1036
36. Vincent S, Vonesch JL, Giangrande A (1996) Glide directs glial fate commitment and cell fate switch between neurones and glia. Development 122(1): 131-139
37. Klaes A, Menne T, Stollewerk et al. (1994) The Ets transcription factors encoded by the Drosophila gene *pointed* direct glial cell differentiation in the embryonic CNS. Cell 78: 149-160
38. Giesen K, Hummel T, Stollewerk A et al. (1997) Glial development in the Drosophila CNS requires concomitant activation of glial and repression of neuronal differentiation genes. Development 124(12): 2307-2316
39. Brown JL, Sonoda S, Ueda H et al. (1991) Repression of the Drosophila *fushi tarazu* *(ftz)* segmentation gene. EMBO J 10: 665-674
40. Tanabe Y, Jessell TM (1996) Diversity and pattern in the developing spinal cord. Science 274: 1115-1123
41. Hynes M, Poulsen K, Tessier-Lavigne M, Rosenthal A (1995) Control of neuronal diversity by the floor plate: contact-mediated induction of midbrain dopaminergic neurons. Cell 80(1): 95-101
42. Ye W, Shimamura K, Rubenstein JL et al. (1998) FGF and Shh signals control dopaminergic and serotonergic cell fate in the anterior neural plate. Cell 93(5): 755-766
43. Noll E, Miller RH (1993) Oligodendrocyte precursors originate at the ventral ventricular zone dorsal to the ventral midline region in the embryonic rat spinal cord. Development 118(2): 563-573
44. Raff MC, Williams BP, Miller RH (1984) The in vitro differentiation of a bipotential progenitor cell. EMBO J 3: 1857-1864
45. Ono K, Bansal R, Payne J et al. (1995) Early development and dispersal of oligodendrocyte precursors in the embryonic chick spinal cord. Development 121(6): 1743-1754
46. Hajihosseini M, Tham TN, Dubois-Dalcq M (1996) Origin of oligodendrocytes within the human spinal cord. J Neurosci 16(24): 7981-7994
47. Pringle NP, Yu WP, Guthrie S et al. (1996) Determination of neuroepithelial cell fate: induction of the oligodendrocyte lineage by ventral midline cells and sonic hedgehog. Dev Biol 177(1): 30-42

48. Poncet C, Soula C, Trousse F et al. (1996) Induction of oligodendrocyte progenitors in the trunk neural tube by ventralizing signals: effects of notochord and floor plate grafts, and of sonic hedgehog. Mech Dev 60: 13-32
49. Pringle NP, Guthrie S, Lumsden A, Richardson WD (1998) Dorsal spinal cord neuroepithelium generates astrocytes but not oligodendrocytes. Neuron 20: 883-893
50. Spassky N, Goujet-Zalc C, Parmantier E et al. (1998) Multiple restricted origin of oligodendrocytes. J Neurosci 18: 8331-8343
51. Cameron-Curry P, Le Douarin NM (1995) Oligodendrocyte precursors originate from both the dorsal and the ventral parts of the spinal cord. Neuron 15: 1299-1310
52. Raff MC, Miller RH, Noble M (1983) A glial progenitor cell that develops in vitro into an astrocyte or an oligodendrocyte depending on culture medium. Nature 303(5916): 390-396
53. Mc Kinnon RD, Dubois-Dalcq M (1995) Cytokines and growth factors in the development and regeneration of oligodendrocytes. In: R Benveniste (ed) Cytokines and the CNS: development, defense and disease, CRC. Boca-Raton, Florida
54. Pfeiffer S, Warrington AE, Bansal R (1994) The oligodendrocyte and its many cellular processes. Trends Cell Biol 3: 191-197
55. Grinspan JB, Franceschini B (1995) Platelet-derived growth factor is a survival factor for PSA-NCAM$^+$ oligodendrocyte pre-progenitor cells. J Neurosci Res 41(4): 540-551
56. Goldman JE, Hirano M, Yu RK, Seyfried TN (1984) GD3 ganglioside is a glycolipid characteristic of immature neuroectodermal cells. J Neuroimmunol 7(2-3): 179-192
57. Eisenbarth GS, Walsh FS, Nirenberg M (1979) Monoclonal antibody to a plasma membrane antigen of neurons. Proc Natl Acad Sci USA 76(10): 4913-4917
58. Sommer I, Schachner M (1981) Monoclonal antibodies (O1 to O4) to oligodendrocyte cell surfaces: an immunocytological study in the central nervous system. Dev Biol 83(2): 311-327
59. Bansal R, Warrington AE, Gard AL et al. (1989) Multiple and novel specificities of monoclonal antibodies O1, O4, and R-mAb used in the analysis of oligodendrocyte development. J Neurosci Res 24(4): 548-557
60. McMorris FA, McKinnon RD (1996) Regulation of oligodendrocyte development and CNS myelination by growth factors: prospects for therapy of demyelinating disease. Brain Pathol 6(3): 313-329
61. Bansal R, Kumar M, Murray K et al. (1996) Regulation of FGF receptors in the oligodendrocyte lineage. Mol Cell Neurosci 7(4): 263-275
62. Weiss S, Reynolds BA, Vescovi A et al. (1996) Is there a neural stem cell in the mammalian forebrain? Trends Neurosci 19(9): 387-393
63. Barres BA, Lazar MA, Raff MC (1994b) A novel role for thyroid hormone, glucocorticoids and retinoic acid in timing oligodendrocyte development. Development 120(5): 1097-1108
64. Baas D, Bourbeau D, Sarlieve LL et al. (1997) Oligodendrocyte maturation and progenitor cell proliferation are independently regulated by thyroid hormone. Glia 19(4): 324-332
65. Noble M, Murray K, Stroobant P et al. (1988) Platelet-derived growth factor promotes division and motility and inhibits premature differentiation of the oligodendrocyte/type-2 astrocyte progenitor cell. Nature 333(6173): 560-562
66. Richardson WD, Pringle N, Mosley MJ et al. (1988) A role for platelet-derived growth factor in normal gliogenesis in the central nervous system. Cell 53(2): 309-319
67. Armstrong RC, Harvath L, Dubois-Dalcq ME (1990) Type 1 astrocytes and oligodendrocyte-type 2 astrocyte glial progenitors migrate toward distinct molecules. J Neurosci Res 27(3): 400-407

68. Bögler O, Wren D, Barnett SC et al. (1990) Cooperation between two growth factors promotes extended self-renewal and inhibits differentiation of oligodendrocyte-type-2 astrocyte (O-2A) progenitor cells. Proc Natl Acad Sci USA 87(16): 6368-6372
69. McMorris FA, Dubois-Dalcq M (1988) Insulin-like growth factor I promotes cell proliferation and oligodendroglial commitment in rat glial progenitor cells developing in vitro. J Neurosci Res 21(2-4): 199-209
70. Barres BA, Raff MC, Gaese F et al. (1994a) A crucial role for neurotrophin-3 in oligodendrocyte development. Nature 367(6461): 371-375
71. Canoll PD, Musacchio JM, Hardy R et al. (1996) GGF/neuregulin is a neuronal signal that promotes the proliferation and survival and inhibits the differentiation of oligodendrocyte progenitors. Neuron 17(2): 229-243
72. Kramer R, Bucay N, Kane DJ et al. (1996) Neuregulins with an Ig-like domain are essential for mouse myocardial and neuronal development. Proc Natl Acad Sci USA 93(10): 4833-4838
73. Beck KD, Powell-Braxton L, Widmer HR et al. (1995) Igf1 gene disruption results in reduced brain size, CNS hypomyelination, and loss of hippocampal granule and striatal parvalbumin-containing neurons. Neuron 14(4): 717-730
74. Soriano P (1997) The PDGF alpha receptor is required for neural crest cell development and for normal patterning of the somites. Development 124(14): 2691-2700
75. Calver A, Hall A, Yu W et al. (1998) Oligodendrocyte population dynamics and the role of PDGF in vivo. Neuron 20: 869-882
76. Barres BA, Raff MC (1994) Control of oligodendrocyte number in the developing rat optic nerve. Neuron 12(5): 935-942
77. Kreider BQ, Grinspan JB, Waterstone MB et al. (1995) Partial purification of a novel mitogen for oligodendroglia. J Neurosci Res 40(1): 44-53
78. Giulian D, Johnson B, Krebs JF, Tapscott MJ et al. (1991) A growth factor from neuronal cell lines stimulates myelin protein synthesis in mammalian brain. J Neurosci 11(2): 327-336
79. Hunter SF, Bottenstein JE (1990) Growth factor responses of enriched bipotential glial progenitors. Brain Res Dev Brain Res 54(2): 235-248
80. Hardy R, Reynolds R (1993) Neuron-oligodendroglial interactions during central nervous system development. J Neurosci Res 36(2): 121-126
81. Avellana-Adalid V, Nait-Oumesmar B, Lachapelle F, Baron-Van Evercooren A (1996) Expansion of rat oligodendrocyte progenitors into proliferative "oligospheres" that retain differentiation potential. J Neurosci Res 45(5): 558-570
82. Zhang S, Lundberg C, Lipsitz D (1998) Generation of oligodendroglial progenitors from neural stem cells. J Neurocytol (*in press*)
83. Asakura K, Hunter SF, Rodriguez M (1997) Effects of transforming growth factor-beta and platelet-derived growth factor on oligodendrocyte precursors: insights gained from a neuronal cell line. J Neurochem 68(6): 2281-2290
84. Franklin R, ffrench-Constant C (1996) Transplantation and repair in multiple sclerosis. In: the molecular biology of multiple sclerosis. Russel W (ed) John Wiley and Sons, Cambridge, pp 231-242
85. Lachapelle F, Nait-Oumesmar B, Avellana-Adalid V et al. (1996) FGF-2, EGF and PDGF differentially activate in vivo the potential of grafted cells derived from adult subventricular zone to generate myeling-forming oligodendrocytes. J Neurosci 390: 13
86. Nait-Oumesmar B, Lachapelle F, Vignais L et al. (1996) Potential repair of adult subventricular zone following chemically induced demyelination. J Neurosci 390: 13
87. Jacobson M (1991) Developmental Neurobiology. Plenum, New York, London, p 101
88. Jacobson M (1991) Developmental Neurobiology. Plenum, New York, London, p 45

Chapter 12

Genomic Screening in Multiple Sclerosis

P. MOMIGLIANO RICHIARDI, on behalf of the Italian Group for the Study of the Genetics of Multiple Sclerosis

The Genetic Component of Multiple Sclerosis

Multiple sclerosis (MS) is caused by an interplay of environmental and genetic factors. Their relative weight can be evaluated, as in all diseases, by three approaches: population epidemiology, twin concordance and family aggregation studies. Epidemiological studies point to environmental factors, likely one or more infectious agents, playing a major role as demonstrated by alteration of MS risk consequent to migration from high to low risk areas and viceversa and by occasional "epidemics" in small communities after contact with groups of individuals from high risk areas [1]. However, they also demonstrate the importance of genetic factors in that some ethnic groups maintain their relative resistance to MS even when they reside in areas where MS is common (e.g. Gypsies in Hungary, Blacks and Asians in USA, Maoris in New Zealand, and Lapps in Scandinavia [2]). Twin studies clearly demonstrate the role of genetic factors since monozygotic (MZ) concordance is substantially above dizygotic (DZ) concordance (25%-30% versus 3%), but also show the importance of the environment since the concordance level in MZ twins is well below 100%.

Remarkably, familial aggregation is largely controlled by genetic factors since the disease risk in relatives closely follows the degree of genetic similarity with the proband according to a polygenic inheritance model [3]. More direct evidence comes from adoption [4] and half-sib [5] studies in Canadian families. The current interpretation is that although MS is heavily influenced by environmental factors, they are not family-specific. Hence the study of multiple case families is most likely to provide information on genetic factors. The rather high level (20-40) of the relative risk of disease recurrence in a sibling (λs), compares favorably with that of type 1 diabetes (λs=15) in which linkage studies have been largely successful. Moreover, in MS the low contribution of the major histocompatibility complex (MHC) as compared with type 1 diabetes leaves more "room" for the contribution of other genes.

Chair of Human Genetics, Department of Medical Sciences, University of Eastern Piedmont A. Avogadro - 28100 Novara, Italy. e-mail: momiglia@scimed1.med.no.unipmn.it

Search for MS Loci by Genome Screens

Prompted by this promising background, a systematic search for MS genes was undertaken in several populations by whole genome linkage screens. They were based on testing widely available and highly informative microsatellite polymorphisms in families with affected sib pairs (ASP). ASP allow the analysis of linkage by non-parametric tests that do not need specification on a genetic model and are not influenced by penetrance, which in MS is generally low and hinders classical lod score analysis. Actually, ASP analysis considers only sibs with overt forms of disease and evaluates whether they share alleles in more than the expected 0.50 proportion. Unfortunately, the sensitivity of ASP analysis is comparatively low and large numbers of these families are needed. There is an inverse relationship between the effect of a gene and the number of families needed for its detection. Linkage of genes with very low effect cannot be detected even with thousands of sib pair families. This bottleneck is illustrated by the number of families tested in the studies published so far (Table 1). Even if all families could be analyzed jointly, which is not the case because of practical problems in a meta-analysis, they would not provide enough power to detect linkage of genes with small effect. This problem has been theoretically treated by Risch and Merikangas [6] and has been the subject of a strong debate [7]. The conclusions from genome screens were that no predominant susceptibility gene is involved, no locus (not even the human MHC, HLA) can be consistently shown to be MS-linked in all screens, and the presence of minor loci (with $\lambda s < 1.5$) cannot be excluded in any sizable portion of the genome.

Nevertheless for some chromosome regions, multiple indications of possible linkage have accumulated, particularly on 2p, 5p, 5q, 17q and 19q [8, 9]. The evidence is weak, in no case reaching a level of "suggestive evidence" as defined by

Table 1. Families utilized in MS linkage studies

Study location	Reference	Set	Families (n)	Pairs (n)
UK	Sawcer et al. [13]	1	129	143
		2	98	128
USA-France	Haines et al. [14]	1	52	81
		2	23	45
Canada	Ebers et al. [15]	1	61	100
		2	42	44
		3	72	78
Finland	Kuokkanen et al. [9]	1	21	24
Italy	Present study	Continental	41	39
		Sardinian	28	28
Total		All	567	710

Table 2. Comparison of published genome screens

Chromosome	Cytogenetic band	Marker	UK (MLS)	USA/France (Lod score)	Canada (MLS)	Finland (Lod score)
1	p36.12	D1S199	1.2			
	p34.3	D1S201	0.93		0.95	
	p22.2	D1S216	1.0			
2	p16.3	D2S119			1.24	
	p11.2	D2S169	1.3			
	p11.1	D2S139	1.4			
3	p21.2	D3S1289	1.2			
	p14.3	D3S1300	1.3			
	p14.1	D3S1285	1.1			
	p13	D3S1261	0.8		0.99	
	q22.2	D3S1309			1.01	
4	q35.2	D4S426	1.4			
5	p15.33	D5S417			1.8	
	p15.31	D5S406			4.2	
	p15.2	D5S416				3.4
	p13.1	D5S477				1.5
	p12	D5S455				2.2
	p11	D5S1968				2.0
	q11.1	D5S427	2.5			
	q12.1	D5S424	1.3		0.42	
	q14.1	D5S428	1.0			
	q14.3	D5S815		1.14		
7	p15.3	D7S629	1.6			
	p15.2	D7S484	1.2		0.38	
	q11.23	D7S524	0.5		0.7	
	q22.1	D7S554		2.86		
	q31.31	D7S523		1.11		
12	p13.31	D12S77	0.8			
	p12.3	D12S364	1.0			
	p12.3	D12S62	2.2			
	p12.3	D12S310	1.8			
	p11.1	D12S87	0.6			
	q23.2	PAH		1.56		
	q24.33	D12S392		1.71		
14	q32.33	D14S292	1.4			
17	p11.2	D17S953	1.5			
	p11.1	D17S798	1.5			
	q21.33	D17S807	2.7			2.8
19	q13.12	D19S251	1.1			
	q13.12	D19S225	1.1			
	q13.13	D19S220	1.2			
	q13.32	D19S219		1.13		
	q13.32	APO-C2	1.1	1.47		
	q13.33	D19S246	1.5			
22	q13.1	D22S283	1.1			
	q13.1	PDGFB	1.4			
	q13.1	D22S274	1.2			
X	p11.4	DXS1068	0.3		1.85	
	p11.21	DXS991	1.8			
	q21.33	DXS990	1.2			
	q23	DXS1059	1.7			

PAH, Phenylalanine hydroxylase; *APO-C2*, Apolipoprotein C2; *PDGFB*, Platelet-derived growth factor B.

Lander and Kruglyak [10], and does not necessarily concern the same marker in the different screens. However, in some cases the loci coincide with regions of homology to EAE-linked genes in the mouse [11] and rat [12]. Thus, these regions are a logical starting point for further analyses and replication attempts. Markers showing some evidence of linkage in the British [13], American/French [14], Canadian [15] and Finnish [9, 16] screens are reported in Table 2.

A Linkage Study in Italian Populations

The Italian Group for the Study of the Genetics of Multiple Sclerosis performed an analysis of markers in selected genome regions and single candidate genes in Southern European populations. What is the rationale for this study? All linkage studies so far have been done on subjects of Northern European (UK, Finland, France) or North American (US and Canada) descent. However, linkage scores are very sensitive to the disease gene frequency. For a gene of moderate to low effect, 10-100 fold less sib pairs may be needed to show linkage in a population where the frequency of the disease allele is intermediate as compared to one where it is either high or low [6]. Therefore the choice of the population can be quite critical. For instance, the CTLA-4 gene region (IDDM12 locus) yielded significant linkage data with type 1 diabetes in Southern Caucasoids while weak or no evidence of linkage was reported in Northern Caucasoids [17, 18].

This study included a panel of multiplex families from continental Italy (central and southern regions) and from Sardinia. The latter location is especially meaningful because Sardinians represent a rather homogeneous and isolated population with a genetic background substantially different from that of other populations [19]. These characteristics are particularly relevant in view of subsequent association studies. Moreover, the incidence of MS in Sardinia is higher than in other Italian regions and is closer to that of northern Europe [20].

For this study, markers in candidate regions were selected on the basis of previous linkage data (Table 2). In addition, markers corresponding to a few candidate genes, namely those of HLA-DRB1, CTLA-4, IL-9, APO-E, BCL-2 and TNFR2 were also considered. HLA was chosen as the only locus that showed genetic association and, although not in all studies, genetic linkage with MS. Cytotoxic T lymphocyte antigen-4 (CTLA-4) has been associated with insulin-dependent diabetes mellitus (IDDM) [17] and other autoimmune diseases [21], and is involved in the regulation of lymphocyte activation as are B cell lymphoma-2 (BCL-2) and tumor necrosis factor receptor-2 (TNFR2). IL-9 is representative of the interleukin (IL) gene cluster that resides within an experimental allergic encephalomyelitis (EAE)-linked region in the rat genome [12]. Apolipoprotein E (APO-E) gene is located in a region showing some evidence of linkage to MS in previous studies and its variation has been prominent in other neurologic diseases.

Affected siblings with a diagnosis of definite multiple sclerosis according to Poser et al. [22] were enrolled in the Hospitals of Bari, Cagliari, Catanzaro, Chieti, Florence and Rome Universities. Each patient was submitted to clinical evaluation

Table 3. Composition of tested multiplex families

Families (n)	Continental Italy	Sardinia
Total	41	28
Sib pairs	37	24
Sib trios	1	2
Families with		
Two parents	22	13
One parent	9	8
Without parents	7	5
Other relatives affected	3[a]	2[b]
Individuals affected	87	57
Individuals analyzed	186	158

[a] One uncle/ nephew pair and two first cousin pairs.
[b] First cousins.

by a trained neurologist and to cerebrospinal fluid and magnetic resonance imaging analysis. Relatives of index cases and affected siblings were recruited in the same centers as part of a program for collection of genomic material from Italian MS multiplex families sponsored by the Italian Multiple Sclerosis Foundation (FISM). Details of the families, are shown in Table 3. Enrollment of patients and relatives followed their informed consent. DNA samples were analyzed for 67 microsatellites with a preference to the markers showing some suggestion of linkage in the previous genome screens (Table 2).

Linkage Analysis

The data were analyzed for linkage by two different non-parametric programs.
1. The GENEHUNTER program [23] was adopted because it allows extraction of linkage information from all relatives, even in families with affected members other than sibs and lacking one or both parents. Non-parametric linkage (NPL) scores were determined. Single point analysis was performed for all markers. The program calculates p values based on the asymptotically normal statistic. NPL scores for X-linked markers were calculated by the xgh version of the program. Multipoint analysis was also performed in the chromosomal regions where several closely located markers were tested. Genetic distances between the markers were calculated by the LIK2P module of the GAS 2.0 package and were in substantial agreement with the Genethon map distances from CEPH families.
2. SimIBD [24] uses a conditional simulation approach to produce an empirical null distribution and empirical p values. Like GENEHUNTER, it uses all available genotypes in unaffected individuals to measure identity-by-descent

Table 4. Single point linkage scores in Sardinians, continental Italians and combined sets

Chromosome	Cytogenetic band	Marker	Distance[a] (cM)	GENEHUNTER (NPL)		
				Total	Sardinian	Continental
1	p36.22	TNFR2	24.4	0.10	-0.94	0.91
	p36.12	D1S199	22.7	-0.62	-0.96	-0.01
	p34.3	D1S201	>50[b]	-0.67	0.18	-1.03
	p22.2	D1S216		-0.01	-0.21	0.16
2	p16.3	D2S119	41[c]	-0.98	-0.26	-1.07
	p11.2	D2S169	2.6	0.57	0.22	0.56
	p11.1	D2S139	>50[b]	**0.98**	0.92	0.53
	q33.1	CTLA-4		0.67	**1.04**	0.01
3	p21.2	D3S1289	9	0.18	-0.41	0.58
	p14.3	D3S1300	10	0.13	-0.30	0.41
	p14.1	D3S1285	6	-0.03	0.14	-0.16
	p13	D3S1261	18	-0.98	-1.00	-0.48
	p11.1	D3S1595	16	0.43	-0.49	**0.96**
	q21.1	D3S1278	14	**1.10**	0.85	0.73
	q21.3	D3S3607	14	-0.25	-0.32	-0.06
	q22.2	D3S1309		-0.21	0.01	-0.29
4	q32.1	D4S415	27[c]	-1.07	-1.27	-0.32
	q35.2	D4S426		-0.40	-0.19	-0.37
5	p15.33	D5S417	6	0.93	-0.00	**1.24**
	p15.31	D5S406	17	-0.27	-0.21	0.18
	p15.2	D5S416	7	-1.06	-0.83	-0.68
	p15.1	D5S655	12	-0.85	-0.23	-0.93
	p13.1	D5S477	7	-0.41	0.06	-0.60
	p12	D5S455	16	-0.29	-0.67	0.20
	p11	D5S1968	8	0.55	-0.33	1.03
	q11.1	D5S427	15	0.50	0.56	0.18
	q12.1	D5S424	13[c]	0.51	**1.00**	-0.19
	q14.1	D5S428	4	-1.27	0.11	-1.80
	q14.3	D5S815	47[c]	-0.55	**1.15**	-1.75
	q31.1	IL-9	1.5	0.33	0.70	-0.16
	q31.1	D5S393		-0.13	0.10	-0.27
6	p21.31	HLA-DRB1		**1.26**	0.56	1.17
7	p15.3	D7S629	19	-0.14	0.36	-0.48
	p15.2	D7S484	>50[b]	**1.66**	**1.60**	0.84
	q11.23	D7S524	10	0.50	-0.37	0.95
	q22.1	D7S554	13	0.13	0.39	-0.15
	q31.31	D7S523		0.14	0.76	-0.44
12	p13.31	D12S374	10	-0.19	-0.04	-0.22
	p13.31	D12S77	8.8	-0.38	0.11	-0.61
	p12.3	D12S364	5.6	-0.32	0.06	-0.50
	p12.3	D12S62	3	-1.82	-1.41	-1.19
	p12.3	D12S310	10	-1.91	-0.81	-1.84
	p11.1	D12S87	>50[b]	-0.72	-0.28	-0.72
	q23.2	PAH	>50[b]	-1.28	-0.64	-1.14
	q24.33	D12S392		0.33	-0.01	0.45
14	q32.33	D14S292		-0.06	-0.04	-0.04

Table 4. (continued)

Chromosome	Cytogenetic band	Marker	Distance[a] (cM)	GENEHUNTER (NPL)		
				Total	Sardinian	Continental
17	p11.2	D17S953	17	0.36	0.25	0.26
	p11.1	D17S798	15	0.60	-0.07	0.86
	q12	D17S250	26	0.55	**1.16**	-0.28
	q21.33	D17S807		0.93	0.50	0.80
18	q21.3	BCL-2		0.27	-0.19	0.57
19	q13.11	D19S49	7	-0.04	-0.26	0.17
	q13.12	D19S251	6	0.28	-0.01	0.38
	q13.12	D19S225	9	-0.02	-0.04	0.00
	q13.13	D19S220	7	0.29	0.31	0.13
	q13.2	D19S217	4	-0.44	-0.04	-0.55
	q13.32	D19S219	7	-0.01	-0.59	0.48
	q13.32	D19S606	7	-0.42	-0.04	-0.50
	q13.32	APO-C2	5[b]	-0.37	-0.19	-0.32
	q13.32	APO-E	3[b]	-0.27	-0.37	-0.65
	q13.33	D19S246		-0.57	0.84	-1.43
22	q13.1	D22S283	12	0.92	**1.01**	0.36
	q13.1	PDGFB	14	-0.02	0.18	-0.18
	q13.1	D22S274		0.01	-0.86	0.73
X	p11.4	DXS1068	13	-0.84	0.07	-1.19
	p11.3	MAOB	18	-0.33	**1.04**	-1.37
	p11.21	DXS991	17[c]	-0.09	-0.85	0.63
	q21.33	DXS990	19[c]	0.01	-0.83	0.7
	q23	DXS1059		-0.06	0.28	-0.33

[a] Calculated distances; [b] LDB distances; [c] Genethon map distances.
TNFR2, Tumor necrosis factor receptor-2 gene; *CTLA-4*, Cytotoxic T lymphocyte antigen-4 gene; *IL-9*, Interleukin-9 gene; *HLA-DRB1*, Human leukocyte antigen DRB1 gene; *PAH*, Phenylalanine hydroxylase; *BCL-2*, B cell lymphoma-2; *APO*, Apolipoprotein C2 or E gene; *PDGF*B, Platelet-derived growth factor B gene; *MAOB*, Monoamine oxidase B gene.

(IBD) sharing. In some instances, it possesses a power of extracting linkage information higher than that of GENEHUNTER-All statistics [25]. The 1/sqrt(p) weighting function was used, where p is the population frequency of a given allele, with a number of replicates of 1000.

Results

A summary of the data is provided in Table 4 that shows: the tested markers with their cytogenetic localization, as derived from the location database (LDB) integrated map [26]; the genetic distances between the markers; the NLP scores obtained by GENEHUNTER. Linkage data are reported for each of the two sets of families, Sardinian and continental Italians and for both sets combined. Only the single point NPL statistics are shown. For simplicity, SimIBP p values are not shown, since in only one case (D2S169) did they reach significance while the NPL score was low.

NPL probability levels were in all cases >0.05 except for D7S484 in the combined set (NPL 1.66, $p=0.048$). Slight hints of linkage, (i.e. NPL scores equal to or higher than 1.0 and/or SimIBD p values < 0.05 in at least one family set) were found in 2p11, 2q33 (CTLA-4), 3p11, 3q21, 5p15, 5p11, 5q12, 5q14, 6p21 (HLA), 7p15, 17q12, 22q13, and Xp11. These are highlighted in Table 4. No evidence of linkage was observed in chromosomes 1p, 3p, 4q, 7q, 12p, 12q, 14q, 17p, 19q, and Xq, nor for the individual candidate genes TNFR2, IL-9, BCL-2 and APO-E. The D17S807 marker in 17q21 was particularly interesting because it showed evidence of linkage both in the Finnish [9] and in the UK [13] screens. In the present study, the NPL score was about 0.9, lower than the relatively high scores of the previous studies, but close to the arbitrary threshold of 1.0.

Multipoint NPL analysis was applied in the regions where multiple markers were tested but did not lead to a substantial increase of the linkage scores, despite acceptable levels of extraction of information content. Low or negative scores were given by all the chromosome 5 regions, even though this chromosome appeared the most promising as judged from data of linkage in the other studies and from the homology of the p14-p12 region with the EAE2 mouse region.

Outlook for Future Studies

In general, our data provide some additional, although weak, evidence of linkage in regions that were singled out by previous studies. The linkage scores are in general low, yet positive and not dissimilar from those expected from genes of low effect studied in populations of limited sizes. Actually, even the contribution of HLA, supported by association studies in several populations, was not detected by linkage in all studies nor, within each study, in all sets [13-16]. Likewise, we observed an NPL score >1 with HLA-DRB1 in continental Italians and a lower, although still positive score in Sardinians, even though a significant HLA association was observed in both populations (C. Ballerini, unpublished results and [27]).

The problem of replication is strictly connected with the magnitude of the genetic effect to be detected. The chance of replication decreases not linearly, but according to the square of the effect of the gene. All linkage studies so far indicate the absence of any gene with a predominant effect in MS. However, the evidence for the presence of MS genes and for them accounting for most of the family aggregation is overwhelming [3-5]. Assuming that 10 (or more) epistatic genes of equal effect contribute to the λs of about 30 attributed to MS, each of them would have a $\lambda s = 1.4$ (or less). Under these conditions, linkage analysis is very unlikely to provide significant lod scores with any practical number of multiplex families [6] and any low level lod score that may be obtained is unlikely to be replicated in independent studies or sets of patients.

Why is it important, then, to record "weak linkage data" despite the seemingly disproportionate effort required both in finding multiplex families and in performing microsatellite analysis? The hopes of finding "major genes" have disappeared in MS as well as in other complex diseases. The underlying reasons may be, besides the low sensitivity of the analytical tools, also the presence of strong and difficult to unravel epistatic interactions between different genes. However, there is a recognizable trend toward accumulation of this kind of weak linkage evidence on some chromosomal regions and not in others, with some sharing between different autoimmune diseases [28]. These "tips of icebergs" that in some cases coincide with EAE homology regions may lead to regions in which to perform association studies. The low-degree HLA linkage provides a good example. Historically, since HLA was an obvious candidate locus, linkage studies were preceded by association studies. The rather low and scattered linkage scores obtained in the different sets of families, including those of the present study, do not prove involvement of HLA in disease. However, should we decide that they provide a sufficient basis for performing an association study, retrospectively this turn out to be successful. One can hope that similar weak evidence of linkage in other regions underlies the presence of some disease gene that can be detected by association studies. The latter are *per se* more sensitive and experimentally easier since they require single-case (simplex) families. In this regard, some regions stand out as good candidates from the present work, providing additional evidence for the presence of linkage to MS, i.e. 2p11, marked by microsatellites D2S169 and D2S139, 3q21 (microsatellite D3S1278), 7p15 (microsatellite D7S484), 17q21.3 (microsatellite D17S807) and 22q13 (microsatellite D22S283). The 7p15 is especially appealing since it is homologous to a region harboring a rat EAE locus [12] and coincides with a region where markers linked to several autoimmune diseases (MS, Crohn's disease and asthma) are clustered [28].

Perspectives of an Association Study on Chromosome 5

Even though the present data do not add strong evidence for linkage in any portion of chromosome 5, this region remains the best candidate for harboring MS genes, taking into account the current available data from different sources. Data from a complete genome screen in EAE mice show strong linkage to a locus, EAE2, on a region of the murine chromosome 15 [11] that is homologous to human 5p14-12. Accordingly, in a study carried out in Finnish multiplex families, a substantial lod score was found with markers on chromosome 5p14-12 [16]. Finally, data obtained in whole genome linkage studies in several populations [13-15] showed a weak but convergent evidence for linkage in a rather extended portion of chromosome 5, from 5p15 to 5q14 (Table 2).

Thus, it seems worthwhile to search for MS genes in this region by a genetic association study. Studies involving the non-random association between genetic markers and disease alleles are based on the presence of linkage disequilibrium between the two. This depends on several factors, mainly related to the history of

Table 5. Localization of available polymorphic STS in the interval 5 ptel-q13

cM[a]	Framework microsatellite marker[b]	Polymorphic STS Present study[c]	Polymorphic STS Wang et al. [33][d]	Type
ptel	D5S678		EST285652	No information
			EST117976	UT
			WI-6093	Random STS
10.7	D5S406	A001X45		UT
13.8	D5S635		WI-18821	Human steroid 5 alpha-reductase
18.6	D5S630			
19.7			A003B29	UT
		SGC32541	SGC32541	UT
			SGC35145	UT
		SHGC-12556		DAP-1
		SHGC-12958		T complex protein 1epsilon subunit
		WI-6722		UT
21.4	D5S478	stSG9739		UT
27.4	D5S1954			
35.6	D5S655		WI-11163	Random STS
38.9	D5S502		WI-5826	Random STS
		stSG10348		UT
			WI-4325	Random STS
44.6	D5S477	stSG9937		UT
46.6	D5S651	A001V01		Threonyl-tRNA synthetase
		A004148		UT
		WI-12996		UT
		WI-13029		UT
51.6	D5S426		SGC333368	UT
			stSG9574	No information
		stSG8574		UT
		stSG9574		UT
52.2	D5S395		ESTC22	Complement C6 precursor
			ESTD-C6	Complement component C6
		C9 exon11		Complement component C9
54.4	D5S2021	stSG1444		UT
		SHGC-13615		UT
		WI-16394		UT
61.1	D5S628		SGC31091	UT
			SGC30610	UT
		SGC33636		UT
63.9	D5S474		WI-8827	UT
65.0	D5S407		U28413	CSA protein
67.2	D5S491	SGC31768		UT
69.6	D5S427		WI-6915	UT
		IB482		UT
74	D5S2048			
74.7	D5S647	A007J09		UT
		SGC36907		UT
76.5	D5S637			
79.2	D5S650		WI-19224	SMA
		WI-11362		Similar to Ras GTPase Activating-like protein
		IB320		KIAA0264
82.3			X74070	Transcription factor BTF3
			WI-17114	UT
82.8	D5S424		STS-R41876	UT
		WI-6272	WI-6272	Antiquitin

Table 5. (continued)

cM[a]	Framework microsatellite marker[b]	Polymorphic STS		Type
		Present study[c]	Wang et al. [33][d]	
84.8	–	–	ESTD-F2	Human prothrombin (F2)
85	–	–	ESTD-ARSB	Human arylsulfatase B
		WI-13020	WI-13020	UT
		–	WI-20907	UT
		–	WI-10775	Random STS
		–	WI-4719	Random STS
		SHGC-11568	–	UT
923	D5S641	–	–	–

[a] Cumulative genetic distances in cM, starting from ptel.
[b] Markers mapped by Genethon. Information on their position from http://www.genome.wi.mit.edu/cgi-bin/SNP/human/sts_info;
[c] Genetic map positions of the STSs are referred to framework microsatellite markers and are available at http:// www.ncbi.nlm.nih.gov/SCIENCE96.
[d] Genetic map positions of the STSs from http://www.genome.wi.mit.edu/SNP/human/maps/Chr5.ALL.html
UT, Unidentified transcript; *DAP-1*, Death-associated protein-1; *CSA*, Cockayne syndrome type A; *SMA³*, Spinal muscular atrophy.

the population under test, and it is hard to predict for each specific region of the genome. However, it has been suggested that linkage disequilibrium between alleles <0.5 cM apart is the "normal" expectation given several assumptions, such as recent population expansion and relatively recent breakage of barriers to gene flow [29].

On this basis, the most important requirement for association studies is the availability of a dense array of markers located in the region of interest at a distance of not more than 1 cM from each other. For this reason, the Italian Group for the Study of Genetics of MS has set up a search for single nucleotide polymorphisms (SNPs) mapping in the 90 cM interval between 5ptel and 5q13 using denaturing high performance liquid chromatography (DHPLC) [30]. DHPLC is an innovative method for DNA variant detection based on the capability of ion-pair reverse-phase liquid chromatography to resolve homoduplex from heteroduplex molecules under conditions of partial denaturation.

Polymorphisms were looked for in expressed genes exploiting the map of >16 000 sequence tagged sites (STS) in the 3' UTR (untranslated regions) of cDNA clones (ESTs) [31]. This approach offers several advantages. First, these STS are present in transcribed genes. Second, since 3' UTR are not expected to undergo a strong selective pressure, single nucleotide polymorphisms (SNPs) should occur rather frequently, at least 1 per 1000 nucleotides [32]. Third, the analysis can be performed directly on genomic DNA under uniform conditions, and primers are commercially available at low cost. Fourth, their mapping is already defined with reference to framework microsatellite markers.

On this basis, 124 STS were screened in the region of interest, totalling about 30000 bp; 30 SNPs were expected on the basis of the reported estimates of about 1 SNP per 1000 bp [32]. This work led to the development of an SNPs map in the region 5ptel-q13 with 28 markers distributed over 90 cM (Table 5). When this set of markers was integrated with the SNP map generated by Wang et al. [33] in the same chromosomal region, it was increased to a total of 55 polymorphic STS (Table 5), including 60 SNPs (5 STS carrying 2 SNPs). Thus, a map of 55 markers at a mean spacing of 1 cM, suitable for an association study, is now available in the 5ptel-q13 region. The Italian Group for the Study of the Genetics of MS is currently accumulating a large set of MS simplex families from Sardinia and continental Italy in which to test these SNPs for MS association.

Acknowledgements: This work was supported by FISM (Italian Foundation for Multiple Sclerosis) and by grants of the Italian Institute of Health and of the University Ministry (MURST). M. Giordano was supported by a fellowship from FISM.

The Italian Group for the Study of the Genetics of Multiple Sclerosis is composed of:

S. *D'Alfonso*, D. *Bocchio*, M. *Giordano*, P. *Zavattari*, Department of Medical Sciences, University of Eastern Piedmont "A. Avogadro", Novara;

L. *Nisticò*, R. *Tosi*, Institute of Cell Biology, CNR, Rome;

M. *Marrosu*, M. *Lai*, R. *Murru*, Section of Neurophysiopathology, Department of Neuroscience, University of Cagliari, Cagliari;

L. *Massacesi*, C. *Ballerini*, D. *Gestri*, Department of Neurological and Psychiatric Sciences, University of Florence, Florence;

M. *Salvetti*, R. *Bomprezzi*, G. *Ristori*, Department of Neurological Sciences, University of Rome "La Sapienza", Rome;

M. *Trojano*, M. *Liguori*, Department of Neurological Science, University of Bari, Bari;

D. *Gambi*: Department of Clinical Neurology, University of Chieti "G.D'Annunzio", Chieti;

A. *Quattrone*, Department of Medical Sciences, University of Catanzaro and Institute of Experimental Medicine and Biotechnology, CNR, Catanzaro;

F. *Cucca*, Institute of Clinics and Biology of the Growing Age, University of Cagliari, Cagliari.

References

1. Kurtzke JF (1995) MS epidemiology world wide. One view of current status. Acta Neurol Scand S161: 23-33
2. Ebers GC, Sadovnick AD (1993) The geographic distribution of multiple sclerosis: a review. Neuroepidemiology 12: 1-5
3. Robertson NP, Fraser M, Deans J et al. (1996) Age-adjusted recurrence risks for relatives of patients with MS. Brain 119: 449-455
4. Ebers GC, Sadovnick AD, Risch NJ and the Canadian Collaborative study Group (1995) Familial aggregation in MS is genetic. Nature 377: 150-151
5. Sadovnick AD, Ebers GC, Dyment DA, Risch NJ and the Canadian Collaborative Study Group (1996) Evidence for genetic basis of MS. Lancet 347: 1728-1730
6. Risch N, Merikangas K (1996) The future of genetic studies of complex human diseases. Science 273: 1516-1517
7. Scott WK, Pericak-Vance MA, Bell DA et al. (1997) Science 275: 1327-1330
8. Dyment DA, Sadnovich AD, Ebers GC (1997) Genetics of MS. Hum Mol Genet 6: 1693-1698

9. Kuokkanen S, Gschwend M, Rioux JD et al. (1997) Genomewide scan of MS in Finnish multiplex families. Am J Hum Genet 61: 1379-1387
10. Lander ES, Kruglyak L (1995) Genetic dissection of complex traits: Guidelines for interpreting and reporting linkage results. Nat Genet 11: 241-247
11. Sundvall M, Jirholt J, Yang HT et al. (1995) Identification of murine loci associated with susceptibility to chronic experimental autoimmune encephalomyelitis. Nat Genet 10: 313-317
12. Roth MP, Viratelle C, Dolbois L et al. (1998) A genome-wide search identifies susceptibility loci for experimental autoimmune encephalomyelitis on rat chromosomes 4 and 10. J Immunol (in press)
13. Sawcer S, Jones HB, Feakes R et al. (1996) A genome screen in multiple sclerosis reveals susceptibility loci on chromosome 6p21 and 17q22. Nat Genet 13: 464-468
14. Haines JL, for The Multiple Sclerosis Genetics Group (1996) A complete genomic screen for multiple sclerosis underscores a role for the major histocompatibility complex. Nat Genet 13: 469-471
15. Ebers GC, Kukay K, Bulman DE et al. (1996) A full genome screen in multiple sclerosis. Nat Genet 13: 472-476
16. Kuokkanen S, Sundvall M, Terwilliger JD et al. (1996) A putative vulnerability locus to multiple sclerosis maps to 5p14-p12 in a region syntenic to the murine locus Eae2. Nat Genet 13: 477-480
17. Nisticò L, Buzzetti R, Pritchard LE et al. (1996) The CTLA-4 gene region of chromosome 2q33 is linked to, and associated with, type 1 diabetes. Hum Mol Genet 5: 1075-1080
18. Marron MP, Raffel LJ, Garchon HJ et al. (1997) Insulin-dependent diabetes mellitus (IDDM) is associated with CTLA-4 polymorphisms in multiple ethnic groups. Hum Mol Genet 6: 1275-1282
19. Piazza A, Mayr WR, Contu L et al. (1985) Genetic and population structure of four Sardinian villages. Ann Hum Genet 49: 47-63
20. Rosati G, Aiello I, Pirastu MI et al. (1996) Epidemiology of MS in north-western Sardinia: further evidence for higher frequency in Sardinians compared to other Italians. Neuroepidemiology 15: 10-19
21. Kotsa K, Watson PF, Weetman AP (1997) A CTLA-4 gene polymorphism is associated with both Graves' disease and autoimmune hypothyroidism. Clin Endocrinol 46: 551-554
22. Poser CM, Paty DW, Scheinberg L (1983) New diagnostic criteria for multiple sclerosis: Guidelines for research protocols. Ann Neurol 13: 227-231
23. Kruglyak L, Daly MJ, Reeve-Daly MP, Lander ES (1996) Parametric and non-parametric linkage analysis: a unified multipoint approach. Am J Hum Genet 58: 1347-1363
24. Davis S, Schroeder M, Goldin LR, Weeks DE (1996) Nonparametric simulation-based statistics for detecting linkage in general pedigrees. Am J Hum Genet. 58: 867-880
25. Davis S, Goldin LR, Weeks DE (1997) SimIBD: A powerful robust non-parametric method for detecting linkage in general pedigrees. In: Pawlowitzki I-H, Edwards JH, Thompson EA (eds) Genetic mapping of disease genes. Academic, San Diego, pp 189-204
26. Collins A, Frezal J, Teague J, Morton NE (1996) A metric map of humans: 23,500 loci in 850 bands. Proc Natl Acad Sci USA 93: 14771-14775 (http://cedar.genetics.soton.ac.uk/public_html/gmap.html)
27. Marrosu MG, Murru MR, Costa G et al. (1997) Multiple sclerosis in Sardinia is associated and in linkage disequilibrium with HLA-DR3 and -DR4 alleles. Am J Hum Genet 61: 454-457
28. Becker KG, Simon RM, Bailey-Wilson JE et al. (1998) Clustering of non-major histocompatibility complex susceptibility candidate loci in human autoimmune diseases. Proc Natl Acad Sci USA 95: 9979-9984

29. Thompson EA, Neel JV (1997) Allelic disequilibrium and allele frequency distribution as a function of social and demographic history. Am J Hum Genet 60: 197-204
30. Oefner PJ, Underhill PA (1998) DNA mutation detection using denaturing high performance liquid chromatography (DHPLC). In: Current protocols in human genetics. John Wiley & Sons, New York (*in press*)
31. HudsonTJ, Stein LD, Gerety SS et al. (1995) An STS-based map of the human genome. Science 270: 1945-1954
32. Kruglyak L (1997) The use of a genetic map of biallelic markers in linkage studies. Nat Genet 17: 21-24
33. Wang DG, Fan JB, Siao CJ et al. (1998) Large-scale identification, mapping, and genotyping of single-nucleotide polymorphisms in the human genome. Science 280: 1077-1082

Chapter 13

MHC and Multiple Sclerosis

M.G. Marrosu

Introduction

The major histocompatibility complex (MHC, also called HLA or human leukocyte antigen) and its products have a central role in the immune response. Sharing of the MHC class II region is important for CD4$^+$ (helper) T cell interactions, and sharing of the MHC class I region is important for CD8$^+$ (cytotoxic) T cell interactions. This phenomenon is called MHC restriction of antigen-specific T cell responses.

The MHC consists in a region of about 4 Mbp DNA located on the short arm of chromosome 6p21. The region contains more than 100 genes, the majority of which have immunological relevance in antigen processing and presentation. The MHC is divided into three regions, named class I, class II and class III. Class I region is located telomerically and contains HLA-A, -B and -C genes, involved in transplantation response. Class II region, situated centromerically, contains HLA-DP, -DQ and -DR genes, mainly involved in antigen presentation. Class I and class II loci are functionally related, because they are all members of the immunoglobulin superfamily.

Structure of the HLA Region

HLA-class II region encompasses approximately 1000 kb DNA. DP, DQ and DR subregions contain at least one pair of expressed genes encoding a pair of functionally expressed α and β chains. For example, the DQ region contains DQA1 and DQB1 genes, which code for DQα and DQβ chains, respectively. The α and β chains form an α-β heterodimer expressed on the surface of the antigen-presenting cell, restricted to helper or suppressor T lymphocytes. Some haplotypes contain up to 14 class II loci, but limitation in the expression of permissive α-β dimers restricts the expressed repertoire to four class II molecules per haplotype: DPα-β, DQα-β, DRα-β1, DRα-β3, 4, 5. Each class II haplotype includes DPα-β, DQα-β, DRα-β1, DRα-β3 or 4 or 5, the latter being determined by the haplotype considered.

Neurophysiopathology Section, Department of Neuroscience, University of Cagliari, Via Ospedale 119 - 09124 Cagliari, Italy. e-mail: gmarrosu@unica.it

Whereas DRα chains are essentially invariant, DRβ, DQα, DQβ, DPα and DPβ chains exhibit sequence variants that are responsible for polymorphisms of the class II region characteristic of these molecules. Such polymorphism is determined by the germline genetic repertoire of the class II region with multiple alleles at most class II loci. Initially, the nomenclature of HLA was referred to on the basis of serologic specificity, such as DR2, DR3 and DR4. This specificity does not distinguish the sequence of alleles. On the contrary, molecular analysis provides a high degree of gene-specific and haplotype-specific information.

Polymorphism of the MHC system, in particular polymorphic residues situated in and around the peptide-binding cleft, regulates the developmental selection of T cell receptor (TCR) specificities during the process of T-cell differentiation and maturation in the thymus. The primary sequence of the amino-terminal polymorphic domain is critical for proper presentation of antigen to the TCR. Both α and β chains contain two external domains, the membrane-external domain being the major site of variation. The foundation for genetic restriction of foreign antigen recognition lies in the formation of a peptide-binding groove, the structure of which has been hypothesized by analogy with class I crystallographic structure, with the exception that it is more open at both ends allowing the bound peptide to protrude out of the cleft [1].

Differences among HLA-DR molecules have been defined within three major regions of hypervariability and two regions of more limited variability. The peptide binding site of HLA-DR is generated by the first domain of the DRα and DRβ chains. The primary anchor residue is accommodated by a hydrophobic pocket, the size of which is controlled by a diallelic polymorphism determining the possibility of accommodating a large hydrophobic residue or an aromatic residue of the foreign peptide in this region [2]. It has been demonstrated that endogenous peptides can bind multiple DR alleles and that this capability must be dependent on the composition and location of several key amino acids within the primary structure [3].

Function of the HLA Region in Disease Association

HLA associations with various autoimmune diseases have been recognized over 25 years ago, but the exact mechanism underlying such an association is not yet clarified. With the introduction of molecular biology techniques, alleles associated with specific diseases have been identified [4], and the sequences of such alleles have been examined [5].

Association in terms of both susceptibility to and protection from autoimmune and infectious diseases (e.g. severe malaria) has been reported, referring as "protective" a negative association and "susceptible" a positive one. Mechanisms hypothesized in the association of HLA with diseases include:
- Peptide presentation
- Presentation of self-peptide
- Receptor theory
- Linkage disequilibrium

- Altered self-peptide presentation
- Superantigenic stimulation
- Thymic selection of TCR repertoire

Peptide Presentation

HLA class I molecules bind short peptides (8-12 amino acids) derived from intracellular proteins, so they are important in the immune response to viruses and other intracellular pathogens. HLA class II molecules bind longer peptides (10-34 amino acids), mainly derived from extracellular and cell-surface proteins.

Polymorphism of class II alleles is crucial to immune activation events. Polymorphic amino acids determine whether or not specific antigenic peptides will bind and will be presented on the surface of antigen-presenting cells to T lymphocytes. The HLA association with many autoimmune diseases can stem from differences in the capacity to bind different sets of peptides. Diseases can occur if a given HLA allele presents a pathogenic peptide either as a function of the allele-specific peptide-binding motif or by access to peptides derives from unusual (self) sites. The penetrance of the disease trait may depend by competition between the affinity of the pathogenic peptide to a susceptible and a protective HLA allele. An example of such a mechanism is the hierarchy of HLA association in insulin-dependent diabetes mellitus (IDDM). Protective and susceptible HLA alleles can be codified in *cis* and in *trans*, so distinct peptide-binding properties may be encoded by different haplotypes, as occurs in coeliac disease.

Altered Self-Peptide Presentation

Central and peripheral mechanisms devoted to tolerance maintain unresponsive HLA molecules to self-peptide. Alteration of self-peptide can occur via exposure of cryptic epitopes or post-translational modification of proteins.

Presentation of Self-Peptide

HLA antigens present not only molecules derived from extracellular organisms but also many self-proteins. The majority of self-peptides are themselves derived from other MHC class I and II molecules, so HLA molecules can regulate immune response also by acting as a source of presented peptides.

Superantigenic Stimulation

Superantigens are proteins produced by virus and bacteria able to stimulate T cells by cross-linking their TCRs with the MHC molecule, resulting in the simultaneous activation of a large number of T cells expressing the particular TCR Vβ family. This stimulation, usually followed by clonal delection, could break self-tolerance.

Receptor Theory

This theory suggests that some microorganisms use MHC molecules as a vehicle to enter cells. There are several examples of such a mechanism in the blood-group system.

Thymic Selection of TCR Repertoire

Thymic TCR repertoire formation is driven by MHC molecules presenting self-peptide, which acts as a template. Immature T cells entering the thymus and expressing TCR which fail to recognize self-MHC undergo apoptosis, while high avidity TCR-self-MHC-peptide interactions produce cell death (negative selection). A proper avidity of TCR with self-MHC and peptide permits positive selection so that the cell can egress into the periphery. Low avidity of certain HLA haplotypes to self-peptide can result in selection of auto-reactive or cross-reactive T cells.

Linkage Disequilibrium

The association between HLA alleles and diseases may result from the strong linkage disequilibrium between genes located in this region. Linkage disequilibrium refers to the presence of two alleles at different loci occurring together more frequently than would be expected by chance, in relation to the physical distance and the recombination of the two loci involved. Some DR and DQ alleles are inherited together on a chromosome as a single genetic unit termed "haplotype" (haploid genotype). In this case, when a particular HLA allele is associated with a disease, it is not immediately obvious whether the gene itself or some linked gene on the same haplotype is primarily responsible for the disease association. For example, a single allele such as DQB1*0201 may be found in different haplotypes, associated with different DQA and DRB genes. Analysis of the extended haplotypes by fine genetic mapping may help to clarify the locus-specific and allele-specific contribution to susceptibility.

Methodology in HLA-Disease Association Studies

Most HLA-associated diseases have genetic components which do not display simple Mendelian patterns. Family studies can reveal the number of genes involved, the inheritance pattern, and the penetrance of the trait. The most important epidemiological parameter is the relative risk, λ_r, which specifies the recurrence risk of a relative (R) of the proband compared to the incidence of the disease in the population. For siblings (S), therefore, λ_s = risk of recurrence in sibling of proband/risk of disease in general population. Values of λ_s in multiple sclerosis (MS) are 20-40, i.e. the genetic component accounts for 20%-40% of the disease as whole. The contribution of the HLA locus itself can be estimated from the ratio of the expected (0.25) and the observed proportions of sib pairs sharing zero alleles identical by descent [6]. In MS, λ_s HLA = 2.4, a value which is lower than in other autoimmune diseases, such as IDDM and rheumatoid arthritis (RA).

The molecular genetic approach to matters concerning HLA and disease associations involves indentification of:
- disease-predisposing haplotype(s),
- individual genes included in haplotypes (in both cis and trans) responsible for the association,
- relationship between HLA-associated alleles and immunopathogenetic mechanisms.

Identification of Disease-predisposing Haplotypes

Methods in detecting the HLA predisposing haplotype include association studies from both population-based samples and nuclear families, affected sib pair analysis and lod score linkage analysis.

Population-based Association Studies. The method consists in comparing marker allele frequencies in ethnically matched patient and control populations. To avoid bias due to ethnic mixing, an artificial control population may be constructed using the non-transmitted parental haplotype as control. This approach is termed the AFBAC (affected family-based controls) method [7]. However, the association of a marker with disease implies either that the marker itself influences disease susceptibility or that it is in linkage disequilibrium with the disease-predisposing locus.

Association Studies on Nuclear Families. The transmission disequilibrium method tests deviation from the random 50:50 of allele transmission from parents to affected offspring [8]. In this test, polymorphic markers are examined in a large set of nuclear families, which include both parents and one affected child. Significant deviation from the expected 50:50 ratio defines linkage in presence of association.

Affected Sib Pair Analysis. In sib pairs, random expectation is 1/4, 2/4 and 1/4 for affected sib pairs which share 2, 1 or 0 identical-by-descent (IBD) parental alleles. Deviation from random has been used to confirm linkage between a marker and a disease. This method has been a powerful test in confirming HLA linkage in IDDM. However, in MS deviation from expected has not been detected, perhaps reflecting a high frequency of the disease-predisposing allele or heterogeneity in disease predisposition.

Lod Score Linkage Analysis. This tool tests modes of inheritance. A model specifies location of the disease-association allele (locus), allele frequency and allele penetrance. The observed genotypes and phenotypes in a pedigree are used to test a variety of models (M) against a null hypothesis (M0) which assume no linkage to a susceptible gene in a region of interest. The analysis works well in monogenic diseases, but in some cases can also be used in polygenic diseases. In MS, analysis of a Finnish pedigree showed linkage with both MHC and myelin basic protein [9].

Identification of Individual Genes

Unlike the cases for IDDM, a role for individual HLA alleles in MS has not been demonstrated. Different HLA allele associations in different populations have been reported. In Caucasians, DRB1*1501 (DR2) allele has been associated with MS [10], while consistent analyses in both population-based studies [11, 12] and using transmission disequilibrium test [13] showed association of MS with both DRB1*0405 (DR4) and DRB1*0301 (DR3) alleles in Sardinians. No sequence motif common to these associated alleles has been found. The trans-ethnic analysis of associated and not associated haplotypes in Caucasians and Sardinians suggests that HLA may not be the primarily involved gene but merely a marker in linkage disequilibrium with MS [13]. Another hypothesis in explaining differences in the HLA-MS association is the possibility that the disease is heterogeneous in etiopathogenetic mechanism. The polymorphism of HLA molecules is selected on the basis of their immune function. This selection happens in two ways: selective advantage of heterozygous (overdominant selection), in which a heterozygous individual responds quantitatively better than either homozygote, and rare allele advantage (frequency-dependent selection), in which individuals with rare MHC alleles respond better to new pathogen variants that are evolved to evade common MHC alleles. Differences in alleles associated with MS may reflect different immunogenetic mechanisms in reactions between the immune system MHC repertoire and environmental pathogens.

Identification of Relationships Between HLA-Associated Alleles and Immunopathogenetic Mechanisms

Such a step could bridge the gap between strictly genetic analysis and functional questions. Several working hypotheses can be formulated:
- The association is caused by several polymorphic residues critical in peptide binding;
- HLA MS-associated molecules have structural motifs which use charge-charge interactions at points critically involved in MHC-peptide binding. Such a situation has been proposed in RA [14];
- Similarity of viral and bacterial peptides and immunodominant self-peptides can activate autoaggressive T lymphocytes via molecular mimicry with MHC and TCR structural motif-binding sites [15].

References

1. Stern LJ, Brown JH, Jardestzky TS et al. (1994) Crystal structure of the human class II MHC protein HLA-DR1 complexed with an influenza virus peptide. Nature 368: 215-221
2. Krieger JI, Karr RW, Grey HM et al. (1991) Single amino acid changes in DR and antigen define residues crucial for peptide-MHC binding and T cell recognition. J Immunol 146: 2331-2338
3. Chicz RM, Urban RG, Gorga JC et al. (1993) Specificity and promiscuity among naturally processed peptides bound to HLA-DR alleles. J Exp Med 178: 27-47
4. Bell JI, Todd JA, McDevitt HO (1989) The molecular basis of HLA disease association. In: Harris H, Hirschorn K (eds) Advances in human genetics. Plenum, London, pp10-17
5. Marsh SGE, Bodmer G (1991) HLA class II nucleotide sequences, 1991. Tissue Antigens 37: 181-189
6. Risch N (1987) Assessing the role of HLA-linked and unlinked determinants of disease. Am J Hum Genet 40: 1-14
7. Thomson G (1995) Mapping disease genes: family-based association studies. Am J Hum Genet 75: 487-498
8. Spielman RS, McGinnis RE, Ewens WJ (1993) Transmission test for linkage disequilibrium: the insulin gene region and insulin-dependent diabetes mellitus (IDDM) Am J Hum Genet 52: 506-516
9. Tienari PJ, Terwilliger JD, Ott J et al. (1994) Two-locus linkage analysis in multiple sclerosis (MS). Genomics 19: 320-325
10. Hillert J, Olerup O (1993) Multiple sclerosis is associated with genes within or close to the HLA-DR-DQ subregion on a normal DR15, DQ6, Dw2 haplotype. Neurology 43: 163-168
11. Marrosu MG, Muntoni F, Murru MR et al. (1988) Sardinian multiple sclerosis is associated with HLA-DR4: a serologic and molecular analysis. Neurology 38: 1749-1753
12. Marrosu MG, Murru MR, Costa G et al. (1997) Multiple sclerosis in Sardinia is associated and in linkage disequilibrium with HLA-DR3 and -DR4 alleles. Am J Hum Genet 61: 454-457
13. Marrosu MG, Murru MR, Costa G et al. (1998) DRB1-DQA1-DQB1 loci and multiple sclerosis predisposition in the Sardinian population. Hum Mol Genet 7: 1235-1237
14. Hammer J, Gallazzi F, Bono E et al. (1995) Peptide-binding to HLA-DR4 molecules: Correlation to rheumatoid arthritis association. J Exp Med 181: 1847-1855
15. Wucherpfennig KW, Strominger JL (1995) Molecular mimicry in T cell-mediated autoimmunity: Viral peptides activate human T cell clones specific for myelin basic protein. Cell 80: 695-705

Chapter 14

Cytokine Genes in Multiple Sclerosis

F.L. Sciacca, L.M.E. Grimaldi

Introduction

Multiple sclerosis (MS) is an autoimmune/inflammatory disease of the central nervous system (CNS) resulting in polymorphic and unpredictable clinical manifestations. Although described as a clinical and pathological entity by Cruveilheir, Carswell and Charcot more than a century ago, its etiology is still obscure and many questions regarding its complex pathogenesis are still unanswered. Current views credit environmental factors, possibly infectious, for triggering MS in genetically susceptibile individuals. The nature of this genetic influence in MS has been the subject of intense studies: classic genetic observations have excluded the involvement of a single gene with full mendelian inheritance [1], both in susceptibility to and modulation of MS, as recently confirmed by full genome family-based studies [2–5]. Alternative to a monogenic hypothesis, polygenic inheritance was proposed for MS almost 50 years ago [6]. Multiple strategies, including full genome screening and association studies, have been employed to identify the discrete number of genes expected to be involved in the pathogenetic processes leading to MS.

The first approach, entailing the use of hundreds of highly polymorphic markers to scan the entire genome of affected MS family members for linkage with potential susceptibility genes, has been so far reported by four groups [2–5] and has indicated that there is not a single master gene for MS. However, this approach has revealed several chromosomal regions likely to contribute in a limited manner to the susceptibility to MS, the human leukocyte antigens (HLA) region having the strongest association in the majority of the studies. The poor interstudy concordance may be attributed to genetic heterogeneity, lack of detection power, or simply absence of significant common genetic contribution to familial aggregation in MS.

The second approach, so far largely unrevealing, seeks associations of one or more candidate genes with the disease. The selection of these candidate genes has been influenced by the established autoimmune pathogenesis of MS. As such, the focus of this search has been primarily on genes coding for immunorelevant molecules, including T cell receptor subunits, immunoglobulins (Ig), myelin antigens

---

Neuroimmunology Unit, Neuroscience Department, San Raffaele Scientific Institute, Via Olgettina 58 - 20132 Milan, Italy. e-mail: luigi.grimaldi@hsr.it

and cytokines. As already mentioned, the only genetic region for which a positive, although minor, influence on MS susceptibility has been convincingly demonstrated is a locus associated with the HLA class II region on chromosome 6 [7]. However, genetic heterogeneity has also been demonstrated for the HLA locus since different haplotypes have been associated with the disease and with disease progression in different populations [8, 9]. Recently, following several contrasting observations on the association of T cell receptor (TCR) β-chain with MS susceptibility, an epistatic interaction between the HLA allele Drw15 and TCR β-chain locus has been shown for a subgroup of MS patients affected by relapsing-progressive disease course [10]. This observation suggests for the first time that genes or alleles involved in MS susceptibility not only may affect the occurrence of the disease, but also may determine different disease courses.

Cytokines in MS

The cytokine family represents a polymorphic system whose effects are potentially able to interfere with the immunopathological processes occurring in MS patients and whose levels of interaction are wide enough to sustain the obvious complexity of the resulting MS phenotypes (from very mild and slow-progressing to very active and rapidly disabling). Both pro- and anti-inflammatory cytokines have been implicated in the pathogenesis of MS and in its capricious course. A hypothetical cytokine contributing to MS is likely to be initially secreted by CD4$^+$ autoreactive T cells entering the central nervous system and, later on, by recruited inflammatory or activated glial cells. The proinflammatory cytokines that most efficiently induce the recruitment of additional inflammatory cells include interleukin (IL)-1β, interferon (IFN)-γ, lymphotoxin (LT) and tumor necrosis factor (TNF)-α. Upregulation of proinflammatory cytokines such as IL-1β, IL-2, IL-6, IFN-γ, TNF-α and LT and downregulatory cytokines such as IL-4, IL-10 and transforming growth factor (TGF)-β have been detected within or nearby CNS MS lesions [11], as well as in the cerebrospinal fluid (CSF) or in the peripheral circulation of MS patients [12]. Complex interactions between cytokines may foster major clinical changes in the course of MS; the occurrence of clinical relapses, for instance, has been associated with a synchronous increase in IFN-γ, TNF-α and its receptors expression [13, 14], while remission phases of the disease have been related to an increase of secretion of TGF-β, IL-10 and IL-1 receptor antagonist (IL-1Ra) [15,16]. Due to the potential regulatory role of cytokines in determining the onset or the course of MS, cytokine genes have been a favorite target for genetic studies aimed to verify their association with MS occurrence and disease variability. However, despite the established biological role played by several cytokines in MS, most (if not all) association studies on cytokine genes in patients with this disease have provided controversial results. The simplest explanation for this occurrence is that these studies simply did not look for the right cytokine. We believe that other factors have hampered the reproducibility of these association studies. First, ethnic differences among the populations studied exist and provide a strong level of variability in these studies. In the majority of cases, however, the limited number of subjects evaluated (around or less than 100) may have prevented reaching statistical relevance for associations between the various cytokines and MS. Association studies, in fact, need at least several hundreds to one thousand patients to detect the relatively modest contribution of one of the probably numerous genetic determinants of a complex disease such as MS [17]. It derives that multicentric efforts would be needed most of the time to recruit the appropriate number of MS cases and relatives for effective case-control and family-based association studies. This fact clearly limits the detection power of these studies.

Here follows a review of the main cytokines involved in the pathogenesis of MS whose genes have been evaluated as potential contributors to occurrence or clinical variability of MS.

Interferons

The interferons (IFNs) are a family of cytokines initially described as proteins able to interfere with viral infections of target cells. Additional biological effects of these cytokines include modulation of the immune system. The IFN family is divided in type I and type II subgroups. The type I IFNs (including IFN-α and -β) are approximately 20 acid-stable proteins having similar structure and biological activities and sharing a common gene locus on chromosome 9. IFN-γ, or type II IFN, is an acid-labile protein whose gene is located on chromosome 12. IFN-γ is more immunomodulating than type I IFNs and shows in some cases contrasting biological activities: IFN-γ, but not IFN-β, for instance, induces the expression of HLA class II molecules on CNS resident cells [18], stimulates the release of oligodendrocytotoxic cytokines such as TNF-α by macrophages [19], triggers the production of toxic molecules such as nitric oxide metabolites by microglial cells [20], and mediates oligodendrocyte apoptosis [21]. Compared to controls, IFN-γ serum protein or CSF mRNA levels are elevated in MS patients [13, 22] irrespective of their clinical status (exacerbations vs. stability) [23]. In vitro, peripheral $CD4^+$ and $CD8^+$ T cells from MS patients produce more IFN-γ, both in resting [24] and in CD3-stimulated conditions [25], and show effective intracellular Ca^{2+} elevations in response to IFN-γ [26] which precedes by 30 days evidence of clinical activity [27]. Moreover, treatment with IFN-γ causes immediate reactivation of MS progression [28], while part of the beneficial effects of IFN-β in MS patients are mediated by its anti-IFN-γ activity [29].

An initial association study between an MS occurence and a CA dinucleotide repeat polymorphism located in the first intron of the IFN-γ gene found a slightly positive association (LOD score of 0.88, theta=0.01) [30]. Further reports have not confirmed this claim [31], and have also excluded an association between a $(CA)_n$ microsatellite DNA within the proximal half of the IFN-β/α/ω gene cluster and the occurrence of MS. Despite the established role of IFNs in the pathogenesis of MS, genetic influence of these cytokines' genes on the occurrence or clinical heterogeneity of this disease has never been confirmed.

TNF-α and Lymphotoxin

TNF-α and lymphotoxin (LT) are members of a protein family able to trigger proliferating and apoptotic signals. TNF-α is a critical inflammatory mediator for the demyelinating events typical of MS and experimental autoimmune encephalitis (EAE), the animal model of MS. The TNF cytokines may directly damage oligodendrocytes [19, 32] and endothelial cells (EC) [33]. Serum TNF-α levels were significantly higher in MS patients compared to controls [13]. An elevation of TNF-α and its receptors (p55 and p75) consistently precedes the occurence of clinical exacerbations in MS patients [14]. Both TNF-α and LT genes are located within the HLA locus: after an initial report of a different TNF-α (two G/A polymorphisms in the promoter region) and LT (GT dinucleotide repeats in the promoter region) allele distribution between MS patients and healthy controls (HC) [34], it has been shown that this association was secondary to the linkage of the TNF-α gene with the DR2 HLA haplotype predisposing to MS [31, 35]. At present, the TNF-α gene should not be considered as primarily associated with MS.

IL-2

IL-2 is a lymphoproliferative cytokine involved in MS pathogenesis mainly for its potential ability to induce autoimmune T cell clone proliferation. IL-2 has been found within MS brain lesions [36], a strategic location to sustain the proinflammatory cellular phenomena of the relapsing phases of MS. However, circulating blood and CSF protein and mRNA detection, as well as in vitro stimulation experiments, have repeatedly failed to find significant differences between MS patients and controls [37-39]. A microsatellite marker (CA dinucleotide repeats) located in the 3'-untranslated region of IL-2 has been reported not to be associated with the occurrence of MS [30, 31]. At present there is no evidence for a role of this cytokine gene in the genetic predisposition to MS.

IL-4

This cytokine inhibits the production by monocytes of many proinflammatory cytokines and chemokines and promotes B cell proliferation and activation [40]. IL-4 has been detected in the CSF of MS patients [13]. The expression of IL-4 mRNA and the production of its mature protein is increased in peripheral blood mononuclear cells from MS patients compared to controls [41] with little fluctuation over time [24] and an interesting restriction to HLA-Drw2 positive individuals [42]. In addition, IL-4 has shown beneficial effects in the treatment and prevention of EAE [43]. A polymorphic marker [variable number of tandem repeats (VNTR)] located in the third intron of the IL-4 gene has been investigated and found not associated with occurrence of MS [30]. We actually found that the IL-4 gene is involved in the modulation of clinical phenomena in MS, since a VNTR polymorphism is associated with the age at disease onset [44]. The role of IL-4 in the genetic predisposition to MS deserves further attention.

IL-10

IL-10 was first identified as a cytokine synthesis inhibitory factor produced by Th2 clones and acting on Th1 cells via macrophage accessory cells [45, 46]. Subsequent work revealed that IL-10 is a pleiotropic cytokine that can inhibit various functions of T and B lymphocytes, mast cells, monocytes and EC [47, 48], induce anergy in $CD4^+$ T cells [49], and stimulate B cell growth and Ig production [50]. Moreover, it has been shown that IL-10 administration prevents and treats EAE in Lewis rats [43, 51]. Serum levels of IL-10 are not significantly elevated in MS, although several patients have elevated serum levels of IL-10 compared to controls [13]. Interestingly, IL-10 mRNA and protein levels increase in vivo, after glucocorticoid [52] or IFN-β [53] treatment of MS patients and after CD3 stimulation in vitro [25].

A polymorphic marker (CA dinucleotide repeat) in the promoter region of the IL-10 gene has been investigated and found not associated with the occurrence of MS [30].

TGF-β

Transforming growth factor beta (TGF-β) is an exceedingly pleiotropic cytokine involved in a wide variety of biological processes, ranging from inflammation to development, tissue repair and tumorigenesis. It occurs in three isoforms, TGF-β1, -β2, and -β3, which bind to three types of receptors. TGF-β treatment reduces [54-56], while neutralizing antibody to TGFβ-1 enhances, the severity of EAE [57]. A defective production of TGF-β has been found in T cell lines of MS patients and showed some correlation with disease activity [58]. Autoantigens [59], IFN-β [60] or corticosteroids [61] induce TGF-β mRNA expression in mononuclear cells from MS patients.

A polymorphism (CA dinucleotide repeats) located in the 5'-untranslated region of the TGF-β gene has been evaluated for its implication in MS by several authors. All published reports concordantly indicate a lack of association with the disease [30]. At present, the TGF-β gene should be considered not associated with MS.

IL-1

IL-1α and IL-1β are the two prototypical proinflammatory cytokines. The IL-1Ra is a protein structurally related to IL-1β that forms a non-productive complex with the two IL-1 receptors [62]. IL-1Ra occurs in three alternatively spliced forms, two exclusively found in the intracellular compartment (icIL-1Ra type I and II), and one secreted (sIL-1Ra). sIL-1Ra is an effective inhibitor of the IL-1-induced proinflammatory effects [62], and has the potential to inhibit autoimmune inflammatory demyelination in rats with EAE [63]. In MS, both IL-1α and IL-1β have been found within MS lesions [64, 65]. Serum levels of sIL-1Ra normal during remission phases in patients with the relapsing-remitting (RR) form, but increased significantly during exacerbations or in response to IFN-β treatment [16, 66].

The genes for IL-1α, -1β and -1Ra are located in a cluster along the long arm of chromosome 2 (Fig. 1). IL-1α and -1β have several single base polymorphisms, two of which (both C/T dimorphisms) are located in the promoter regions of the cytokines. The IL-1Ra gene is polymorphic in several regions: the gene of the soluble protein is polymorphic in intron 2 and exon 2 and these polymorphisms are completely associated [67], while there is not linkage disequilibrium between the polymorphism of IL-1Ra and those of IL-1α and β. The second intron of the IL-1Ra gene contains an 86 bp VNTR polymorphism. Although five alleles have been described so far (4, 2, 5, 3 and 6 repeats, respectively from A1 to A5), A1 and A2 are the only alleles with a potential clinical relevance, accounting for approximately 99% of all IL-1Ra alleles. Several studies seeking association between these three cytokines and MS occurrence have provided contrasting results [68-70].

We performed a case-control association study based on a large cohort of Italian subjects (339 MS patients and 339 age-matched healthy controls (HC)), and found that, while IL-1α and IL-1β polymorphisms did not show significant differences in the allele and genotype frequency (GF) between MS subjects and HC, the IL-1Ra gene intron 2 polymorphism is moderately associated with the occurrence of the disease. The A1 allele appeared the major predisposing factor for the disease, conferring an odds ratio (OR) of 1.83 in double copy and 1.35 in single copy. A novel finding of our study was the association between IL-1Ra-polymorphism and clinical variables of MS. In our cohort of Italian subjects, in fact, a progressive increase in A1/A1 (and a mirror decrease A2/A2) GF actually paralleled

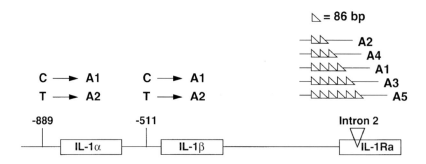

Fig. 1. IL-1 cluster in the chromosome region 2q12-21 includes IL-1α, IL-1β and IL-1Ra genes. Indicated are the polymorphisms studied: two C/T dimorphisms in the promoter region of IL-1α and IL-1β, and the VNTR in the IL-1Ra gene

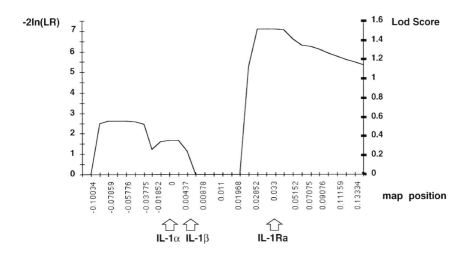

Fig. 2. Linkage analysis of IL-1 polymorphism clusters in multiple sclerosis patients

the clinical outcome of MS, being significantly more pronounced in non-benign (patients that accumulated an expanded disability status scale (EDSS) score > 3 within 10 years from the clinical onset of the disease) than in benign MS patients (EDSS ≤ 3 after at least 10 years from disease onset) and, even more so, in HC. Moreover, the percentage of patients with a more aggressive disease course observed at 3 cut-off time points was always higher within the A1$^+$ genotype groups than within the A2/A2 group. This aggravating effect was demonstrated around a mild disability burden, measured EDSS score ≤ 3, a clinical situation where inflammatory phenomena are still likely to be active and able to influence disease progression. Accordingly with these genetic observations, we found that in peripheral blood mononuclear cells from HC and in patient sera the three polymorphic genotypes were associated with a different production of IL-1Ra: the A1$^+$ genotypes were always associated with a lower production of IL-1Ra compared to the A2/A2 genotype. Since the IL-1Ra is an anti-inflammatory cytokine, an increase in A1$^+$ GF, leading to decreased cytokine production, is likely to foster a sharpening of inflammatory processes, a more aggressive inflammatory/autoimmune disease and, possibly, a greater accumulation of disease clinical burden (as assessed by the EDSS) over the years. The three association studies regarding the three IL-1 cytokines located in the cluster 2q12-21 were combined for a linkage disequilibrium analysis seeking possible association in a chromosomal region of this locus. The results showed that indeed a "hot" region for MS occurence was located nearby the IL-1Ra gene (Fig. 2). Genetical studies of the IL-1 gene family cluster on chromosome 2 may reveal interesting insights into the complex phenomena underlying MS and will have to be confronted with the results of more detailed linkage analyses of the region derived from family based studies.

Conclusions

The large number of potential candidate genes decrease the chances for association studies (both based on single cases or on families) to reveal relatively small effects of cytokine genes on predisposition or modulation of MS. However, at present no better approaches are available to sort out the genetic control of complex diseases. Current genetic models of MS susceptibility suggest the existance of two or more genes, possibly varying among different populations and patients affected by clinical subtypes of the disease. Several "modulatory" genes which could affect the disease clinical variables (age at onset, accumulation of disability, lesion localization, etc.) are likely to exist. Some of these genes might be found among those coding for immunorelevant molecules such as cytokines. Moreover, an indefinite number of environmental (epigenetic) factors might interact with genetic loci predisposing to this disease and influence the resulting clinical burden, an event likely to occur in a disease developing over many years and for which infective, inflammatory and hormonal phenomena modulate complex immunological activities (e.g. cell activation, proliferation, cytokine secretion). Although so far unrewarding and limited by technical problems, genetic studies in MS are already popular. Large series of patients currently gathered for genetic studies will provide the critical mass to achieve the needed statistical significance. Confirmation by independent studies in other populations will be required before accepting as meaningful the results of these studies [71].

We conclude that studies on cytokine gene polymorphisms have the potential to provide relevant insights into the mechanisms of immunomediated CNS demyelination in humans. Polymorphic genes of a great number of immunorelevant molecules should be evaluated to achieve a more complete picture of the complex pathogenesis of MS.

References

1. Bell JI, Lathrop GM (1996) Multiple loci for multiple sclerosis. Nat Genet 13: 377-378
2. Sawcer S, Jones HB, Feakes R et al. (1996) A genome screen in multiple sclerosis reveals susceptibility loci on chromosome 6p21 and 17q22. Nat Genet 13: 464-468
3. Haines JL, Ter-Minassian M, Bazyk A et al. (1996) A complete genome screen for multiple sclerosis underscores a role for the major histocompatibility complex. The multiple sclerosis genetic group. Nat Genet 13: 469-471
4. Ebers GC, Kukai K, Bulman DE et al. (1996) A full genome search in multiple sclerosis. Nat Genet 13: 472-476
5. Kupkkanene S, Gschwend M, Rioux JD et al. (1997) Genomewide scan of multiple sclerosis in Finnish multiplex families. Am J Hum Genet 61: 1379-1387
6. Pratt RTC, Compston ND, McAlpine D (1952) The familial incidence of multiple sclerosis and its significance. Brain 74: 191-232
7. Hillert J, Olerup O (1993) Multiple sclerosis is associated with genes within or close to the HLA-DR-DQ subregion on a normal DR15, DQ6, Dw2 haplotype. Neurology 43: 163-168

8. Eoli M, Pandolfo M, Amoroso A et al. (1995) Evidence of linkage between susceptibility to multiple sclerosis and HLA-class II loci in Italian multiplex families. Eur J Hum Genet 3: 303-311
9. Marrosu MG, Murru MR, Costa G et al. (1997) Multiple sclerosis in Sardinia is associated and in linkage disequilibrium with HLA-DR3 and -DR4 alleles. Am J Hum Genet 61: 454-458
10. Hockertz MK, Paty DW, Beall SS (1998) Susceptibility to relapsing-progressive multiple sclerosis is associated with inheritance of genes linked to the variable region of the TcR beta locus: use of affected family-based controls. Am J Hum Genet 62: 373-385
11. Cannella B, Raine CS (1995) The adhesion molecule and cytokine profile of multiple sclerosis lesions. Ann Neurol 37: 424-435
12. Birebent B, Semana G, Commeurec A et al. (1998) TCR repertoire and cytokine profiles of cerebrospinal fluid- and peripheral blood-derived T lymphocytes from patients with multiple sclerosis. J Neurosci Res 51: 759-770
13. Hohnoki K, Inoue A, Koh CS (1998) Elevated serum levels of IFN-gamma, IL-4 and TNF-alpha/unelevated serum levels of IL-10 in patients with demyelinating diseases during the acute stage. J Neuroimmunol 87: 27-32
14. Martino G, Consiglio A, Franciotta DM et al. (1997) Tumor necrosis factor alpha and its receptors in relapsing-remitting multiple sclerosis. J Neurol Sci 152: 51-61
15. Rieckmann P, Albrecht M, Kitze B et al. (1995) Tumor necrosis factor-alpha messenger RNA expression in patients with relapsing-remitting multiple sclerosis is associated with disease activity. Ann Neurol 37: 82-88
16. Voltz R, Hartmann M, Spuler S et al. (1997) Longitudinal measurement of IL-1 receptor antagonist. J Neurol Neurosurg Psychiatry 62: 200-201
17. Risch N, Merikangas K (1996) The future of genetic studies of complex human diseases. Science 273: 1516-1517
18. Cannella B, Raine CS (1989) Cytokines up-regulate Ia expression in organotypic cultures of central nervous system tissue. J Neuroimmunol 24: 239-248
19. Selmaj K, Raine CS (1988) Tumor necrosis factor mediates myelin damage in organotypic cultures of nervous tissue. Ann N Y Acad Sci 540: 568-570
20. Hartung HP, Jung S, Stoll G et al. (1992) Inflammatory mediators in demyelinating disorders of the CNS and PNS. J Neuroimmunol 40: 197-210
21. Vartanian T, Li Y, Zhao M, Stefansson K (1995) Interferon-gamma-induced oligodendrocyte cell death: implications for the pathogenesis of multiple sclerosis. Mol Med 1: 732-734
22. Monteyne P, Van Laere V, Marichal R, Sindic CJ (1997) Cytokine mRNA expression in CSF and peripheral blood mononuclear cells in multiple sclerosis: detection by RT-PCR without in vitro stimulation. J Neuroimmunol 80: 137-142
23. van Oosten BW, Barkhof F, Scholten PE et al. (1998) Increased production of tumor necrosis factor alpha, and not of interferon gamma, preceding disease activity in patients with multiple sclerosis. Arch Neurol 55: 793-798
24. Lu CZ, Jensen MA, Arnason BG (1993) Interferon gamma- and interleukin-4-secreting cells in multiple sclerosis. J Neuroimmunol 46: 123-128
25. Brod SA, Nelson LD, Khan M, Wolinsky JS (1997) Increased in vitro induced CD4+ and CD8+ T cell IFN-gamma and CD4+ T cell IL-10 production in stable relapsing multiple sclerosis. Int J Neurosci 90: 187-202
26. Martino G, Clementi E, Brambilla E et al. (1994) Gamma interferon activates a previously undescribed Ca^{2+} influx in T lymphocytes from patients with multiple sclerosis. Proc Natl Acad Sci USA 91: 4825-4829

27. Martino G, Filippi M, Martinelli V et al. (1997) Interferon-gamma induced increases in intracellular calcium in T lymphocytes from patients with multiple sclerosis precede clinical exacerbations and detection of active lesions on MRI. J Neurol Neurosurg Psychiatry 63: 339-345
28. Panitch HS, Hirsch RL, Haley AS, Johnson KP (1987) Exacerbations of multiple sclerosis in patients treated with gamma interferon. Lancet 1: 893-895
29. Van Weyenbergh J, Lipinski P, Abadie A et al. (1998) Antagonistic action of IFN-beta and IFN-gamma on high affinity Fc gamma receptor expression in healthy controls and multiple sclerosis patients. J Immunol 161: 1568-1574
30. He B, Xu C, Yang B et al. (1998) Linkage and association analysis of genes encoding cytokines and myelin proteins in multiple sclerosis. J Neuroimmunol 86: 13-19
31. Epplen C, Jackel S, Santos EJ et al. (1997) Genetic predisposition to multiple sclerosis as revealed by immunoprinting. Ann Neurol 41: 341-352
32. Selmaj K, Raine CS, Farooq M et al. (1991) Cytokine cytotoxicity against oligodendrocytes. Apoptosis induced by lymphotoxin. J Immunol 147: 1522-1529
33. Estrada C, Gomez C, Martin C et al. (1992) Nitric oxide mediates tumor necrosis factor-alpha cytotoxicity in endothelial cells. Biochem Biophys Res Commun 186: 475-482
34. Kirk CW, Droogan AG, Hawkins SA et al. (1997) Tumour necrosis factor microsatellites show association with multiple sclerosis. J Neurol Sci 147: 21-25
35. Roth MP, Nogueira L, Coppin H et al. (1994) Tumor necrosis factor polymorphism in multiple sclerosis: no additional association independent of HLA. J Neuroimmunol 51: 93-99
36. Woodroofe MN, Cuzner ML (1993) Cytokine mRNA expression in inflammatory multiple sclerosis lesions: detection by non-radioactive in situ hybridization. Cytokine 5: 583-588
37. Sivieri S, Ferrarini AM, Gallo P (1998) Multiple sclerosis: IL-2 and sIL-2R levels in cerebrospinal fluid and serum. Review of literature and critical analysis of ELISA pitfalls. Mult Scler 4: 7-11
38. Musette P, Benveniste O, Lim A et al. (1996) The pattern of production of cytokine mRNAs is markedly altered at the onset of multiple sclerosis. Res Immunol 147: 435-441
39. Crucian B, Dunne P, Friedman H et al. (1996) Detection of altered T helper 1 and T helper 2 cytokine production by peripheral blood mononuclear cells in patients with multiple sclerosis utilizing intracellular cytokine detection by flow cytometry and surface marker analysis. Clin Diagn Lab Immunol 3: 411-416
40. Brown MA, Hural J (1997) Functions of IL-4 and control of its expression. Crit Rev Immunol 17: 1-32
41. Link J, Soderstrom M, Olsson T et al. (1994) Increased transforming growth factor-beta, interleukin-4 and interferon-gamma in multiple sclerosis. Ann Neurol 36: 379-86
42. Soderstrom M, Hillert J, Link J et al. (1995) Expression of IFN-gamma, IL-4, and TGF-beta in multiple sclerosis in relation to HLA-Dw2 phenotype and stage of disease. Mult Scler 1: 173-180
43. Shaw MK, Lorens JB, Dhawan A et al. (1997) Local delivery of interleukin 4 by retrovirus-transduced T lymphocytes ameliorates experimental autoimmune encephalomyelitis. J Exp Med 185: 1711-1714
44. Vandenbroeck K, Martino G, Marrosu MG et al. (1997) Occurence and clinical relevance of an interleukin-4 gene polymorphism in patients with multiple sclerosis. J Neuroimmunol 76: 189-192
45. Fiorentino DF, Bond MW, Mosmann TR (1989) Two types of mouse T helper cell. IV. Th2 clones secrete a factor that inhibits cytokine production by Th1 clones. J Exp Med 170: 2081-2095

46. Mosmann TR, Schumacher JH, Street NF et al. (1991) Diversity of cytokine synthesis and function of mouse CD4+ T cells. Immunol Rev 123: 209-229
47. Zlotnik A, Moore KW (1991) Interleukin 10. Cytokine 3: 366-371
48. Sironi M, Munoz C, Pollicino T et al. (1993) Divergent effects of interleukin-10 on cytokine production by mononuclear phagocytes and endothelial cells. Eur J Immunol 23: 2692-2695
49. Groux H, Bigler M, de Vries JE, Roncarolo MG (1996) Interleukin-10 induces a long-term antigen-specific anergic state in human CD4$^+$ T cells. J Exp Med 184: 19-29
50. Itoh K, Hirohata S (1995) The role of IL-10 in human B cell activation, proliferation, and differentiation. J Immunol 154: 4341-4350
51. Rott O, Fleischer B, Cash E (1994) Interleukin-10 prevents experimental allergic encephalomyelitis in rats. Eur J Immunol 24: 1434-1440
52. Gayo A, Mozo L, Suarez A et al. (1998) Glucocorticoids increase IL-10 expression in multiple sclerosis patients with acute relapse. J Neuroimmunol 85: 122-130
53. Rudick RA, Ransohoff RM, Peppler R et al. (1996) Interferon beta induces inter-leukin-10 expression: relevance to multiple sclerosis. Ann Neurol 40: 618-627
54. Racke MK, Dhib-Jalbut S, Cannella B et al. (1991) Prevention and treatment of chronic relapsing experimental allergic encephalomyelitis by transforming growth factor-beta 1. J Immunol 146: 3012-3017
55. Stevens DB, Gould KE, Swanborg RH (1994) Transforming growth factor-beta 1 inhibits tumor necrosis factor-alpha/lymphotoxin production and adoptive transfer of disease by effector cells of autoimmune encephalomyelitis. J Neuroimmunol 51: 77-83
56. Fabry Z, Topham DJ, Fee D et al. (1995) TGF-beta 2 decreases migration of lymphocytes in vitro and homing of cells into the central nervous system in vivo. J Immunol 155: 325-332
57. Johns LD, Sriram S (1993) Experimental allergic encephalomyelitis: neutralizing antibody to TGF beta 1 enhances the clinical severity of the disease. J Neuroimmunol 47: 1-7
58. Mokhtarian F, Shi Y, Shirazian D et al. (1994) Defective production of anti-inflammatory cytokine, TGF-beta, by T cell lines of patients with active multiple sclerosis. J Immunol 152: 6003-6010
59. Link J, Fredrikson S, Soderstrom M et al. (1994) Organ-specific autoantigens induce transforming growth factor-beta-1 mRNA expression in mononuclear cells in multiple sclerosis patients and myastenia gravis. Ann Neurol 35: 197-203
60. Nicoletti F, Di Marco R, Patti F et al. (1998) Blood levels of transforming growth factor-beta-1 (TGF-beta1) are elevated in both relapsing remitting and chronic progressive multiple sclerosis (MS) patients and further augmented by treatment with interferon-beta 1b (IFN-beta1b). Clin Exp Immunol 113: 96-99
61. Ossege LM, Sindern E, Voss B, Malin JP (1998) Corticosteroids induce expression of transforming growth factor-beta-1 mRNA in peripheral blood mononuclear cells of patients with multiple sclerosis. J Neuroimmunol 84: 1-6
62. Dinarello CA (1996) Biological basis for interleukin-1 in disease. Blood 87: 2095-2147
63. Martin D, Near SL (1995) Protective effect of the Interleukin-1 receptor antagonist (IL-1ra) on experimental encephalomyelitis in rats. J Neuroimmunol 61: 241-245
64. Brosnan CF, Cannella B, Battistini L, Raine CS (1995) Cytokine localization in multiple sclerosis lesions: correlation with adhesion molecule expression and reactive nitrogen species. Neurology 45: S16-21
65. Wucherpfennig KW, Newcombe J, Li H et al. (1992) T cell receptor V alpha-V beta repertoire and cytokine gene expression in active multiple sclerosis lesions. J Exp Med 175: 993-1002

66. Nicoletti F, Patti F, Di Marco R et al. (1996) Circulating serum levels of IL-1ra in patients with relapsing remitting multiple sclerosis are normal during remission phases but significantly increased either during exacerbations or in response to IFN-beta treatment. Cytokine 8: 395-400
67. Clay FE, Tarlow JK, Cork MJ et al. (1996) Novel interleukin 1 receptor antagonist exon polymorphisms and their use in allele specific mRNA assessment. Hum Genet 97: 723-726
68. Crusius JBA, Peña AS, vanOsten BW et al. (1995) Interleukin 1 receptor antagonist gene polymorphism and multiple sclerosis. Lancet 346: 979-980
69. Huang WX, He B, Hillert J (1996) An interleukin 1-receptor-antagonist gene polymorphism is not associated with multiple sclerosis. Neuroimmunol 67: 143-144
70. Semana G, Yaouanq J, Alizadeh M et al. (1997) Interleukin-1 receptor antagonist gene in multiple sclerosis. Lancet 349: 476
71. Rosenthal N, Schwartz RS (1998) In search of perverse polymorphisms. N Engl J Med 338: 122-124

Chapter 15

Adhesion Molecules and the Blood-Brain Barrier in Multiple Sclerosis

J.J. Archelos, H.-P. Hartung

Introduction

Multiple sclerosis (MS) is an immune-mediated disease of the central nervous system (CNS) and constitutes a major cause of transient and permanent neurological disability in the adult. The etiology and pathogenesis of MS are only partially understood. On a cellular level, focal mononuclear cell infiltration with demyelination and eventual axonal loss is a crucial pathogenetic event leading to inflammation and subsequent dysfunction. Here we review evidence that adhesion molecules (AM) expressed at the blood-brain barrier (BBB) and on T cells play a central role in immune cell recruitment to the CNS. Therapeutic targeting of AM has been very successful in the corresponding animal model of experimental autoimmune encephalomyelitis and holds promise as a novel treatment strategy to combat human immune-mediated disorders of the CNS.

Anatomy of the BBB

The BBB separates the intravascular and CNS compartments. It constitutes one of the tightest blood-organ barriers known in the human body. The BBB consists of a vascular endothelial cell lining with tight junctions, a basement membrane, the pericytes and the glia limitans formed by astrocyte processes linked by gap-junctions and, occasionally, microglial cells (Fig. 1). The intact BBB is a major barrier for large molecules and for cells [1]. Only activated immune cells can pass this barrier [2]. The latter observation is of special relevance for immune-mediated disorders of the CNS such as multiple sclerosis. There, the presence of T and B lymphocytes and monocytes in the perivascular demyelinating lesion is a constant histopathological feature. There are several lines of evidence to suggest an important pathogenetic role of these immune cells in this chronic inflammatory demyelinating disease of the CNS. Understanding the mechanisms involved in lymphocyte and monocyte accumulation in MS lesions could lead to new therapeutic options. Studies of the molecules involved in the transendothe-

---

Department of Neurology, Karl-Franzens-Universität, Auenbruggerplatz 22 - 8036 Graz, Austria. e-mail: hans.hartung@kfunigraz.ac.at

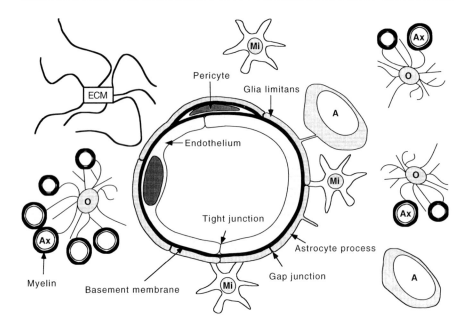

Fig. 1. The intact BBB and the extracellular compartments in the CNS. In immune-mediated disorders of the CNS such as multiple sclerosis, activated T cells enter the CNS by crossing the BBB, the interface between the capillary compartment and the neural environment. The cellular components of the white matter are embedded in the extracellular matrix (*ECM*). In the CNS, specialized anatomical structures such as vascular basement membrane (BM) and the white matter profoundly differ in the composition of their ECM. The BM is composed of classic ECM proteins, such as collagen type IV, fibronectin, laminins, vitronectin, entactin, and heparan sulfate. The ECM of the white matter comprises hyaluronic acid, heparan sulfate proteoglycans, lecticans such as brevican, neurocan, versican, and other chondroitin sulfate proteoglycans, glial hyaluronic acid-binding protein, thrombospondin and tenascin-R, and does not contain classic ECM proteins. The ECM is mainly produced by astrocytes, and oligodendrocytes; the BM is also produced by endothelial cells. *A*, Astrocyte; *Ax*, Axon; *Mi*, Microglial cell; *O*, myelin-forming oligodendrocyte

lial migration of T cells across the BBB have relied on both observations in MS patients and in the animal model of MS, experimental autoimmune encephalomyelitis (EAE). On a molecular basis, adhesion molecules on both T cells and endothelial cells seem to be of crucial importance for the active extravasation of T cells and monocytes [3]. This chapter will briefly summarize this evidence in MS and EAE.

Adhesion Molecules

On a molecular level, adhesion molecules (AM) are critically involved in all steps of the immune response [4, 5]. Based on their structure, three main classes of AM can be distinguished: (1) members of the immunoglobulin superfamily, (2) inte-

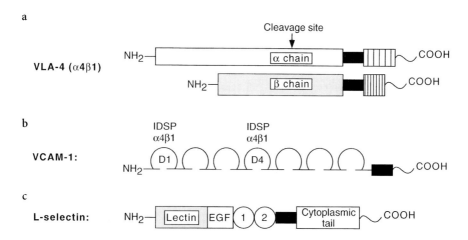

Fig. 2. a-c. *Structural characteristics of adhesion molecules.* Based on their structure, three main classes of AM can be distinguished: integrins, members of the immunoglobulin superfamily, and selectins. **a** Integrins such as VLA-4 ($\alpha 4\beta 1$) are transmembrane, noncovalently linked heterodimers formed by two variable chains with the common structure $\alpha_x \beta_y$. They bind to members of the Ig-superfamily and to a variety of adhesive glycoproteins on cells and in the extracellular matrix. Integrins mediate high-strength interactions with ligands and their affinity is mainly regulated by conformational changes in the heterodimer. **b** Members of the immunoglobulin superfamily such as VCAM-1, ICAM-1 or PECAM-1 are transmembrane molecules characterized by an extracellular region composed of immunoglobulin (Ig)-like domains. This notoriously stable Ig-backbone interacts with other members of the Ig family and with integrins. **c** Selectins such as L-selectin have a common modular structure consisting of an amino-terminal lectin domain, an epidermal growth-factor (*EGF*)-like region, a variable number of short consensus-repeats homologous to a complement-binding domain, a transmembrane region, and a short cytoplasmic tail. Despite their structural similarities, selectins have distinct patterns of expression. The lectin domain is directly involved in binding carbohydrate ligands on a variety of molecules including mucins. Selectins have a low intrinsic binding affinity but a high avidity for their ligands

grins, and (3) selectins (Fig. 2). Table 1 gives an overview of the best characterized members of each group and their ligands [3].

The Blood-Brain Barrier in EAE

Expression of Adhesion Molecules

In naive EAE-susceptible mice, intercellular cell adhesion molecule-1 (ICAM-1), vascular cell adhesion molecule-1 (VCAM-1) and platelet/endothelial cell adhesion molecule-1 (PECAM-1) are present at low constitutive levels on endothelial cells of large venules in the spinal cord and brain. Mucosal addressin cell adhesion molecule-1 (MAdCAM-1), P-selectin and E-selectin are not detected in the CNS of these animals [6-8]. In addition, a variety of endothelial integrins have been described at the BBB in the normal Lewis rat [9, 10].

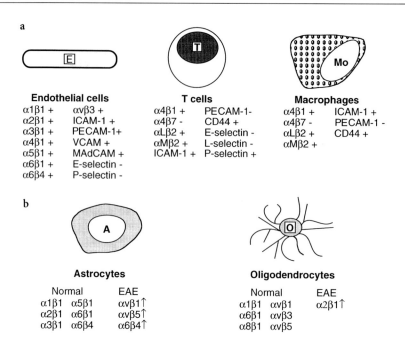

Fig. 3 a, b. *Expression of adhesion molecules and integrins in EAE.* **a** In rat and mouse EAE, endothelial cells of the CNS, infiltrating T cells and macrophages have an adhesion molecule phenotype characteristic of T-helper-1-mediated disease. **b** Astrocytes and oligodendrocytes express a variety of integrins in normal rats. In acute EAE, there is upregulation of the vitronectin and laminin receptors on astrocytes and neo-expression of α2β1 integrin on oligodendrocytes, both mediated in vitro by tumor necrosis factor α, a central proinflammatory cytokine in EAE

In the acute phase of EAE both ICAM-1 and VCAM-1 are upregulated on all blood vessels (irrespective of lesion location), PECAM-1 is redistributed and enriched at the endothelial tight junctions without an increase of expression, and MAdCAM-1, P-selectin and E-selectin are not found at any stage of the disease [7, 8, 11, 12]. Endothelial integrins are modulated in EAE and exhibit a defined spatiotemporal expression pattern at the BBB [9, 10] (Fig.3). Clinical relapses and parenchymal infiltration are paralleled by upregulation of ICAM-1 and VCAM-1 on blood vessels [13].

The presence of a variety of AM on mononuclear cells in EAE (Fig. 3) and their upregulation on endothelial cells at the BBB during active disease suggest a pathophysiological role of AM in the initiation of CNS inflammation.

AM and Transendothelial Migration at the BBB

The transendothelial migration of T cells at the BBB is of paramount importance in the pathogenesis of immune-mediated disorders of the CNS. Systemic autoreactivity only results in local autoaggression when activated T cells penetrate an intact BBB, whose first and strongest component is the endothelial lining. Blockade of the egress of T cells from blood vessels could prevent inflammation and subsequent demyelination in the CNS.

A cascade of sequentially interacting pairs of AM seems to be responsible for the crucial transendothelial migration of T cells into the target organ (Fig. 4) [3, 14, 15]. In vitro T cells predominantly use ICAM-1/lymphocyte function-associated molecule-1 (LFA-1) pathways if exposed to unstimulated endothelium [16]. By contrast, very late antigen-4 (VLA-4)/VCAM-1 is crucial for their transmigration through stimulated endothelium [17]. CD44, a transmembrane glycoprotein expressed on T cells, macrophages and other non-hematopoietic cells such as astrocytes, is involved in T cell migration into inflamed sites in several animal models, although its exact role in T cell recruitment to the CNS remains to be defined [18].

It is noteworthy that the kinetics of endothelial AM expression are regulated differentially. ICAM-1, VCAM-1 and E-selectin are expressed at low baseline levels and are strongly upregulated upon cytokine stimulation. In contrast, endothelial PECAM-1 and mucins exhibit a high level of constitutive expression and are upregulated much less upon cytokine challenge [19].

Treatment of EAE with Antibodies Acting at the BBB

To obtain more conclusive evidence for a pathogenic role of AM in EAE, therapeutic manipulation of EAE with monoclonal antibodies (mAbs) has been undertaken. Antibodies to AM involved in transendothelial migration at the BBB prevented clinical disease and reduced mononuclear infiltration and demyelination. The mechanisms underlying suppression of EAE achieved with such intervention vary considerably depending on which AM is targeted. Thus, antibodies directed to ICAM-1, LFA-1, CD2, CTLA-4/CD28, B7-1 and B7-2 primarily inhibit the process of antigen presentation that takes place in the lymph nodes draining the immunization site during the induction phase of EAE or in the CNS where local antigen presentation amplifies the incipient immune reaction. By contrast, antibodies to VLA-4, VCAM-1, LFA-1, macrophage glycoprotein associated with complement receptor function (Mac-1), or L-selectin interfere with transendothelial migration of lymphocytes and monocytes in EAE [20]. Antibodies to Mac-1 involved in transendothelial migration of monocytes and to the chemoattractant macrophage inflammatory protein-1α (MIP-1α) [21] activating integrin-dependent adhesion of T cells to endothelium markedly diminished severity of murine EAE [21].

In rodent models of EAE, not all antibodies masking a specific AM are equipotent: some prevent clinical signs completely while others only retard disease; even worsening of EAE with anti-adhesion monoclonal antibody treatment was observed [22]. It is difficult to interpret these conflicting results. A single AM member can exhibit many different functions, each of which may be "localized" to different domains of the molecule. On T cells, ICAM-1, LFA-1 and PECAM-1

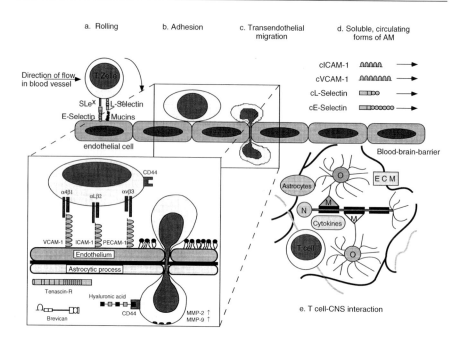

Fig. 4. Transendothelial migration of activated T cells across the BBB in EAE is mediated by adhesion molecules. Through the action of selectins – a family of cell adhesion molecules – T cells establish a loose reversible contact with endothelial cells. This initial rolling is followed by a firm irreversible adhesion mediated by integrins $\alpha 4\beta 1$ (VLA-4), $\alpha L\beta 2$ (LFA-1) and possibly $\alpha v\beta 3$ (vitronectin receptor) on T cells and their Ig-like receptors on endothelium (VCAM-1, ICAM-1, PECAM-1). Locally released chemokines (*black circles*) bound to endothelial glycoproteins (glycocalyx) increase integrin adhesiveness, and induce directed movement of T cells and monocytes. The subsequent transendothelial migration of T cells through the BBB is thought to be mediated mainly by $\alpha 4\beta 1$/VCAM-1. Engagement of $\alpha 4\beta 1$ with VCAM-1 activates matrix metalloproteinases (MMP) which degrade certain ECM proteins. The extracellular domain of VCAM-1 and ICAM-1 is shed from the cell surface after T-cell-endothelial-cell interaction, and soluble forms circulate in the blood and cerebrospinal fluid (cVCAM-1, cICAM-1). Physiological levels of circulating forms of integrin ligands increase with pathology. T cells that have migrated into the CNS interact with local cellular and ECM components such as hyaluronic acid via CD44 and possibly with tenascin-R/lectican networks via $\beta 1$ integrins. The nature of this complex interaction determines the outcome of the immune response in the target tissue

are the best characterized AM exhibiting different domain-dependent functions in one molecule. Depending on the epitope of AM recognized, antibodies to LFA-1 can provide either stimulatory or inhibitory signals to T cells, or even mimic the binding of a ligand with subsequent conformational change of this integrin resulting in activation [23]. For PECAM-1, domains 1 and 2 seem to be responsible for transendothelial migration of monocytes in response to chemokines and cytokines, and domain 6 appears to mediate the migration of macrophages in the ECM once these have left the vessel [24]. Severe side effects of antibodies to AM have been described. Antibodies to ICAM-1 induced focal spleen and liver necrosis, a generalized state of immunosuppression [25], and cerebral bleeding [26], whereas antibodies to L-selectin precipitated severe lymphopenia in the Lewis rat EAE model [27].

Potentially opposing actions and even severe side effects noted with these antibodies underscore the complexity of the biological role of cell-adhesion molecules and emphasize the need for careful experimentation in appropriate animal models prior to initiating clinical trials of antibodies to AM in MS. To understand the multitude of individual functions and to be able to target the most appropriate ones with high specificity, further studies are needed using antibodies with well-characterized specificity or even engineered to react with defined epitopes/domains.

Summarizing in vitro and in vivo experimentation, the following roles of membrane-anchored AM in the immune response emerge: integrins and members of the immunoglobulin superfamily are involved in the migration of T cells and monocytes/macrophages across vessel walls. Selectins are necessary for T cell recirculation and for transendothelial migration (Table 1). Based on the studies published to date, the α4 chain of VLA-4, CTLA-4, and ICAM-1 appear to be the most promising targets for intervention.

Disease Activity and MRI in EAE

In EAE, disease activity has been monitored and quantified with a semiquantitative clinical score and immunohistochemistry. Recent technical improvements offer the possibility to monitor disease activity and to observe a breakdown of the BBB by magnetic resonance imaging (MRI) [28, 29]. Similar to the situation in MS, Gd-enhancing lesions indicate a breakdown of the BBB in EAE and correlate with mononuclear cell infiltrations in situ [30, 31].

The Blood-Brain Barrier in MS

AM Expression on T Cells and at the BBB in MS

Do the mechanisms of transendothelial migration studied in EAE by antibody blockade also govern the immune response in MS? Most studies of the role of AM during MS have relied on immunohistochemical analyses of AM expression during different disease stages on autopsic CNS material or on blood or cerebrospinal fluid (CSF)-derived lymphocytes. In addition, circulating adhesion molecules (cAM) have been measured in blood and CSF in MS patients.

Adhesion Molecule Expression in the CNS. In acute and chronic active lesions, ICAM-1, VCAM-1 and E-selectin are upregulated on endothelial cells [10]. Similar to the situation in EAE, certain integrins on endothelial cells are modulated at the BBB [32]. The immune molecule expression patterns of infiltrating

Table 1. Adhesion molecules and their ligands. Only cellular distributions and functions of adhesion molecules with potential relevance for the pathogenesis of MS are given

Adhesion molecule (alternative name)	Cellular expression	Ligands
Immunoglobulin superfamily		
ICAM-1 (CD54) [a]	E, T, B, Ma, Mi, A	LFA-1, Mac-1, CD43
VCAM-1 (CD106) [b]	E, Ma, Pe	VLA-4, $\alpha_4\beta_7$
PECAM-1 (CD31) [c]	E, T (naive), Ma, NK	PECAM-1, $\alpha_v\beta_3$
LFA-2 (CD2) [d]	T, NK	LFA-3
LFA-3 (CD58) [d]	E, T, B, Ma, Mi	LFA-2
CD28 [e]	T (resting, activated)	B7-1, B7-2
CTLA-4 (CD152) [f]	T (activated)	B7-1, B7-2
B7-1 (CD80) [d]	T, B, Ma, Mi, A	CTLA-4, CD28
B7-2 (CD86) [d]	T, B, Ma, Mi, A	CTLA-4, CD28
Integrins		
VLA-4 ($\alpha_4\beta_1$) (CD49d/CD29)	T, B, Ma	VCAM-1, FN, VLA-4, $\alpha_4\beta_7$
LPAM-1 ($\alpha_4\beta_7$) (CD49d/CD104)	T (memory)	VCAM-1, FN,
1 ($\alpha_M\beta_2$) (CD11b/CD18)	Ma, Mi, NK	ICAM-1, FN, C3bi
Selectins		
E-selectin (CD62E)	E	sLewis^x, ESL-1
L-selectin (CD62L)	T, B, Ma	sLewis^x, mucins [g]
P-selectin (CD62P)	E	sLewis^x, PSGL-1
Other		
CD40	B, MA, E	CD154 (CD40L)
CD44	T, B, Ma, A, O	ECMproteins [h]

A, Astrocytes; *B*, B cells; *E*, Endothelial cells; *Ma*, Macrophages or monocytes; *Mi*, Microglia; *NK*, Natural killer cells; *Pe*, Pericytes, *O*, Oligodendrocytes; *T*, T cells (naive or memory); *C3bi*, Complement factor 3bi; *CTLA-4*, Cytotoxic T lymphocyte antigen-4; *ESL-1*, E-selectin ligand-1; *FN*, Fibronectin; *ICAM-1*, Intercellular cell adhesion molecule-1; *LFA-1*, Lymphocyte function-associated molecule-1; *LPAM-1*, Lymphocyte Peyer´s patch adhesion molecule-1; *Mac-1*, Macrophage glycoprotein associated with complement receptor function; *MAdCAM-1*, Mucosal addressin cell adhesion molecule-1; *PECAM-1*, Platelet/endothelial cell adhesion molecule-1; *PSGL-1*, P-selectin glycoprotein ligand-1; *VCAM-1*, Vascular cell adhesion molecule-1; *VLA-4*, Very late antigen-4

[a] 5 Ig domains
[b] 7 Ig domains
[c] 6 Ig domains
[d] 2 Ig domains
[e] 1 Ig domain
[f] 1 Ig domain (dimer)
[g] Includes CD34, MAdCAM, PSGL-1 and glycosylation-dependent cell adhesion molecule-1 (GlyCAM-1)
[h] Includes hyaluronate, collagen, fibronectin and laminin.

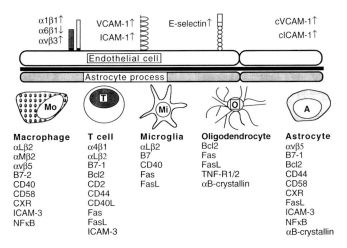

Fig. 5. Expression of integrins and immune molecules in active MS lesions. Integrins are expressed and regulated on endothelial, neural and infiltrating mononuclear cells in MS. Together with other immunomodulating molecules present on these cells, they contribute to the development and maintenance of the inflammatory lesion. cICAM-1 and cVCAM-1 (circulating integrin ligands) are elevated in remitting-relapsing MS and correlate with disease activity on cranial MRI. Only expression patterns described in MS lesions are shown. Data from EAE, although they may apply to MS, are not included. *B7*, B cell activation antigen B7; *CD40L (CD154)*, ligand of CD40; *CXR*, Chemokine receptors; *FasL*, Fas ligand; *ICAM-3 (CD50)*, Intercellular cell adhesion molecule-3; *LFA-3 (CD58)*, Lymphocyte function-associated molecule-3; *MMP*, Matrix metalloproteinase; *NFκB*, Nuclear factor kappa B; *TNF-R*, Tumor necrosis factor receptor; *VCAM-1 (CD106)*, Vascular cell adhesion molecule 1

T cells and resident neural cells have been described in the MS lesion in situ (Fig. 5). It is noteworthy that the ICAM-1/LFA-1 pair was expressed at relatively high levels in plaques of all ages. In contrast, expression of the VCAM-1/VLA-4 pair was lowest in acute lesions and increased in chronic lesions. The presence of certain integrin subunits has been associated with a chronic breakdown of the BBB in MS [32].

AM Expression on T Cells in Blood and Cerebrospinal Fluid. Increased levels of LFA-1, ICAM-1, LFA-3, CD2 and CD44 but decreased expression of VLA-4 and VLA-5 on the surface of blood-derived T cells of MS patients have been reported by some authors [33] but not confirmed by others [34]. Adhesion of T cells to endothelial cells derived from MS patients was enhanced [33].

Adhesion molecule expression patterns on peripheral T cells have been analyzed in a longitudinal study: levels of CD28, VLA-4 and LFA-1 did not change significantly during a 6-month period, although patients showed disease activity on MRI [33]. Treatment with interferon-β-1a downregulated expression of VLA-4 but not of LFA-1, ICAM-1, L-selectin, or CD44 [35]. Treatment with methylprednisolone transiently reduced LFA-1 expression on T cells, had no effect on L-selectin and VLA-4 on T cells, but significantly altered the expression pattern of

L-selectin, Mac-1 and VLA-4 on circulating monocytes [34-36]. T cells isolated from CSF showed an increased expression of LFA-1, LFA-3, ICAM-1, VLA-3, VLA-4, VLA-5, VLA-6, CD2, B7-1, and CD44 [37, 38].

Circulating Adhesion Molecules in MS

AM are not only present at the BBB as anchored transmembrane molecules but, after interaction with the corresponding ligand, some are cleaved or shed and circulate as soluble forms in body fluids [39]. Areas of interest are the cellular origin of the circulating forms, the agents inducing cleavage and shedding, and the physiological function of circulating adhesion molecules (cAM). In addition, in MS an easy-to-measure disease activity marker is much needed: cAM could be promising candidates.

In MS, circulating forms of ICAM-1, ICAM-3, VCAM-1, E-selectin, and L-selectin have been studied [20]. Circulating ICAM-1 (cICAM-1) is the best characterized cAM in MS. Serum levels are consistently elevated in active relapsing-remitting MS (RR MS) and correlate with the presence of Gd-enhancing lesions in MRI indicating an open BBB. cVCAM-1 has a similar profile, although less data are available and – compared to cICAM-1 – MRI findings are less consistent. Data for cE-selectin are less convincing for RR MS, but it could be a useful marker for monitoring chronic progressive MS. Of course, more studies are needed. The few studies published on cL-selectin indicate an elevation in RR MS which is correlated with the presence of Gd-positive lesions in MRI. Currently available assays measure both neutrophil-derived and lymphocyte-derived L-selectin. It would be interesting to quantitate only cL-selectin shed from lymphocytes by use of specific assays. This may considerably improve the utility of this marker in MS (for a review see [36]).

Research on cAM needs to be intensified. More longitudinal studies are warranted to evaluate the usefulness of cAM as easy-to-measure and reliable markers of disease activity in MS. cAM in CSF, although of pathophysiological relevance, will most likely not become a routine marker of disease activity. However, cAM should not only be studied as indicators of disease; the pathobiological functions of cAM in MS should also be elucidated. Most soluble AM circulate in an active form as monomers, although cICAM-1 also circulates as a homodimer. Most cAM are committed to bind their ligands either at their release site or systemically. Some cAM have been shown to compete with membrane-anchored AM for their ligands. Although ligand-bound cAM are not detected in serum or CSF, they may possess relevant biological function. Illumination of the pathophysiological role of cAM in MS may also have future therapeutic implications.

Conclusions

The blood-brain barrier as a defined anatomical structure is intimately involved in the pathogenesis of immune-mediated diseases of the CNS. Adhesion molecules expressed at the BBB and their corresponding ligands present on immune cells are of major importance for the transendothelial migration of T cells and monocytes across the BBB. This step initiates the local immune response in the target tissue and leads to mononuclear infiltration, demyelination and eventual axonal loss with the associated neurological symptoms. The mechanisms operating at the BBB during migration of T cells into the CNS have been elaborated in experimental autoimmune encephalomyelitis - an animal model of multiple sclerosis. These findings may have considerable therapeutic implications in the near future. A better understanding of the BBB in physiology and pathology will likely enlarge the therapeutic options in immune-mediated diseases of the central nervous system.

References

1. Pardridge WM (1997) Drug delivery to the brain. J Cereb Blood Flow Metab 17: 713-731
2. Hickey WF, Hsu BL, Kimura H (1991) T-lymphocyte entry into the central nervous system. J Neurosci Res 28: 254-260
3. Archelos JJ, Hartung H-P (1997) The role of adhesion molecules in multiple sclerosis: Biology, pathogenesis and therapeutic implications. Mol Med Today 3: 310-321
4. Oppenheimer-Marks N, Lipsky PE (1996) Adhesion molecules as targets for the treatment of autoimmune diseases. Clin Immunol Immunopathol 79: 203-210
5. Springer TA (1994) Traffic signals for lymphocyte recirculation and leukocyte emigration: The multistep paradigm. Cell 76: 301-314
6. Raine CS (1994) Multiple sclerosis: immune system molecule expression in the central nervous system. J Neuropathol Exp Neurol 53: 328-337
7. Engelhardt B, Vestweber D, Hallmann R, Schulz M (1997) E- and P-selectin are not involved in the recruitment of inflammatory cells across the blood-brain barrier in experimental autoimmune encephalomyelitis. Blood 90: 4459-4472
8. Williams KC, Zhao RW, Ueno K, Hickey WF (1996) PECAM-1 (CD31) expression in the central nervous system and its role in experimental allergic encephalomyelitis in the rat. J Neurosci Res 45: 747-757
9. Previtali S, Archelos JJ, Hartung H-P (1997) Modulation of the expression of integrins on glial cells during experimental autoimmune encephalomyelitis. A central role for TNF-α. Am J Pathol 151: 1425-1435
10. Archelos JJ, Previtali SC, Hartung H-P (1999) The role of integrins in immune-mediated diseases of the nervous system. Trends Neurosci 22: 30-38
11. Dopp JM, Brenemann SM, Olschowka JA (1994) Expression of ICAM-1, VCAM-1, L-selectin, and leukosialin in the mouse central nervous system during the induction and remission stages of experimental allergic encephalomyelitis. J Neuroimmunol 54: 129-144
12. Lindsey JW, Steinman L (1993) Competitive PCR quantification of CD4, CD8, ICAM-1, VCAM-1, and MHC class II mRNA in the central nervous system during development and resolution of experimental allergic encephalomyelitis. J Neuroimmunol 48: 227-234
13. Cannella B, Cross AH, Raine CS (1990) Upregulation and coexpression of adhesion molecules correlate with relapsing autoimmune demyelination in the central nervous system. J Exp Med 172: 1521-1524

14. Butcher E, Picker LJ (1996) Lymphocyte homing and homeostasis. Science 272: 60-66
15. Springer TA (1994) Traffic signals for lymphocyte recirculation and leukocyte emigration: the multistep paradigm. Cell 76: 301-314
16. Reiss Y, Hoch G, Deutsch U, Engelhardt B (1998) T cell interaction with ICAM-1 deficient endothelium in vitro: essential role for ICAM-1 and ICAM-2 in transendothelial migration of T cells. Eur J Immunol 28: 3086-3099
17. Pryce G, Male D, Campbell I, Greenwood J (1997) Factors contolling T-cell migration across rat cerebral endothelium in vitro. J Neuroimmunol 75: 84-94
18. DeGrendele HC, Estess P, Siegelman MH (1997) Requirement for CD44 in activated T cell extravasation into an inflammatory site. Science 278: 672-675
19. Bevilacqua MP, Nelson RM, Mannori G, Cecconi O (1994) Endothelial-leukocyte adhesion molecules in human disease. Annu Rev Med 45: 361-378
20. Archelos JJ, Hartung H-P (1999) Adhesion molecules in multiple sclerosis: A review. In: Siva A, Thompson A, Kesselring J (eds) Frontiers in multiple sclerosis II. Martin Dunitz, London (*in press*)
21. Karpus WJ, Lukacs NW, McRae BL et al. (1995) An important role for the chemokine macrophage inflammatory protein-1α in the pathogenesis of the T cell-mediated autoimmune disease, experimental autoimmune encephalomyelitis. J Immunol 155: 5003-5010
22. Cannella B, Cross AH, Raine CS (1993) Anti-adhesion molecule therapy in experimental autoimmune encephalomyelitis. J Neuroimmunol 46: 43-56
23. Hogg N, Berlin C (1995) Structure and function of adhesion receptors in leukocyte trafficking. Immunol Today 16: 327-330
24. Newman PJ (1997) The Biology of PECAM-1. J Clin Invest 99: 3-8
25. Archelos JJ, Jung S, Mäurer M et al. (1993) Inhibition of experimental autoimmune encephalomyelitis by an antibody to the intercellular adhesion molecule ICAM-1, Ann Neurol 34: 145-154
26. Soilu-Hänninen M, Roytta M, Salmi A, Salonen R (1997) Therapy with antibody against leukocyte integrin VLA-4 (CD49d) is effective and safe in virus-facilitated experimental allergic encephalomyelits. J Neuroimmunol 72: 95-105
27. Archelos JJ, Jung S, Rinner W et al. (1998) Role of leukocyte adhesion molecule L-selectin in experimental autoimmune encephalomyelitis. J Neurol Sci 159: 127-134
28. Seeldrayers PA, Syha J, Morrissey SP et al. (1993) Magnetic resonance imaging investigation of blood-brain barrier damage in adoptive transfer experimental autoimmune encephalomyelitis. J Neuroimmunol 46: 199-206
29. Namer IJ, Steibel J, Piddlesden SJ et al. (1994) Magnetic resonance imaging of antibody-mediated demyelinating experimental allergic encephalomyelits. J Neuroimmunol 54: 41-50
30. Hawkins CP, Munroe PMG, Mackenzie F et al. (1990) Duration and selectivity of blood-brain barrier breakdown in chronic relapsing experimental allergic encephalomyelitis studied by gadolinium-DTPA and protein markers. Brain 113: 365-367
31. Morrissey SP, Stodal H, Zettl U et al. (1996) In vivo MRI and its histological correlates in acute adoptive transfer experimental allergic encephalomyelits. Quantification of inflammation and oedema. Brain 119: 239-248
32. Sobel RA, Hinojoza JR, Maeda A, Chen M (1998) Endothelial cell integrin laminin receptor expression in multiple sclerosis lesions. Am J Pathol 153: 405-415
33. Stüber A, Martin R, Stone LA et al. (1996) Expression pattern of activation and adhesion molecules on peripheral blood CD4+ T-lymphocytes in relapsing-remitting multiple sclerosis patients: a serial analysis. J Neuroimmunol 66: 147-151

34. Droogan AG, Crockard AD, McMillan SA, Hawkins SA (1998) Effects of intravenous methylprednisolone therapy on leukocyte and soluble adhesion molecule expression in MS. Neurology 50: 224-229
35. Soilu-Hänninen M, Salmi A, Salonen R (1996) Interferon-beta downregulates expression of VLA-4 antigen and antagonizes interferon-gamma-induced expression of HLA-DQ on peripheral blood monocytes. J Neuroimmunol 60: 99-106
36. Pitzalis C, Sharrack B, Gray IA et al. (1997) Comparison of the effects of oral versus intravenous methylprednisolon regimens on peripheral blood T lymphocyte adhesion molecule expression, T cell subsets distribution and TNF alpha concentrations in multiple sclerosis. J Neuroimmunol 74: 62-68
37. Svenningsson A, Hansson GK, Andersen O et al. (1993) Adhesion molecule expression on cerebrospinal fluid T lymphocytes: evidence for common recruitment mechanisms in multiple sclerosis, aseptic meningitis, and normal controls. Ann Neurol 34: 155-161
38. Svenningsson A, Dotevall L, Stemme S, Andersen O (1997) Increased expression of B2-7 costimulatory molecule on cerebrospinal fluid cells of patients with multiple sclerosis and infectious central nervous system disease. J Neuroimmunol 75: 59-68
39. Hartung HP, Archelos JJ, Zielasek J et al. (1995) Circulating adhesion molecules and inflammatory mediators in demyelination. Neurology 45(Suppl 6): S22-S32

Chapter 16

Non-Myelin Antigen Autoreactivity in Multiple Sclerosis

G. Ristori[1], C. Montesperelli[1], C. Buttinelli[1], L. Battistini[2], S. Cannoni[1], G. Borsellino[2], R. Bomprezzi[1], A. Perna[1], M. Salvetti[1]

Introduction

Immune-mediated damage is probably the most relevant event among those that contribute to the pathogenesis of multiple sclerosis (MS). The search for autoantigens inducing T cell responses in MS as well as in other immune-mediated, organ-specific diseases was initated as soon as efficient techniques for the isolation and expansion of antigen-specific T cell lines became available [1].

Myelin antigens were obviously considered first (discussed by M. Vergelli [2]), because of the demyelinating nature of the disease and because there was formal proof that the animal model of MS - experimental allergic encepahalomyelitis (EAE) - could be induced through a T cell response against such antigens (discussed by A. Uccelli [3]). However, comparison between patients and controls of precursor frequency, fine specificity, and functional characteristics of T cells specific for myelin antigens did not provide decisive results [4]. Contemporarily, cells of the central nervous system - particularly oligodendrocytes - were shown to suffer from "bystander damage" as a result of toxic mediators produced during inflammatory responses. These toxic mediators include macrophage-derived factors such as tumor necrosis factor (TNF)-α [5-7], complement, eicosanoids, and in some cases nitric oxide, whose synthesis by activated glial cells can be induced by the proinflammatory cytokines TNF-α, interleukin (IL)-1 and interferon (IFN)-γ [8-12].

Hence, after a few years of research on myelin antigens, investigations on potential autoantigens in MS were extended to non-myelin constituents of the central nervous system (CNS). (Note that the adjective "non-myelin" may not be fully correct in that the majority of these constituents are expressed by different CNS cells, including oligodendrocytes; this term is therefore used as the contrary to "myelin antigens" that refers to proteins expressed exclusively on the myelin sheath.)

---

[1] Department of Neurological Sciences, University of Rome La Sapienza, Viale dell'Università 30 - 00185 Rome, Italy. e-mail: md0914@mclink.it
[2] IRCCS Santa Lucia, Rome, Italy

Evidence from Animal Models

Recently, the implication of non-myelin antigens in autoimmunity received support from the demonstration that axonal loss is a conspicuous event in MS lesions [13, 14]. However, the first evidence for this association came from experiments in which EAE, the animal model of MS, was induced in rodents with a non-myelin antigen. In these experiments, the S-100β protein in complete Freund's adjuvant induced CNS inflammation in the cerebral cortex, retina and uvea in addition to the typical white matter lesions [15]. The lesions were particularly rich in infiltrating T cells (in contrast to those of myelin basic protein-induced EAE in which activated macrophages predominate). The other non-myelin antigen so far shown to be potentially encephalitogenic is αB-crystallin, a small heat shock protein (Hsp) having encephalitogenic potential in Biozzi AB/H mice [16].

Introduction of the novel concept of "determinant spreading" in the pathogenesis of organ-specific, immune-mediated diseases further fostered the interest in T cell responses directed against non-myelin constituents of the CNS [17, 18]. Determinant spreading defines the recruitment of T cells specific for an increasingly broad set of self-determinants as disease progresses. The spreading may involve different epitopes of the same autoantigen (intramolecular spreading) or different autoantigens (intermolecular spreading) as a result of secondary sensitization of T cells against determinants that, in normal conditions, are not available for antigen presentation. This event has now been described in several models of autoimmune disease and is generally regarded as a promoter of disease progression. However, recent reports have suggested a protective role in animal models if the sensitization of T cells against new determinants involves lymphocytes with a Th2 cytokine profile [19]. The relevance of determinant spreading for MS progression is unclear. So far only two examples of intramolecular spreading are available, one within the proteolipid protein (PLP) (reported as pathogenetically relevant) [20] and the other within the myelin basic protein (MBP) sequence (considered to be unrelated to disease progression) (Ristori et al., submitted). Intermolecular spreading has not yet been documented in MS.

The concept of determinant spreading may even be extended, in view of recent examples of molecular mimicry at the T cell receptor (TCR) level. TCR binding to antigenic peptides is more degenerate than previously thought [21, 22]. Hence, the probability of molecular mimicry between different epitopes is relatively high [23]. These findings may explain, at the epitope level, why pre-immunization not only with autoantigens but also with apparently non-pathogenic antigens confers disease protection in animal models of autoimmune disease [24]. If, as a result of epitope spreading, there are several encephalitogenic epitopes on different autoantigens, there are increased probabilities that pre-immunization procedures with non-encephalitogenic antigens may lead to the induction of specific unresponsiveness or deletion of autoreactive cells through molecular mimicry. Both epitope spreading and molecular mimicry have important implications for future therapeutic strategies and indicate that current attempts at specific immunotherapies may have limited efficacy.

Evidence from Pathology

An essential requisite for an autoantigen is its expression in the target organ. This is obvious for myelin antigens but not for non-myelin ones. Among the potential non-myelin autoantigens in MS, Hsp have received the greatest attention as far as their CNS expression is concerned.

The rationale for investigating Hsp as potential autoantigens in MS is the same as in other immune-mediated disorders [25]. Hsp are the most conserved molecules known to date. Furthermore, they are relevant immune targets during infection. Hence, there is a theoretically high chance of cross-reactive responses to epitopes shared by host and microbial Hsp. This suggests the possibility that Hsp trigger an autoaggressive response through a mechanism of molecular mimicry. Moreover, as stress-induced proteins, Hsp antigens are upregulated during inflammatory processes and are therefore ideal candidates for sustaining the immunopathological loop of determinant spreading. It has recently been shown that, within a proinflammatory milieu, material from damaged cells is engulfed, processed and presented by dendritic cells to resting autoreactive T lymphocytes that become activated. This bystander activation has been demonstrated in EAE and in insulin-dependent diabetes mellitus in nonobese diabetic (NOD) mice [26-28].

The best studied families of stress proteins are the 60 kDa (Hsp60), 70 kDa (Hsp70) and small Hsp (particularly αB-crystallin). Much is known about Hsp60 and Hsp70 from the field of basic immunology [29, 30]. (Hsp60 and Hsp70 are major immunogens. Hsp60 is potentially involved in the pathogenesis of other immune-mediated diseases such as insulin-dependent diabetes mellitus and rheumatoid arthritis. Sequences from self-Hsp70 are among the motifs eluted at high frequency from MHC class II molecules. Hsp70 is the most conserved protein known to date [31], and thus has increased potential to elicit cross-reactive responses). T cell responses to Hsp in EAE and MS suggest that reactivity against these antigens is dysregulated (see following section). Hsp expression in the brain is compatible with these proteins being target antigens in a disease of the cerebral white matter [32]. Nonetheless, given the complex events that regulate Hsp expression in different cell types, depending on the intensity and duration of various stress events, it was important to verify Hsp induction in glial cells following proinflammatory stimuli and during an autoimmune inflammatory condition of the brain. Hsp70 and Hsp27 can be variably induced in oligodendrocytes and, according to some reports, in astrocytes following exposure to the cytokines IL-1, IFN-γ and TNF-α [33-36].

During the acute onset of EAE in SJL mice, Hsp60 immunoreactivity revealed increased expression of the protein in inflammatory cells at lesion sites. In more chronic phases, Hsp60 expression was detected in oligodendrocytes and astrocytes in areas adjacent to the inflammatory infiltrates. In lesioned areas, Hsp60 was found in mitochondria, in the cytosol, and on the surface of inflammatory cells [36-38]. These findings may have implications in terms of modulation of the inflammatory response. During immune responses to infectious agents, leukocytes express Hsp that become targets of cytotoxic cells [39]. Cytotoxic reactivity to Hsp60 on the cell surface of infiltrating leukocytes may help limit the autoaggressive process. In EAE induced in Lewis rats [37], increased heat shock cognate 70 (Hsc70) mRNA levels were predominantly localized in neurons, while some reactivity was noted in glia (probably astrocytes). At variance with data on Hsp60 expression in mouse EAE, inflammatory cells were not responsible for the notable increase in Hsc70 message, indicating that in this case a role in the downregulation of inflammation is unlikely.

Apart from some exceptions that go beyond the scope of this review, the expression patterns of Hsp60, Hsp70 and αB-crystallin in the human brain during MS are similar to that in EAE [36, 40, 41]. In active demyelinating lesions, expression of these protein is maximal. Cell types involved are inflammatory cells, oligodendrocytes (including proliferating oligodendrocytes at the lesion edge), reactive astrocytes, endothelial cells and microglia.

Evidence from Functional Studies on Lymphocytes

Hsp60, Hsp70 and αB-crystallin are the most thoroughly studied proteins among non-myelin autoantigens for their expression in tissues. In a passive transfer model of EAE obtained using MBP-specific encephalitogenic T cell clones, Hsp60-reactive T cells were recruited as the disease progressed [42], implicating such cells in sustaining the disease process. Determinant spreading or mimicry between Hsp60 and myelin constituents are both suitable explanations for this phenomenon. The latter possibility may be supported by the work of Birnbaum and colleagues [43] who recently identified an *M. Leprae* Hsp60 peptide that is cross-reactive with the myelin protein 2',3' cyclic nucleotide 3' phosphodiesterase (CNP).

In MS, a dysregulated T cell response to Hsp70 but not to Hsp60 was reported [44] and then confirmed in a larger series of patients [45]. Interestingly, the reactivity against Hsp70 appeared to be directed against conserved regions of the protein, with frequent events of cross-reactivity between mycobacterial and human Hsp70 sequences. Dysregulation of the T cell response to Hsp70 in MS was also suggested by Stinissen et al. [46] who screened γδ T cell clones derived from peripheral blood and cerebrospinal fluid (CSF) for their proliferative response to recombinant mycobacterial Hsp65 and Hsp70.

Van Noort and colleagues [47] identified αB-crystallin as the low molecular weight fraction from MS-affected brain tissue which stimulated mononuclear peripheral blood cells of patients but not of controls. The same group then showed that Epstein-Barr virus infection induces the expression of αB-crystallin in peripheral blood leukocytes. This may activate αB-crystallin-specific Th1 T cells, supporting a model in which viral infections trigger myelin-targeted autoimmunity [48].

The search for non-myelin autoantigens has not been restricted to Hsp. Among other proteins of potential interest are transaldolase (TAL) and CNP. TAL is a rate-limiting enzyme of the pentose phosphate pathway. Immunohistochemical analyses of human brain sections and primary murine brain cell cultures demonstrated that TAL is expressed selectively in oligodendrocytes at high levels. High-affinity autoantibodies to recombinant TAL in serum and CSF were detected in 15 of 20 patients with MS but not in normal individuals or patients with other autoimmune and neurological diseases [49]. These results were confirmed in a subsequent study in which TAL was shown to be a more potent immunogen than MBP in MS [50].

CNP, a protein associated with oligodendrocyte/myelin membranes, is also present in lymphocytes and retina. It constitutes one major target for the humoral response in MS [51]. A persistent humoral (IgM) response to CNP was detected in sera of 74% of MS patients. The antibodies were also present in high titer in CSF. The persistence of the antibody response is consistent with systemic immune activation and ongoing antigenic stimulation. CNP binds the C3 complement, raising the possibility of C3 receptor-mediated opsonization of myelin membrane CNP. Moreover, phagocytosis of CNP-Ig immune complexes may be mediated by membrane Ig Fc receptors of macrophages and CNS microglia. In retina, CNP is expressed principally in photoreceptors at about the same level as in CNS white matter [52]. This may explain the occurrence, in MS, of a periphlebitis reaction in the retina [53], a location where myelin is absent.

Two new lines of research deserve to be mentioned. The first is the possibility that T cell responses to non-proteinaceous moieties contribute to organ damage in MS. This is highly speculative since, so far, it is inferred from (1) the presence in MS lesions of $\gamma\delta$ T cells [54] specific to nonpeptide phosphoantigens (in particular a class of compounds known as the prenyl phosphates), and (2) the upregulated expression in MS lesions of CD1b [55] (the CD1 lineage of antigen-presenting molecules which have evolved to bind and present non-protein lipid and glycolipid self and foreign antigens). Nonetheless, the existence and pathogenetic importance of T cells that cross-recognize self and foreign non-proteinaceous moieties need to be assessed.

The second line of research stems from the identification, in humans, of T cell responses directed against *N*-formylated peptides of mitochondrial origin (Ristori et al., manuscript in preparation). Given the endosymbiotic origin of mitochondria, the probability that T cells specific for *N*-formylated peptides of mitochondrial origin cross-react with *N*-formylated peptides of microbial derivation is, in hypothetical terms, high. Both research fields will be among the many aspects of MS to be explored in the future.

References

1. Pette M, Fujita K, Kitze B et al. (1990) Myelin basic protein-specific T lymphocyte lines from MS patients and healthy individuals. Neurology 40(11): 1770-1776
2. Vergelli M (1999) Myelin antigen autoreactivity in multiple sclerosis. In: Martino G, Adorini L (eds) From basic immunology to immune-mediated demyelination (Topics in neuroscience, vol 2). Springer, Milan, pp 170-184

3. Uccelli A (1999) Animal models of Demyelination of the central nervous system. In: Martino G, Adorini L (eds) From basic immunology to immune-mediated demyelination (Topics in neuroscience, vol 2). Springer, Milan, pp 233-245
4. Hohlfeld R (1997) Biotechnological agents for the immunotherapy of multiple sclerosis. Principles, problems and perspectives. Brain 120: 865-916
5. Brosnan CF, Selmaj K, Raine CS (1988) Hypothesis: a role for tumor necrosis factor in immune-mediated demyelination and its revelance to multiple sclerosis. J Neuroimmunol 18: 87-94
6. Selmaj KW, Raine CS (1988) Tumor necrosis factors mediates myelin and oligodendrocyte damage in vitro. Ann Neurol 23: 339-46
7. Selmaj K, Raine CS, Farooq M et al. (1991) Cytokine cytotoxicity against oligodendrocytes. Apoptosis induced by lymphotoxin. J Immunol 147: 1522-1529
8. Lowenstein CJ, Snyder SH (1992) Nitric oxide, a novel biologic messenger. Cell 70: 705-707
9. Choi DW (1993) Nitric oxide: Foe or a friend to the injured brain? Proc Natl Acad Sci USA 90: 9741-9743
10. Liu J, Zhao M-L, Brosnan CF, Lee SC (1996) Expression of the type II nitric oxide synthase in primary human astrocytes and microglia: Role of IL-1β and IL-1 receptor antagonist. J Immunol 157: 3569-3576
11. Selmaj K, Raine CS, Cannella B, Brosnan CF (1991) Identification of lymphotoxin and tumor necrosis factor in multiple sclerosis lesions. J Clin Invest 87: 949-954
12. Cannella B, Raine CS (1995) The adhesion molecule and cytokine profile of multiple sclerosis lesions. Ann Neurol 37: 424-435
13. Scolding N, Franklin R (1998) Axon loss in multiple sclerosis. Lancet 352: 340-341
14. Trapp BD, Peterson J, Ransohoff RM et al. (1998) Axonal transection in the lesions of multiple sclerosis. N Engl J Med 338: 278-285
15. Kojima K, Berger T, Lassmann H et al. (1994) Experimental autoimmune panencephalitis and uveoretinitis transferred to the Lewis rat by T lymphocytes specific for the S-100β molecule, a calcium binding protein of astroglia. J Exp Med 180: 817-829
16. Amor S, Baker D, Layward L et al. (1997) Multiple sclerosis: variations on a theme. Immunol Today 18: 368-371
17. Sercarz EE, Lenhmann PV, Ametani A et al. (1993) Dominance and crypticity of T cell antigenic determinants. Annu Rev Immunol 11: 729-766
18. Miller SD, McRae BL, Vanderlugt CL et al. (1995) Evolution of the T-cell repertoire during the course of experimental immune-mediated demyelinating diseases. Immunol Rev 144: 225-244
19. Tian J, Lehmann PV, Kaufman DL (1997) Determinant spreading of T helper cell 2 (Th2) responses to pancreatic islet autoantigens. J Exp Med 186(12): 2039-2043
20. Tuohy VK, Yu M, Weinstock-Guttman B, Kinkel RP (1997) Diversity and plasticity of self recognition during the development of multiple sclerosis. J Clin Invest (7): 1682-1690
21. Wucherpfennig KW, Strominger JL (1995) Molecular mimicry in T cell-mediated autoimmunity: viral peptides activate human T cell clones specific for myelin basic protein. Cell 80(5): 695-705
22. Hemmer B, Fleckenstein BT, Vergelli M et al. (1997) Identification of high potency microbial and self ligands for a human autoreactive class II-restricted T cell clone. J Exp Med 185(9): 1651-1659
23. Mason D (1998) A very high level of cross reactivity is an essential feature of the T-cell receptor. Immunol today 19: 395-404

24. Fiori P, Ristori G, Cacciani A et al. (1997) Down-regulation of cell-surface CD4 co-receptor expression and modulation of experimental allergic encephalomyelitis. Int Immunol 9(4): 541-545
25. Young RA (1990) Stress proteins and immunology. Annu Rev Immunol 8: 401-420
26. Wekerle H (1998) The viral triggering of autoimmune disease. Nat Med 4(7): 770-771
27. Horwitz MS, Bradley LM, Harbertson J et al. (1998) Coxsackie virus-induced diabet: Initation by bystander damage and not molecular mimicry. Nat Med 4: 781-785
28. Albert ML, Sauter B, Bhardwaj N (1998) Dendritic cells acquire antigen from apoptotic cells and induce class I restricted CTLs. Nature 392: 86-89
29. van Eden W, van der Zee R, Paul AG et al. (1998) Do heat shock proteins control the balance of T-cell regulation in inflammatory diseases? Immunol Today 19(7): 303-307
30. Newcomb JR, Cresswell P (1993) Characterization of endogenous peptides bound to purified HLA-DR molecules and their absence from invariant chain-associated alpha beta dimers. J Immunol 150(2): 499-507
31. Gupta RS, Golding GB (1993) Evolution of HSP70 gene and its implications regarding relationships between archaebacteria, eubacteria, and eukaryotes. J Mol Evol 37(6): 573-582
32. Mayer RJ, Brown IR (1994) Heat shock proteins in the nervous system. Academic Press, London
33. D'Souza SD, Antel JP, Freedman MS (1994) Cytokine induction of heat shock protein expression in human oligodendrocytes: An interleukin-1 mediated mechanism. J Neuroimmunol 50: 17-24
34. Satoh J, Kim SU (1995) Constitutive and inducible expression of heat shock protein HSP72 in oligodendrocytes in culture. Neuroreport 6: 1081-1084
35. Aquino DA, Lopez C, Farooq M (1996) Antisense oligonucleotide to the 70-kDa heat shock protein inhibits synthesis of myelin basic protein. Neurochem Res 21: 417-422
36. Brosnan CF, Battistini L, Gao YL et al. (1996) Heat shock proteins and multiple sclerosis. J Neuropathol Exp Neurol 55: 389-402
37. Aquino DA, Klipfel AA, Brosnan CF, Norton WT (1993) The 70 kDa heat shock cognate protein is a major constituent of the central nervous system and is upregulated only at the mRNA level in acute experimental autoimmune encephalomyelitis. J Neurochem 61: 1340-1348
38. Gao YL, Brosnan CF, Raine CS (1995) Experimental autoimmune encephalomyelitis. Qualitative and quantitative differences in heat shock protein 60 expression in the central nervous system. J Immunol 154: 3448-3456
39. Koga T, Wand-Wurttenberger A, de Bruyn J et al. (1989) T cells against a bacterial heat shock protein recognize stressed macrophages. Science 246: 1112-1115
40. Aquino DA, Capello E, Weisstein J et al. (1997) Multiple sclerosis: altered expression of 70- and 27-kDa heat shock proteins in lesions and myelin. J Neuropathol Exp Neurol 56: 664-672
41. Bajramovic JJ, Lassmann H, van Noort JM (1997) Expression of αB-crystallin in glia cells during lesional development in multiple sclerosis. J Neuroimmunol 78: 143-151
42. Mor F, Cohen IR (1992) T cells in the lesion of experimental autoimmune encephalomyelitis. Enrichment for reactivities to myelin basic protein and to heat shock protein. J Clin Invest 90: 2447-2455
43. Birnbaum G, Kotilinek L, Schlievert P et al. (1996) Heat shock proteins and experimental autoimmune encephalomyelitis (EAE): I. Immunization with a peptide of the myelin protein 2', 3' cyclic nucleotide 3' phosphodiesterase that is cross-reactive with a heat shock protein alters the course of EAE. J Neurosci Res 44(4): 381-396

44. Salvetti M, Buttinelli C, Ristori G et al. (1992) T-lymphocyte reactivity to the recombinant mycobacterial 65- and 70-kD heat shock proteins in multiple sclerosis. J Autoimmunity 5: 691-702
45. Salvetti M, Ristori G, Buttinelli C et al. (1996) The immune response to mycobacterial 70-kDa heat shock proteins frequently involves autoreactive T cells and is quantitatively disregulated in multiple sclerosis. J Neuroimmunol 65: 143-153
46. Stinissen P, Vandevyver C, Medaer R et al. (1995) Increased frequency of γδ T cells in the cerebrospinal fluid and peripheral blood of patients with multiple sclerosis: reactivity, cytotoxicity and T cell receptor V gene rearrangements. J Immunol 154: 4883-4894
47. van Noort JM, van Sechel AC, Bajramovic JJ et al. (1995) The small heat shock protein alpha-B crystallin as a candidate autoantigen in multiple sclerosis. Nature 375: 798-801
48. van Sechel AC, van Stipdonk MJB, Persoon-Deen C, van Noort JM (1998) Epstein-Barr virus induced expression and HLA-DR restricted presentation of αB-crystallin, a major human myelin antigen. J Neuroimmunol 90: 38 (abstract)
49. Banki K, Colombo E, Sia F et al. (1994) Oligodendrocyte-specific expression and autoantigenicity of transaldolase in multiple sclerosis. J Exp Med 180: 1649-1663
50. Colombo E, Banki K, Tatum AH et al. (1997) Comparative analysis of antibody and cell-mediated autoimmunity to transaldolase and myelin basic protein in patients with multiple sclerosis. J Clin Invest 99: 1238-1250
51. Walsh MJ, Murray JM (1998) Dual implication of 2',3'-cyclic nucleotide 3' phosphodiesterase as major autoantigen and C3 complement-binding protein in the pathogenesis of multiple sclerosis. J Clin Invest 101(9): 1923-1931
52. Giulian D, Moore S (1980) Identification of 2',3' cyclic nucleotide 3'-phosphodiesterase in the vertebrate retina. J Biol Chem 255: 5993-5995
53. Tola MR, Granieri E, Casetta I et al. (1993) Retinal periphlebitis in multiple sclerosis: a marker of disease activity. Eur Neurol 33: 93-96
54. Selmaj K, Brosnan CF, Raine CS (1991) Colocalization of lymphocytes bearing the γδ T cell receptor and heat shock protein 65+ oligodendrocytes in multiple sclerosis. Proc Natl Acad Sci USA 88: 6452-6456
55. Battistini L, Fischer FR, Raine CS, Brosnan CF (1996) CD1b is expressed in multiple sclerosis lesions. J Neuroimmunol 67(2): 145-151

Chapter 17

Myelin Antigen Autoreactivity in Multiple Sclerosis

M. Vergelli

Introduction

It is currently believed that disturbances in the regulation of autoantigen-specific T cells, with a breakdown of immunological tolerance to myelin components, is crucial for the pathogenesis of multiple sclerosis (MS). However, the target autoantigens of this immune reaction are yet unknown. In MS the immune reactivity to myelin proteins has been extensively investigated. To date, this work has primarily focused on the two major myelin proteins, proteolipid protein (PLP) and myelin basic protein (MBP), but there is increasing interest in reactivity to minor myelin components such as myelin oligodendrocyte glycoprotein (MOG). For these three myelin proteins, encephalitogenicity has been demonstrated upon injection with immune adjuvants in susceptible animal strains, and T cells specific for these antigens are sufficient to cause experimental autoimmune encephalomyelitis (EAE).

T Cell Reactivity Against Myelin Basic Protein

Cellular immune response against MBP has been studied in greatest detail probably because MBP is the easier to isolate among myelin components. Although several questions have remained unanswered, data are available regarding the frequency, epitope specificity, mayor histocompatibility complex (MHC) restriction, T cell receptor (TCR) usage, and functional profile of MBP-specific T cell lines generated from MS patients and from normal donors.

Frequency of MBP-Specific T Cells

Although this finding was not expected at the beginning, MBP-specific, potentially autoreactive T cells are present in the peripheral blood (PB) of MS patients as well as in normal individuals [1-5]. Similarly, autoreactive T cells were found in the peripheral lymphoid organs of experimental animals and, upon in vitro activation, these cells mediated autoimmune damage to the central nervous system (CNS) [6,

---

Department of Neurological and Psychiatric Sciences, University of Florence, Viale Morgagni 85 - 50134 Florence, Italy. e-mail: vergelli@cesit1.unifi.it

7]. These findings indicate that autoantigen-reactive T cells escape the negative selection process in the thymus and belong to the "normal" T cell repertoire.

A wide number of studies have addressed the issue of whether the frequency of T cells with specificity to MBP differs between MS patients and controls. These investigations have resulted in controversial findings. First of all, the absolute frequency of MBP-reactive T cells in the peripheral blood varied enormously depending upon the methods used for its determination (i.e. antigen-specific proliferation [2-4], cytokine production after stimulation with antigen [8], or polymerase chain reaction (PCR) approach to identify MBP-specific T cells [9]). Second, although several studies reported higher frequencies of MBP-specific T cells in MS patients [2,8,10] other investigations failed to confirm these results [4]. These controversial results can partly be explained by the fact that patients and controls have not been carefully stratified for MHC-type, age and gender. However, results from family and twin studies, where the immunogenetic differences were minimized or completely eliminated, indicated that the responsiveness to MBP tends to be similar within members of the same family or twin set irrespective of disease status [11, 12]. These data suggest that the capability to mount a T cell response to MBP is linked to the immunogenetic background of the individual rather than to the disease process itself. Another possibility to explain the conflicting results is that the peripheral blood was isolated from MS patients during phases with different disease activity. However, stratification for disease status was not performed either in the studies showing differences or in those which failed to show them. The evolution over time of T cell responses to MBP in relationship with disease activity will be discussed in a following section.

Further evidence for the involvement of MBP-specific T cells in MS is provided by studies investigating in vivo activated peripheral blood T cells. The demonstration of MBP-reactive T cells in a pool of hypoxanthine-guanine-phosphoribosyltransferase (hprt) mutant T lymphocytes (somatic mutations at hprt gene is an index of T cell amplification in vivo) cultured from the peripheral blood of MS patients but not from normal individuals suggests that such cells were pre-activated in vivo either by MBP itself or by foreign antigens with sequence or structural similarity to MBP [13]. In addition, the frequency of MBP-reactive but not tetanus toxoid-reactive T cells generated after primary recombinant interleukin-2 (rIL-2) stimulation (to selectively expand those cells that had been activated in vivo) was significantly higher in MS patients as compared with control individuals [14]. These results provide definitive in vitro evidence of an absolute difference in the activation state of myelin-reactive T cells in the peripheral blood of patients with MS and suggest a pathogenic role of these autoreactive T cells in the disease.

Finally, two independent reports have recently shown that MBP-reactive T cells derived from MS patients have a decreased dependence on CD28/B7-mediated costimulation [15, 16]. These data suggest that MBP-responsive $CD4^+$ T cells are more likely to have been activated and or differentiated into memory T cells in vivo in MS patients compared with controls, and provide further indirect evidence for a role of these cells in the pathogenesis of MS.

Fine Specificity of MBP-Specific T Cells

With respect to the epitope specificity in MBP recognition, a number of studies failed to show preferential T cell recognition of any particular region of MBP in MS patients compared to normal donors. Considerable heterogeneity exists in the fine specificity of human MBP-specific T cells. In most MS patients and normal donors, the anti-MBP response uses a large variety of T cell clones recognizing a broad spectrum of peptide epitopes. Only in few cases the T cell response is focused to restricted segments of the MBP molecule. The spectrum of MBP epitopes recognized by T cells is much more diverse in humans than in inbred rodents. T cell epitopes can be located virtually in all regions of the human MBP sequence. At least three immunodominant areas have been identified in the context of different MHC class II restriction elements [2, 3, 17-20], and immunodominance in T cell responses has been correlated to the binding affinity of the different MBP peptides to the specific HLA-DR molecules [21]. These immunodominant areas are located in the N terminus, middle portion (amino acids 80-99) and C terminus (amino acids 140-170) of the MBP sequence. A novel immunodominant region recognized by HLA-DR4-restricted T cells has been recently described [22]. It is interesting that the immunodominant regions of MBP in humans correspond to sequences which are encephalitogenic in different animal strains susceptible to the induction of EAE [23, 24]. In particular, the middle region of MBP is encephalitogenic in SJL mice as well as in Lewis rats. The potential importance of this portion of MBP is further supported by the observation that the peptide MBP 83-99 (also MBP 84-102 and MBP 87-106) represents the immunodominant region in the context of the MS-associated HLA-DR alleles [2, 18, 19]. MBP 83-99 binds with high affinity to these HLA-class II alleles and in particular to both polymorphic alleles coexpressed in the HLA-DR15 Dw2 haplotype[1], the (MHC) class II molecule with the strongest association with MS in Caucasians [21, 25-27]. In addition, studies of (TCR) expressed in the brains of HLA-DR15 Dw2-positive MS patients revealed a restricted group of CDR3 motifs [28]. Similar motifs have been found in in vivo-activated MBP-specific T cell clones (TCC) derived from MS patients [29], in encephalitogenic TCC isolated from Lewis rats [30], and also in a cytotoxic, HLA-DR2a-restricted MBP 87-99-specific TCC derived from an MS patient [19]. These findings support the hypothesis of a pathogenic role of MBP 87-99-specific T cells, at least in a subgroup of HLA-DR15 Dw2-positive patients [31].

Since T cell responses against the immunodominant portions of MBP represent a potential therapeutic target in MS, characterization of MBP 83-99-specific T cells is considered critical. In a recent study, by using a large panel of peptides with single amino acid substitutions, considerable heterogeneity was found with

---

[1] HLA-DRa paired with the product of HLA-DRB1*1501 = DR2b; HLA-DRa paired with HLA-DRB5*0101 = DR2a.

respect to recognition of the epitope (i.e. individual TCC were differently sensitive to single amino acid exchanges within the peptide sequence) [32]. This heterogeneity was demonstrated even for TCC recognizing the epitope in the context of the same HLA class II restriction molecule. Considerable variability was also observed with respect both to TCR Vα and Vβ usage and to TCR binding affinity for the peptide/MHC complex. Again, even T cells restricted by the same HLA-DR molecule showed a high degree of heterogeneity with respect to TCR expression. Thus, extensive heterogeneity exists within the T cell response to a well-defined immunodominant region, and single TCC recognize the peptide epitope in different manners [32].

In summary no major differences were found in terms of fine specificity of MBP-specific T cells between MS patients and controls. In addition, although immunodominant areas of MBP have clearly been demonstrated, it is evident that the T cell response even to a restricted portion of MBP is extremely complex and heterogeneous.

TCR Usage of MBP-Specific T Cells

Following the description of restricted TCR variable chain usage in encephalitogenic T cells in EAE, numerous investigations were directed to analyze TCR gene expression in human MBP-specific T cells. Although first studies emphasized that TCR V gene usage is restricted in MS patients [33, 34], other investigations did not confirm these findings [3, 20]. In addition, different TCR V gene products can be used to recognize the same MBP peptide in the context of the same restriction element [3, 32], while TCC expressing the same V family may recognize different peptides (i.e. V gene usage does not correlate with T cell epitope specificity or MHC restriction). Furthermore, T cell clones with identical specificity and V gene usage do not necessarily express identical junctional region sequences, indicating that they may recognize the peptide epitope in different manners. Finally, striking sequence homologies were found among TCC recognizing the same MBP peptide/MHC complex in spite of different V gene usage [35].

In spite of this complexity in the recognition of MBP and even of single MBP epitopes by specific TCR, several groups have observed MBP-reactive T cells with identical Vα and Vβ junctional regions in the peripheral blood and CNS of MS patients, suggesting that autoreactive T cell reponses may in some cases be mounted by a limited set of expanded clones rather than by a large polyclonal T cell population [36, 37].

Phenotype and Functions of MBP-Reactive T Cells

It is evident from EAE that the presence of MBP-reactive T cells is not sufficient for the disease. Encephalitogenic T cells must be activated and express the appropriate functional phenotype both in terms of cytokine production and expression of adhesion molecules [38-40]. In EAE, encephalitogenic T cells primarily belong to the Th1 subpopulation of CD4$^+$ T cells producing large amounts of the proinflammatory cytokines interferon gamma(IFN-γ), IL-2, and tumor necrosis factor (TNF-α). Several pieces of experimental evidence support the idea that Th1-like T cells play a major role in MS as well. It is well known that the therapeutical application of IFN-γ resulted in an increased exacerbation rate in MS patients [41]. In addition increased levels of IL-2 have been observed [42]. Finally, several reports demonstrated elevated TNF-α levels in the blood and cerebrospinal fluid (CSF) of MS patients and, more importantly these increased levels have been related to disease activity [43, 44].

When the phenotype and function of MBP-reactive T cells were studied, it became clear that essentially all T cell lines generated with MBP belonged to the CD4+ subpopulation [45]. This observation is clearly biased by the methodological approach used to generate the T cell lines, since exogenous proteins are processed by the endocytic pathway and presented in the context of MHC class II molecules for recognition by CD4+ T lymphocytes. Recent advances in understanding the basic mechanisms of interaction between the TCR of CD8+ T lymphocytes and their specific ligands (MHC class I molecule complexed with the antigenic peptide) have made it possible to investigate CD8-mediated T cell responses to MBP and other myelin proteins by using synthetic peptides carrying the epitopes necessary for binding MHC class I molecules [46]. However, only few data are currently available about CD8-mediated T cell responses to myelin protein in MS. Although CD8+ T cells have been detected in MS plaques, there is much less evidence for a role of CD8+ T lymphocytes in the pathogenesis of MS.

The functional analysis of MBP-reactive CD4+ T cells has definitely shown a wide heterogeneity in both MS patients and normal donors. Although first reports suggested a similarity to encephalitogenic T cells in EAE in terms of cytokine production (i.e. Th1-like phenotype) [47], it is now clear that MBP-specific T cells can secrete all possible combinations of cytokines [32, 48]. TCC were identified which produced IFN-γ and TNF-α but not IL-4 or IL-5 (Th1-like), as well as others that produced high amounts of IL-4 and IL-5 and little or not IFN-γ and TNF-α/β (Th2-like T cells). Most TCC produced cytokines characteristic of both Th1 and Th2 (Th0-like). The cytokine profile of Th0 TCC ranged from being predominantly Th1-like to being primarly Th2-like. Most TCC produced IL-10 and TNF-α, irrespective of their release of IFN-γ, and IL-4. No correlation was found between IL-4 (or IL-5) and IL-10, nor between IFN-γ (or TNF-β) and TNF-α, suggesting that IL-10 and TNF-α are not useful to discriminate Th1-like from Th2-like cells. Interestingly, heterogeneous cytokine secretion patterns were also observed among the TCC generated from single individuals [32].

Another function of CD4+ T lymphocytes that has been correlated to the encephalitogenic potential in EAE is cytotoxicity. It has been shown that a considerable proportion of human MBP-specific CD4+ T cells mediates HLA class II-restricted target cell lysis [49-51]. When cytotoxicity of CD4+ MBP-reactive T cells was analyzed in more detail, a heterogeneous pattern emerged. MBP-specific T cells can be non-cytotoxic or mediate cytolytic activity by two distinct mechanisms: perforin-mediated killing or Fas/Fas ligand-mediated cytotoxicity. These two cytolytic pathways are not mutually exclusive. MBP-specific T cells which mediate highly efficient, perforin-dependent killing and also express high levels of functional Fas ligand have been isolated [52].

In conclusion, the functional analysis of MBP-specific T cells has demonstrated a high degree of complexity even among T cells obtained from single donors. The number of TCC examined so far does not permit definitive conclusions as to whether any functional phenotype is more common in patients affected by MS compared to normal donors.

Evolution Over Time of T Cell Response to MBP

As mentioned previously, T cell responses to MBP have been investigated widely in MS patients. However, little is known about the evolution over time of this response in individual patients. Salvetti and colleagues found that T cell lines derived from three patients reacted to only one or two MBP peptides in spite of a relatively long disease history [53]. A repetition of this assay on T cells from one of these patients eight months later showed the same restricted epitope recognition even though clinical activity in this patient had increased. Similarly, a study of T cell clones derived from two HLA-DR2-positive patients showed dominant reactivity to MBP 84-102 and MBP 143-168. These reactivities were due to clonal expansions which persisted for several months [36]. Finally a recent investigation carried out in MS patients with benign disease showed a predominant T cell response to single MBP epitopes that was stable over time and mediated by clonally expanded populations [54]. Taken together, these studies suggest that the repertoire of autoreactive T cells is stable over time and argues against the hypothesis of extensive determinant spreading during the course of MS.

On the other hand, changes in the pattern of cytokine secretion by autoreactive myelin-specific T cell clones have been described [55]. These changes correlated with the disease status in patients with MS. PLP-specific T cell clones generated during acute attacks of MS were characterized by a Th1-like profile whereas T cell clones raised from the same donors during remissions showed a pattern of cytokine production resembling that of murine Th0 and Th2 subsets.

In a recent study, my colleagues and I have serially investigated the frequency, peptide specificity, and cytokine profile of MBP-specific T cells in patients affected by MS (M. Vergelli et al., manuscript submitted). We aimed to determine if changes in the frequency and functional characteristics of MBP-reactive T cells are associated with inflammatory activity in the CNS. To this purpose, MBP-specific T cell lines (TCL) were generated from the peripheral blood of MS patients who underwent a serial gadolinium-enhanced magnetic resonance imaging (MRI) study (a currently accepted measure of inflammatory activity within the CNS). These T cell lines were expanded in culture and functionally characterized at different times. Episodic changes in the frequency, peptide specificity and functional repertoire of MBP-specific T cells were observed in each patient, suggesting that different sets of T cell clones are used for the recognition of MBP by individual patients at different times. In some patients an increased frequency of MBP-specific T cells was observed at time points corresponding to re-activation of inflammatory activity in the CNS. These increases in the frequency of MBP-specific T cells were associated with changes in the panel of epitopes recognized by T cells, suggesting that specific sub-populations of MBP-specific T cells had been expanded. In other patients similar episodic expansions of MBP-specific sub-populations were detected even if the analysis of MRI images revealed neither new lesions nor enlargement of pre-existing lesions during the entire follow-up. When the functional properties of MBP-specific T cell lines were analyzed, it become obvious that T cells generated during active inflammation within the CNS were more often characterized by a Th0/Th2 phenotype, whereas T cells generated at times without detectable MRI activity produced greater amounts of the Th1 cytokine IFN-γ. These results indicate that the pattern of MBP recognition over time is more complex than previously described. Episodic fluctuations of the number of MBP-specific T cells are probably common events during the course of MS. These fluctuations seem to be related to expansion of MBP-specific T cell populations with restricted specificity and characterized by a certain cytokine profile. These phenomena are probably superimposed on a rather stable pattern of MBP responsiveness.

Degeneracy in Antigen Recognition by MBP-Specific T Cells

As previously mentioned, T cells specific for self myelin antigens can be isolated from the peripheral blood of patients with MS as well as from healthy individuals. The biological meaning of these autoreactive circulating T cells is still unknown. One surprising finding is the high frequency of MBP-reactive T cells that can be isolated from both MS patients and normal individuals by using high antigen concentrations in the primary cultures (A. Vergelli et al., manuscript submitted). When the functional features of these T cells generated with high antigen concentration were examined, it was obvious that they recognized a wider spectrum of MBP epitopes and required higher antigen concentrations to be activated than MBP-specific T cells obtained using low antigen concentrations. The affinity (antigen requirement) of MBP-specific T cells appears therefore to be strictly related to the concentration used in the primary culture: at low antigen concentrations a small number of high affinity T cells was isolated, while the use of high antigen concentration in the primary culture yielded a high number of low-affinity TCL. These findings indicate that a high number of low-affinity autoreactive T cells is part of the "normal" T cell repertoire.

Recent studies have shown a high degree of degeneracy in antigen recognition by MBP-reactive T cells [56, 57]. The identification of single amino acid substitutions in MBP peptides yielding increased T cell responsiveness [58] as well as the proliferative response of MBP-specific T cell clones to combinatorial peptide libraries (consisting of mixtures containing up to 10^{14} different peptides) [59] suggested that the number of productive ligands for a single autoreactive T cell clone is much higher than previously expected. These approaches enabled my colleagues and I to identify cross-reactive peptides inducing T cell responses at lower antigen concentrations than native MBP peptides, suggesting that the "real" specificity of MBP-reactive T cells may be to an antigen different from that used to select and expand of these T cells in culture. The MBP peptide could just happen to be a low-affinity cross-reactive ligand.

The presence of high numbers of circulating low-affinity MBP-reactive T cells is fitting with the hypothesis that, due to the large extent of degeneracy in T cell recognition, self-reactive T cells are commonly activated and expanded during the course of protective immune reactions to foreign agents. However, the low-affinity antigen recognition prevents these autoreactive T cells from triggering autoimmunity even if they encounter the cross-reactive self antigen within the target organ (i.e. the CNS). However, in the case of a concomitant inflammatory process in the CNS causing increased expression of MHC and costimulatory molecules, it is conceivable that the low affinity can be compensated for by the elevated number of available MHC-peptide complexes on the surface of antigen-presenting cells (APC) and by costimulatory factors raising the overall avidity beyond the threshold required for triggering an effector T cell response. Once the inflammatory reaction is established within the CNS, the process can be further amplified by the release of self antigens. In addition, successive cross-stimulation of low-affinity self-reactive T cells in the periphery might result in exacerbations of the inflammatory reaction in the CNS due to the "susceptible" local environment. Therefore, both systemic and organ-related factors are required to trigger autoimmunity in this scenario. Alternatively, a second but not mutually exclusive scenario can be envisioned in which foreign antigens may interact and cross-activate T cells with high-affinity self-reactive TCR. This will lead to expansion and activation of self-reactive T cells that may trigger autoimmunity by themselves. This is probably similar to what happens in experimental autoimmunity when animals are immunized with a self antigen leading to the expansion of high-affinity self-reactive T cells.

T Cell Reactivity Against Other Myelin Proteins

Myelin Oligodendrocyte Glycoprotein

Myelin oligodendrocyte glycoprotein (MOG), a member of the imunoglobulin superfamily, is expressed exclusively in the CNS where it accounts for 0.05% of the total myelin proteins [60, 61]. In spite of its low amount, the immune response to MOG contributes to the autoimmune-mediated demyelination observed in animals immunized with whole CNS homogenates. In addition, MOG by itself is able to generate both an encephalitogenic T cell response and autoantibodies which result in a relapsing-remitting form of EAE characterized by extensive demyelination [62, 63]. Direct demyelinating activity of anti-MOG antibodies has been demonstrated in different in vivo and in vitro systems [61].

These findings have prompted the examination of MOG reactivity in MS even if these studies have been limited by difficulties in purifying sufficient amounts of the protein. Although few data about antibody reactivity to MOG in MS are available [64], two reports demonstrated strong T cell responses against purified MOG in MS [65]. When peripheral blood lymphocytes (PBL) from MS patients and controls were compared for reactivity to different myelin antigens, reactivity to MOG predominated. Interestingly, little reactivity to MOG was demonstrated in controls [65, 66]. These findings have been recently confirmed using a recombinant preparation representing the extracellular Ig-like domain of human MOG [67]. Finally, increased T cell reactivity to a number of synthetic MOG peptides was demonstrated in HLA-DR2(15)-positive MS patients using an Elispot assay to detect the number of T cells secreting IFN-γ in response to the specific antigen [68].

Proteolipid Protein

Proteolipid protein is a highly hydrophobic molecule. Due to its physico-chemical features, PLP is difficult to isolate and to use in cell culture studies because of the tendency to precipitate in aqueous solutions. For these reasons, studies investigating PLP reactivity in MS are fewer than those directed to study MBP. A few investigations documented PLP-reactive T cells in the PB and CSF of patients affected by MS as well as of healthy individuals [69, 70]. In addition, PLP-specific TCL have been established in culture. Nevertheless, the response in culture to the entire protein is usually weak and no differences were detected in the frequency of PLP-reactive T cells between MS patients and normal donors.

Better results in generating PLP-reactive TCL have been obtained using synthetic PLP peptides. Three immunodominant regions of PLP have been identified so far (amino acids 40-60, 95-117 and 185-205) [71-75]. These regions are located in hydrophilic portions of the molecule. Increased proliferative responses to two overlapping PLP peptides (PLP 184-199 and PLP 190-209) were found significantly more frequently in PBL of MS patients than in those of healthy control subjects. TCL generated in vitro with these peptides retained the capability of recognizing the entire PLP, demonstrating that that these peptides can be naturally processed [75]. In a different study, estimated frequency analyses were performed on the PBL of HLA-DR15-positive MS patients and controls using TCL initiated by the three immunodominant peptides of PLP. TCL from HLA-DR15-positive MS subjects recognized PLP 95-117 significantly more frequently than did TCL from control subjects [76]. These results confirm previous investigations by Tabira and collaborators who showed increased responsiveness to this region of PLP in MS patients compared to normal donors, especially in the context of HLA-DR15 [73].

With respect to phenotype and functions, PLP-specific TCL are similar to MBP-specific TCL. They belong to the CD4$^+$ subset, are restricted by HLA-DR molecules, and can mediate cytotoxic activity. Modifications of the cytokine secretion pattern by PLP-specific T cells during the course of MS have been reported, with a predominant Th1-like profile associated with clinical exacerbations. Although it is not completely clear whether there are quantitative or qualitative differences on PLP-specific T cells between MS patients and normal individuals, PLP-specific T cells have been identified in the CSF [70] as well as in the fraction of in vivo-activated PB T cells [14, 77].

Other Proteins

T cell responses have also been demonstrated to other proteins expressed in myelin such as myelin-associated glycoprotein (MAG) [78, 79], or in the myelin forming cells of the CNS (i.e. oligodendrocytes) such as 2'-3'-cyclic nucleotide 3'-phosphodiesterase (CNPase) and transaldolase-H [80, 81]. However it is currently unknown whether these responses differ between MS patients and controls. Several laboratories are involved in the characterization of T cell responsiveness to these alternative autoantigens in humans and in the evaluation of the encephalitogenic potential of these myelin components in animal models.

Conclusions

Besides the well-characterized response against MBP, it is clear that T cell reactivity exists in humans against each myelin component that has been examined so far. Considering that susceptibility to both EAE and MS is under control of the MHC background and that different myelin antigens result in different forms of EAE in different animal strains, it may be speculated that different myelin components may be responsible during the initial stages of MS in individual patients. Finally, little is known about the functional characteristics of the T cell response to myelin antigens different from MBP, in particular with respect to their affinity in antigen recognition and their intrinsic potential for cross-reactivity. These latter points may have important implications if considered in the context of degenerate T cell recognition and its relationship with molecular mimicry. Further studies on well-defined subgroups of MS patients (stratified for MHC background and disease status), in particular if combined with longitudinal evaluation of the fluctuations of disease activity within the CNS, are certainly required to better understand the role of T cell reactivity to myelin antigen in the pathogenesis of MS.

References

1. Burns J, Rosenzweig A, Zweiman B, Lisak RP (1983) Isolation of myelin basic protein-reactive T-cell lines from normal human blood. Cell Immunol 81: 435-440
2. Ota K, Matsui M, Milford EL et al. (1990) T-cell recognition of an immunodominant myelin basic protein epitope in multiple sclerosis. Nature 346: 183-187
3. Martin R, Utz U, Coligan JE et al. (1992) Diversity in fine specificity and T cell receptor usage of the human CD4+ cytotoxic T cell response specific for the immunodominant myelin basic protein peptide 87-106. J Immunol 148: 1359-1366
4. Pette M, Fujita K, Kitze B et al. (1990) Myelin basic protein-specific T lymphocyte lines from MS patients and healthy individuals. Neurology 40: 1770-1776
5. Chou YK, Vainiene M, Whitham R, et al. (1989) Response of human T lymphocyte lines to myelin basic protein: association of dominant epitopes with HLA-class II restriction molecules. J Neurol Sci 23: 207-216

6. Schlüsener H, Wekerle H (1985) Autoaggressive T lymphocyte lines recognize the encephalitogenic region of myelin basic protein; In vitro selection from unprimed rat T lymphocyte populations. J Immunol 135: 3128-3133
7. Genain CP, Lee-Parritz D, Nguyen M-H et al. (1994) In healthy primates, circulating autoreactive T cells mediate autoimmune disease. J Clin Invest 94: 1339-1345
8. Olsson T, Wei Zhi W, Höjeberg B, et al. (1990) Autoreactive T Lymphocytes in multiple sclerosis determined by antigen-induced secretion of interferon-γ. J Clin Invest 86: 981-985
9. Bieganowska KD, Ausbel LJ, Modaber Y et al. (1997) Direct ex vivo analysis of activated, Fas-sensitive autoreactive T cells in human autoimmune disease. J Exp Med 185: 1585-1594
10. Olsson T, Sun J, Hillert J et al. (1992) Increased numbers of T cells recognizing multiple myelin basic protein epitopes in multiple sclerosis. Eur J Immunol 22: 1083-1087
11. Joshi N, Usuku K, Hauser SL (1993) The T-cell response to myelin basic protein in familial multiple sclerosis: Diversity of fine specificity, restricting elements, and T-cell receptor usage. Ann Neurol 34: 385-393
12. Martin R, Voskuhl R, Flerlage M et al. (1993) Myelin basic protein-specific T-cell responses in identical twins discordant or concordant for multiple sclerosis. Ann Neurol 34: 524-535
13. Allegretta M, Nicklas JA, Sriram S, Albertini RJ (1990) T cells responsive to myelin basic protein in patients with multiple sclerosis. Science 247: 718-721
14. Zhang J, Markovic-Plese S, Lacet B et al. (1994) Increased frequency of interleukin 2-responsive T cells specific for myelin basic protein in peripheral blood and cerebrospinal fluid of patients with multiple sclerosis. J Exp Med 179: 973-984
15. Lovett-Racke AE, Trotter JL, Lauber J et al. (1998) Decreased dependence of myelin basic protein-reactive T cells on CD28-mediated costimulation in multiple sclerosis patients. J Clin Invest 101: 725-730
16. Scholz C, Patton KT, Anderson DE et al. (1998) Expansion of autoreactive T cells in multiple sclerosis is independent of exogenous B7 costimulation. J Immunol 160: 1532-1538
17. Richert JR, Reuben-Burnside CA, Deibler GE, Kies MW (1988) Peptide specificities of myelin basic protein-reactive human T-cell clones. Neurology 38: 739-742
18. Pette M, Fujita K, Wilkinson D et al. (1990) Myelin autoreactivity in multiple sclerosis: recognition of myelin basic protein in the context of HLA-DR2 products by T lymphocytes of multiple sclerosis patients and healthy donors. Proc Natl Acad Sci USA 87: 7968-7972
19. Martin R, Howell MD, Jaraquemada D et al. (1991) A myelin basic protein peptide is recognized by cytotoxic T cells in the context of four HLA-DR types associated with multiple sclerosis. J Exp Med 173: 19-24
20. Meinl E, Weber F, Drexler K et al. (1993) Myelin basic protein-specific T lymphocyte repertoire in multiple sclerosis. Complexity of the response and dominance of nested epitopes due to recruitment of multiple T cell clones. J Clin Invest 92: 2633-2643
21. Vergelli M, Kalbus M, Rojo SC et al. (1997) T cell response to myelin basic protein in the context of the multiple sclerosis-associated HLA-DR15 haplotype: peptide binding, immunodominance and effector functions of T cells. J Neuroimmunol 77: 195-203
22. Muraro PA, Vergelli M, Kalbus M et al. (1997) Immunodominance of a low-affinity major histocompatibility complex-binding myelin basic protein epitope (residues 111-129) in HLA-DR4 (B1*0401) subjects is associated with a restricted T cell repertoire. J Clin Invest 100: 339-349

23. Martin R, McFarland HF, McFarlin DE (1992) Immunological aspects of demyelinating diseases. Annu Rev Immunol 10: 153-187
24. Zamvil SS, Steinman L (1990) The T lymphocyte in experimental allergic encephalomyelitis. Annu Rev Immunol 8: 579-621
25. Valli A, Sette A, Kappos L et al. (1993) Binding of myelin basic protein peptides to human histocompatibility leukocyte antigen class II molecules and their recognition by T cells from multiple sclerosis patients. J Clin Invest 91: 616-628
26. Vogt AB, Kropshofer H, Kalbacher H et al. (1994) Ligand motifs of HLA-DRB5*0101 and DRB1*1501 molecules delineated from self-peptides. J Immunol 153: 1665-1673
27. Wucherpfennig KW, Sette A, Southwood S et al. (1994) Structural requirements for binding of an immunodominant myelin basic protein peptide to DR2 isotypes and for its recognition by human T cell clones. J Exp Med 179: 279-290
28. Oksenber JR, Panzara MA, Begovich AB et al. (1993) Selection for T-cell receptor Vβ-Dβ-Jβ gene rearrangements with specificity for a myelin basic protein peptide in brain lesions of multiple sclerosis. Nature 362: 68-70
29. Allegretta M, Albertini RJ, Howell MD et al. (1994) Homologies between T cell receptor junctional sequences unique to multiple sclerosis and T cells mediating experimental allergic encephalomyelitis. J Clin Invest 94: 105-109
30. Gold DP, Offner H, Sun D et al. (1991) Analysis of T cell receptor beta chains in Lewis rats with experimental allergic encephalomyelitis: conserved complementarity determining region 3. J Exp Med 174: 1467-1476
31. Steinman L, Waisman A, Altman D (1995) Major T-cell responses in multiple sclerosis. Mol Med Today 1: 79-83
32. Hemmer B, Vergelli M, Tranquill L et al. (1997) Human T-cell response to myelin basic protein peptide (83-99): extensive heterogeneity in antigen recognition, function and phenotype. Neurology 49: 1116-1126
33. Wucherpfennig KW, Ota K, Endo N et al. (1990) Shared human T cell receptor V beta usage to immunodominant regions of myelin basic protein. Science 248: 1016-1019
34. Kotzin BL, Karuturi S, Chou YK et al. (1991) Preferential T-cell receptor Vβ-chain variable gene use in myelin basic protein-reactive T-cell clones from patients with multiple sclerosis. Proc Natl Acad Sci USA 88: 9161-9165
35. Vergelli M, Hemmer B, Utz U et al. (1996) Differential activation of human autoreactive T cell clones by altered peptide ligands derived from myelin basic protein peptide (87-99). Eur J Immunol 26: 2624-2634
36. Wucherpfennig KW, Zhang J, Witek C et al. (1994) Clonal expansion and persistence of human T cells specific for an immunodominant myelin basic protein peptide. J Immunol 152: 5581-5592
37. Hafler DA, Saadeh MG, Kuchroo V et al. (1996) TCR usage in human and experimental demyelinating diseases. Immunol Today 17: 152-159
38. Ando DG, Clayton J, Kono D et al. (1989) Encephalitogenic T cells in the B10.PL model of experimental allergic encephalomyelitis (EAE) are of the Th-1 lymphokine subtype. Cell Immunol 124: 132-143
39. Baron JL, Madri JA, Ruddle NH et al. (1993) Surface expression of α4 integrin by CD4 T cells is required for their entry into brain parenchyma. J Exp Med 177: 57-68
40. Kuchroo VK, Martin CA, Greer JM et al. (1993) Cytokines and adhesion molecules contribute to the ability of myelin proteolipid protein-specific T cell clones to mediate experimental allergic encephalomyelitis. J Immunol 151: 4371-4382
41. Panitch HS, Hirsch RL, Schindler J, Johnson KP (1987) Treatment of multiple sclerosis with gamma interferon: Exacerbations associated with activation of the immune system. Neurology 37: 1097-1102

42. Hartung HP, Hughes RAC, Taylor WA, et al. (1990) T cell activation in Guillain-Barré syndrome and in MS: elevated serum levels of soluble IL-2 receptors. Neurology 40: 215-218
43. Sharief MK, Hentges R (1991) Association between tumor necrosis factor-α and disease progression in chronic progressive multiple sclerosis. N Engl J Med 325: 467-472
44. Rieckmann P, Albrecht M, Kitze B et al. (1995) Tumor necrosis factor-α messenger RNA expression in patients with relapsing-remitting multiple sclerosis is associated with disease activity. Ann Neurol 37: 82-88
45. Martin R, McFarland HF (1995) Immunological aspects of experimental allergic encephalomyelitis and multiple sclerosis. Crit Rev Clin Lab Sci 32: 121-182
46. Tsuchida T, Parker KC, Turner RV, et al.(1994) Autoreactive CD8$^+$ T-cell responses to human myelin protein-derived peptides. Proc Natl Acad Sci USA 91: 10859-10863
47. Voskuhl RR, Martin R, Bergman C et al. (1993) T helper 1 (TH1) functional phenotype of human myelin basic protein-specific T lymphocytes. Autoimmunity 15: 137-143
48. Hemmer B, Vergelli M, Calabresi P et al. (1996) Cytokine phenotype of human autoreactive T cell clones specific for the immunodominant myelin basic protein peptide (83-99). J Neurosci Res 45: 852-862
49. Jaraquemada D, Martin R, Rosen-Bronson S et al. (1990) HLA-DR2a is the dominant restriction molecule for the cytotoxic T cell response to myelin basic protein in DR2Dw2 individuals. J Immunol 145: 2880-2885
50. Martin R, Jaraquemada D, Flerlage M et al. (1990) Fine specificity and HLA restriction of myelin basic protein-specific cytotoxic T cell lines from multiple sclerosis patients and healthy individuals. J Immunol 145: 540-548
51. Vergelli M, Le H, van Noort JM et al. (1996) A novel population of CD4$^+$CD56$^+$ myelin reactive T cells lyses target cells expressing CD56/neural cell adhesion molecule. J Immunol 157: 679-688
52. Vergelli M, Hemmer B, Muraro P et al. (1997) Human autoreactive CD4$^+$ T cell clones use perforin – or Fas ligand-mediated pathways for target cell lysis. J Immunol 158: 2756-2761
53. Salvetti M, Ristori G, D'Amato M et al. (1993) Predominant and stable T cell responses to regions of myelin basic protein can be detected in individual patients with multiple sclerosis. Eur J Immuol 23: 1232-1239
54. Uccelli A, Giunti D, Salvetti M et al. (1998) A restricted response to myelin basic protein (MBP) is stable in multiple sclerosis (MS) patients. Clin Exp Immunol 111: 186-192
55. Correale J, Gilmore W, McMillan M et al. (1995) Patterns of cytokine secretion by autoreactive proteolipid protein-specific T cell clones during the course of multiple sclerosis. J Immunol 154: 2959-2968
56. Hemmer B, Vergelli M, Pinilla C et al. (1998) Probing degeneracy in T-cell recognition using peptide combinatorial libraries - Importance for T-cell survival and autoimmunity. Immunol Today 19: 163-168
57. Mason D (1998) A very high level of crossreactivity is an essential feature of the T-cell receptor. Immunol Today 19: 395-404
58. Vergelli M, Hemmer B, Kalbus M et al. (1997) Modifications of peptide ligands enhancing T cell responsiveness suggest a broad spectrum of stimulatory ligands for autoreactive T cells. J Immunol 158: 3746-3752
59. Hemmer B, Fleckenstein BT, Vergelli M et al. (1997) Identification of high potency microbial and self ligands for a human autoreactive class II-restricted T cell clone. J Exp Med 185: 1651-1659
60. Abo S, Bernard CCA, Webb M et al. (1993) Preparation of highly purified human myelin oligodendrocyte glycoprotein in quantities sufficient for encephalitogenicity and immunogenicity studies. Biochem Mol Biol Int 30: 945-958

61. Bernard CCA, Johns TG, Slavin A et al. (1997) Myelin oligodendrocyte glycoprotein: a novel candidate autoantigen in multiple sclerosis. J Mol Med 75: 77-88
62. Genain CP, Nguyen MH, Letvin NL et al. (1995) Antibody facilitation of multiple sclerosis-like lesions in a nonhuman primate. J Clin Invest 96: 2966-2974
63. Genain CP, Abel K, Belmar N et al. (1996) Late complication of immune deviation therapy in a nonhuman primate. Science 274: 2054-2057
64. Xiao BG, Linington C, Link H (1991) Antibodies to myelin oligodendrocyte glycoprotein in cerebrospinal fluid from patients with multiple sclerosis and controls. J Neuroimmunol 31: 91-96
65. Kerlero de Rosbo N, Milo R, Lees MB et al. (1993) Reactivity to myelin antigens in multiple sclerosis. Peripheral blood lymphocytes respond predominantly to myelin oligodendrocyte glycoprotein. J Clin Invest 92: 2602-2608
66. Sun J, Link H, Olsson T et al. (1991) T and B cell responses to myelin-oligodendrocyte glycoprotein in multiple sclerosis. J Immunol 146: 1490-1495
67. Kerlero de Rosbo N, Hoffman M, Mendel I et al. (1997) Predominance of the autoimmune response to myelin oligodendrocyte glycoprotein (MOG) in multiple sclerosis: reactivity to the extracellular domain of MOG is directed against three main regions. Eur J Immunol 27: 3059-3069
68. Wallstrom E, Khademl M, Andersson M et al. (1998) Increased reactivity to myelin oligodendrocyte glycoprotein peptides and epitope mapping in HLA-DR2(15)+ multiple sclerosis. Eur J Immunol 28: 3329-3336
69. Trotter JL, Hickey WF, van der Veen RC, Sulze L (1991) Peripheral blood mononuclear cells from multiple sclerosis patients recognize myelin proteolipid protein and selected peptides. J Neuroimmunol 33: 55-62
70. Sun JB, Olsson T, Wang WZ et al. (1991) Autoreactive T and B cells responding to myelin proteolipid protein in multiple sclerosis and controls. Eur J Immunol 21: 1461-1468
71. Pelfrey CM, Trotter JL, Tranquill LR, McFarland HF (1993) Identification of a novel T cell epitope of human proteolipid protein (residues 40-60) recognized by proliferative and cytolytic CD4+ T cells from multiple sclerosis. J Neuroimmunol 46: 33-42
72. Pelfrey CM, Trotter JL, Tranquill LR, McFarland HF (1994) Identification of a second T cell epitope of human proteolipid protein (residues 89-106) recognized by proliferative and cytolytic CD4+ T cells from multiple patients. J Neuroimmunol 53: 153-161
73. Ohashi T, Yamamura T, Inobe J et al. (1995) Analysis of proteolipid protein (PLP)-specific T cells in multiple sclerosis: identification of PLP 95-116 as an HLA-DR2, w15-associated determinant. Int Immunol 7: 1771-1778
74. Markovic-Plese S, Fukaura H, Zhang J et al. (1995) T cell recognition of immunodominant and cryptic proteolipid protein epitopes in humans. J Immunol 155: 982-992
75. Greer JM, Csurhes PA, Cameron KD et al. (1997) Increased immunoreactivity to two overlapping peptides of myelin proteolipid protein in multiple sclerosis. Brain 120: 1447-1460
76. Trotter JL, Pelfrey CM, Trotter AM et al. (1998) T cell recognition of myelin proteolipid protein and myelin proteolipid protein peptides in the peripheral blood of multiple sclerosis and control subjects. J Neuroimmunol 84: 172-178
77. Trotter JL, Damico CA, Cross AH et al. (1997) HPRT mutant T-cell lines from multiple sclerosis patients recognize myelin proteolipid protein peptides. J Neuroimmunol 75: 95-103
78. Zhang YD, Burger D, Saruhan M et al. (1993) The T-lymphocyte response against myelin-associated glycoprotein and myelin basic protein in patients. Neurology 43: 403-407

79. Link H, Sun JB, Wang Z et al. (1992) Virus-specific and autoreactive T cells are accumulated in cerebrospinal fluid in multiple sclerosis. J Neuroimmunol 38: 63-74
80. Banki K, Colombo E, Sia F et al. (1994) Oligodendrocyte-specific expression and autoantigenicity of transaldolase in multiple sclerosis. J Exp Med 180: 1649-1663
81. Rosener M, Muraro P, Riethmuller A et al. (1997) 2',3'-cyclic nucleotide 3'-phosphodiesterase: a novel candidate autoantigen in demyelinating diseases. J Neuroimmunol 75: 28-34

Chapter 18

Inflammation and Multiple Sclerosis: a Close Interplay

G. Martino[1,2], R. Furlan[1], P.L. Poliani[1,2]

Inflammation: Hystorical Highlights

In the first century A.D., Cornelius Celsus first described the inflammatory process by indicating its five cardinal signs: *rubor, tumor, calor, dolor* and *functio laesa*. The intrinsic characteristics of the inflammatory process were subsequently described in 1793 by John Hunter who stated that the inflammatory reaction is not a disease but a specific response that has a "salutary" effect on its host. The pathological aspects of the inflammatory process were first described in the nineteenth century by Julius Cohnheim who claimed that in the inflamed tissue there is an initial vasodilatation associated with increased vascular permeability and changes in the blood flow causing edema and leukocyte extravasation. At the end of the nineteenth century, Elia Metchnikoff suggested that the purpose of the inflammatory process is to bring phagocytic cells to the injured areas to engulf invading bacteria. At the beginning of the twentieth century, Paul Enrich indicated that inflammation involves not only cellular (phagocytosis) but also serum factors (antibodies), and Sir Thomas Lewis concluded that the vascular changes occurring during the inflammatory process are mainly mediated by chemical substances (reviewed in [1]).

The Inflammatory Process

At present, there is a general consensus in defining the inflammatory process as a reaction of blood vessels in vascularized living tissue to a local injury leading to the accumulation of fluid and blood cells [1]. The inflammatory reaction can be further divided in acute or chronic inflammation. The acute reaction has a short duration (few minutes to 1-2 days) and is characterized by exudation of fluid and plasma proteins (edema) along with recruitment of neutrophils. The chronic reaction has a long duration (days to months) and is characterized by the presence in the inflammed tissue of lymphocytes and macrophages and by the prolif-

---

[1] Experimental Neuroimmunology Unit, Department of Neuroscience, San Raffaele Scientific Institute, Via Olgettina 58 - 20132 Milan, Italy. e-mail: g.martino@hsr.it
[2] Department of Neurology, University of Milan, San Raffaele Scientific Institute, Via Olgettina 60 - 20132 Milan, Italy

eration of blood vessels (angiogenesis) and connective tissue (fibrosis). Different stimuli such as hypoxia, physical agents, chemical agents, drugs, infectious agents and immune-mediated processes can induce an inflammatory reaction. The lifespan of the inflammatory reaction depends on the "quality", "quantity" and "persistance" of the inflammatory stimulus.

Inflammatory stimuli induce the production of inflammatory mediators which can be subdivided in plasma and cellular mediators. These factors participate in most of the vascular and cellular responses. Among plasma mediators, the complement system plays a major role along with the kinin system and the clotting/fibrinolytic system. Cellular mediators include preformed chemical substances like histamine, serotonin and lysosomal enzymes as well as newly synthesized substances like prostaglandin, leukotrienes, platelet activating factors, chemokines and cytokines.

Inflammatory Mediators

Inflammatory mediators produced *ex novo* or preformed participate in the inflammatory reaction as a cascade (Fig. 1) [2]. The primary mediators are the

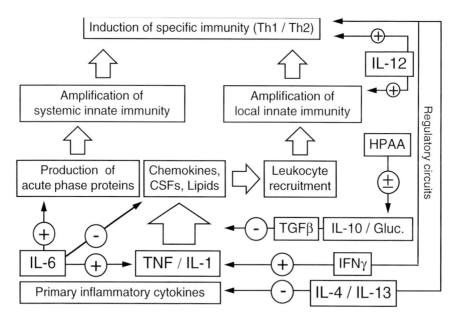

Fig. 1. The role of primary inflammatory cytokines in igniting local and systemic reactions of the innate immune system. The activation of the innate immune system when the inflammatory stimulus is persistent leads to the activation of the specific immune system. Stimulatory as well as inhibitory signals are indicated with plus (+) and minus (-) signs, respectively. *Gluc*, glucocorticoids; *HPAA*, hypothalamic-pituitary-adrenal axis; *CSFs*, colony stimulating factors

so-called primary inflammatory cytokines: interleukin (IL)-1, tumor necrosis factor (TNF)-α, and to some exent IL-6. These cytokines, which are mainly produced by activated macrophages (mostly in response to immune or microbial stimuli) [3], induce the production of (a) secondary mediators which act locally (e.g. in the target tissues) and systemically, and (b) acute phase proteins (by the liver) which in turn (e.g. C-reactive protein) amplify the systemic innate immunity. If the inflammatory stimulus persists the reaction tends to became chronic thus inducing primary inflammatory cytokine-mediated leukocyte recruitment and survival in the target tissue. This recruitment is regulated by the production of lipid and peptide mediators. Among these latter mediators, chemokines (chemo-attractant cytokines) [4] and colony-stimulating factors (CSF) play a major role [5] by setting the stage for the activation of specific immunity such as antigen-dependent activation and differentiation of T lymphocytes. After antigen recognition and activation, $CD4^+$ (and $CD8^+$) T cells (in mouse and human) differentiate into two distinct cell subsets which differ in their cytokine secretion profile. T helper 1 (Th1) cells secrete IL-2, IFN-γ, and TNF-β whereas T helper 2 (Th2) produce IL-4, IL-5, and IL-13 (reviewed in [6]). Thus, Th1 cytokines play a major role in promoting and supporting the inflammatory process via the activation of macrophages secreting primary inflammatory cytokines, but also in regulating delayed-type hypersensitivity (DTH) reactions and in IgG1 (human) or IgG2a (mouse) but not IgE synthesis. Conversely, Th2 cytokines, play a predominant role in immediate-type hypersensitivity (IL-4 is in fact the critical stimulus inducing a switch to IgE antibody production) and in the downregulation of inflammatory responses by inhibiting secretion of primary inflammatory cytokines [6, 7]. The cytokines produced by each Th cell subset are, in fact, inhibitory for the opposite subset. In this complex scenario, glucocorticoid hormones and the hypothalamic-pituitary-adrenal axis (HPAA) regulate the inflammatory process at different levels (Fig. 1).

In conclusion, the induction of specific immunity is therefore one of the primary end points of the inflammatory reaction. This is essential to perpetuate the inflammatory process due to its ability to upregulate the production of primary inflammatory cytokines, which in turn restimulate the critical pathways of the inflammatory process (Fig. 1).

Multiple Sclerosis, Inflammation and Cytokines

Multiple sclerosis (MS) is a demyelinating disease of the central nervous system (CNS) of unknown etiology [8]. Its pathological hallmark is the presence within the CNS of inflammatory infiltrates containing few autoreactive T cells and a multitude of pathogenic nonspecific lymphocytes [9]. Among T cells present in the CNS of MS patients, $CD4^+$ predominate although $CD8^+$ and B cells are also present [10]. The pathological process underlying MS determines the typical patchy CNS demyelination, ranging from demyelination with preservation of oligodendrocytes to complete oligodendrocyte loss and severe glial scarring. In most instances, however, oligodendrocytes or their precursors are morphologically preserved in demyelinating plaques, and remain capable of differentiating and remyelinating [11]. MS is therefore considered an autoimmune organ-specific inflammatory demyelinating disease of the CNS; an immune-mediated reaction triggered by a still unknown stimulus (virus?) ignites a cascade of inflammatory events leading to the activation of the specific immunity which indefinitely perpetuates the inflammatory process. T cells committed against CNS components (e.g. myelin proteins) provide the organ specificity of the pathogenic process. These T cells regulate the recirculation within the CNS of nonantigen-specific lymphomononuclear cells which form the CNS-confined perivascular inflammatory infiltrates characteristic of the disease and in turn act as effector cells by directly destroying oligodendrocytes or by releasing myelinotoxic substances [9]. Studies performed on transgenic animals affected by experimental allergic encephalomyelitis (EAE), the animal model for MS, have confirmed that peripheral polyclonal expansion of lymphocytes driven by nonspecific inflammatory stimuli along with T cells specific for myelin antigens is actually required to obtain CNS perivascular inflammatory infiltration and demyelination [12, 13].

After a CNS-localized immune reaction [8], whose primary trigger as well as the primary target (i.e. myelin components) are still partially unknown, primary inflammatory cytokines are released and represent the starting point of the pathological process occurring in MS. TNF-α, IL-1β as well as IL-6 have been found in MS plaques at the protein as well as the mRNA levels [14]. When the so-called primary inflammatory cytokines are produced within the CNS, the inflammatory MS-specific process procedes via the in situ production of primary inflammatory cytokine-induced mediators. Among them, chemokines [14, 15], colony-stimulating factors [16] and lipids [17-19] have been found in active MS plaques. Due to the combined action of cytokines and chemokines, leukocytes are then recruited within the CNS. The typical pathological aspect of autoptic or brain biopsy material from MS patients indicates that lymphocytes and monocytes predominate in areas of demyelination and perivascular inflammation [11]. The recruitment of leukocytes within the CNS amplifies the local innate immunity via the activation of glial cells (mainly astrocytes and microglial cells) which then actively participate in the ongoing inflammatory process. Activated microglia are the main population of phagocytes in early stage of demyelination [20, 21] and primary inflammatory cytokines are found in astrocytes at the edge of MS plaques [22]. We have recently shown in MS patients that the normal appearing white matter, which subsequently will contain foci of perivascular inflammation, shows magnetic resonance imaging changes early before the appearance of the inflammatory foci [23]. When local innate immunity is amplified, the specific immunity generating T cells against myelin components takes place in the CNS of MS patients. Primary inflammatory cytokines can induce per se a myelin breakdown [14] determining the release of myelin components which then become immunogenic and stimulate a T cell-specific response [8]. In MS patients, T cells against myelin-associated glycoprotein (MAG) [24], myelin basic protein (MBP) [25], proteolipid protein (PLP) [26] and myelin oligodendrocyte glycoprotein (MOG) [27] are present at higher levels compared to healthy subjects either in the blood or in the cerebrospinal fluid (CSF) [28-30]. It has also been shown that these specifc T cells and in particular those reacting against MBP are of the Th1 phenotype and therefore are able to perpetuate the inflammatory process by producing some of the putative primary inflammatory cytokines [31]. A more direct evidence that T cells against myelin components, such as MBP, are present within demyelinating areas comes from a recent study showing that MS brain T cell receptor (TCR) transcripts were shared in part with the CD3 region of a Vβ 5.2-expressing DR2a-restricted MBP87-106-specific cytotoxic T cell line [32]. In MS patients, the local effects due to the inflammatory process are also paralleled by systemic effects which tend to amplify the local innate and specific immunities. Several non-MS specific inflammatory markers have been variably found in blood as well as in CSF samples from MS patients [14]. Among them primary inflammatory cytokines are consistently found and seem to correlate with preclinical phases of disease activity [33]. The presence of inflammatory cytokines in the peripheral circulation of patients with MS could contribute to amplify the CNS-confined inflammatory events. Myelin-specific T cells are in fact present in the peripheral circulation of normal individuals [34] as well as MS patients and can be (re)activated via specific inflammatory stimuli such as primary inflammatory cytokine production [35]. In non-human primates, MBP-specific T cells producing proinflammatory cytokines after in vitro activation are able to induce EAE upon transfer into irradiated autologous recipients [36].

Myelin-reactive T cells are then pivotal in orchestrating and perpetuating the inflammatory events during MS. When the inflammatory process starts, it procedes like a vicious circle. Myelin-specific T cells, in fact, produce Th1-type cytokines including IFN-γ which in turn increase the production and the activity of primary inflammatory cytokines either locally where these cytokines can directly destroy the myelin sheath (e.g. TNF-α), or systemically where these cytokines can activate macrophages as well as T cells in a nonantigen-specific manner [37]. This latter hypothesis is supported by the recent findings that (1) nonantigen-specific priming can be induced in vitro in CD4$^+$ T cells by proinflammatory cytokines (i.e. TNF-α, IL-2, and IL-6) [38] and in CD8$^+$ T cells by IFNs [39] and that (2) T cells (mainly CD4$^+$-memory) from MS patients can be activated in a myelin-independent fashion by the combination of two primary inflammatory cytokines (TNF-α and IL-6), the prototypical Th1 cytokine (e.g. IFN-γ) and a T cell growth factor (e.g. IL-2). We have shown this cytokine-mediated "bystander" effect in MS patients and demonstrated that it is sustained by a cytokine-induced persistent increase of intracellular calcium due to the stimulation of two different intracellular calcium-mediated pathways, one ignited by IFN-γ alone and mediated by protein kinase C [40], and the other ignited by the combination of IL-2, IL-6 and TNF-α and sustained by the increase of inositol-trisphosphate production [35]. This latter pathway mediates nuclear translocation of nuclear factor of activated T cells (NF-AT) to ignite IL-2 gene transcription and therefore promote the T cell activation process [41]. Briefly, using two different antibodies (kindly donated by A. Sica, Milan) recognizing the nuclear or the cytoplasmic form of NF-AT, we found that the combination of IL-2/IL-6/TNF-α induced the nuclear translocation of NF-AT in 22.4% of CD4$^+$CD45RO$^+$ negatively purified T cells while only 3.6% of these cells translocated NF-AT into the nucleus when IL-2 was the only stimulus provided. No differences have been found when IFN-γ was added to the three-cytokine combination. As a positive control [41], we measured NF-AT nuclear translocation in Jurkat T cells stimulated with phorbol myristate acetate (PMA) and ionomycin; nuclear localization of NF-AT was found in 28.3% of the cells (unpublished results). Thus, primary inflammatory cytokines not only participate in the process of myelin antigen-specific T cell activation but may also represent the main mediators of the nonantigen-specific activation of peripheral T cells. The following scenario can then be hypothesized: primary inflammatory cytokines (re)activate myelin-specific resting memory peripheral T cells which act as driver cells for nonantigen-specific effector cells also activated or reactivated in the periphery by the same combination of cytokines. This peripheral mechanism can facilitate the recruitment of nonantigen-specific inflammatory cells (mainly lymphomononocytes) releasing myelinotoxic substances to different CNS areas.

Conclusions

Pathological and immunological findings in MS indicate that the process underlying the disease can be considered to be a T cell-mediated organ-specific inflammatory response leading to the recruitment of effector macrophages to the CNS (Table 1). A possible pathogenic scenario can be therefore depicted (Fig. 2 and

Table 1. Events underlying MS pathogenesis

Etiology
    Still unknown (virus?)

Pathogenesis
    Immune-mediated chronic inflammation

Target antigen(s)
    Myelin components, others?

Immune cells involved
    Regulatory cells (organ specificity)
        Microglia and astrocytes (antigen-presenting cells)
        Myelin antigen-specific T cells (CD4$^+$ Th1)
    Effector cells (myelin toxicity)
        Activated macrophages
        Nonantigen-specific T and B cells

Effector mechanisms
    Proteolytic enzymes and NO$_2$ (glial cells and macrophages)
    Oligodendrotoxic cytokines (glial cells, macrophages, T cells)
    Antibody-dependent cell-mediated cytotoxicity (B cells, macrophages)
    Cytolytic T cells (MHC-unrestricted CD4$^+$, γ/δ T cells)

Inflammation and Multiple Sclerosis: A Close Interplay

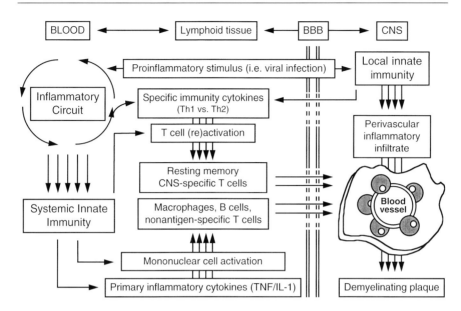

Fig. 2. Possible scenario of the pathogenic process occurring in MS. After an aspecific inflammatory hit in the CNS, T cells reacting against myelin components are generated. These cells, which are of the Th1-type secreting proinflammatory cytokines (e.g. IFN-γ), recruit and activate effector cells (mainly macrophages) from the periphery. When macrophages reach the CNS parenchyma, they secrete primary inflammatory cytokines and ignite the in situ inflammatory process leading to the perivascular accumulation of lymphomononuclear cells. These cells, in turn, release myelinotoxic substances inducing demyelination. The process is self-limiting but can be rekindled whenever a strong-enough nonspecific inflammatory hit in the periphery induces (re)activation of resting memory myelin-specific T cells as well as effector macrophages. *BBB*, blood-brain barrier

Table 1). An initial CNS-confined inflammatory hit, of still unknown origin (virus?), persists within the CNS for a sufficient amount of time to initiate a cascade of inflammatory events overtaking the innate immunity and leading to ignition of the acquired immune system which in turn mounts an antigen-specific response against myelin components. This response is strong enough to induce an autoreactive T cell population which continuously recruits inflammatory effector cells from the periphery. The recruitment is orchestrated by primary inflammatory cytokines secreted either in the CNS by the autoreactive T cells or in the periphery by nonantigen-specific mononuclear inflammatory cells activated by a specific inflammatory stimulus (e.g. upper respiratory tract infections). The peripheral inflammatory stimulus can also activate memory resting myelin-specific peripheral T cells which reach the CNS parenchyma and there secrete proinflammatory cytokines functioning as a stimulus for the perpetuation of the process. When an appropriate number of peripherally driven effector cells reach the CNS parenchyma, a cascade of events leading to myelin breakdown via the secretion of inflammatory myelinotoxic molecules or directly via the action of cytotoxic cells occurs within the white matter. These events lead to the formation of the typical perivascular inflammatory infiltrates causing perivascular demyelination and axonal loss. The accumulation of axonal loss during time causes permanent CNS damage and neurological impairment [10, 11, 42]. However, disease heterogeneity exists. The genetic background of each individual patient can in part explain the disease heterogeneity by contributing (a) to limit vs. amplify the extent of the inflammatory reaction; (b) to the generation of an appropriate number of myelin-specific T cells; and (c) to the net effect of a discrete amount of a certain proinflammatory cytokine on the peripheral and/or CNS-confined immune cells [43]. However, some questions remain unsettled. What is the primary inflammatory event in MS? Which are the predisposing genes? Is the pathogenic mechanism due to a secondary autoimmune phenomenon triggered by a viral infection or is it "tout court" autoimmune? Are myelin antigens the primary target of the immune process or is the autoimmune reaction epiphenomenic and the primary target is represented by oligodendrocytes? What is the immunological mechanism underlying the different disease courses or phases? Moreover, there is some evidence against an exclusive inflammatory pathogenic background underlying MS demyelination: (a) inflammation follows demyelination, (b) in some cases, following immunosuppressive treatments, active demyelination procedes in the absence of perivascular infiltrates and tissue infiltration by T cells, and (c) perivascular T cell infiltration is present in normal CNS and during uncomplicated virus infections.

In conclusion, the dissection of the pathological mechanisms underlying the inflammatory reaction seen in patients with MS is crucial to understand the pathogenic process occurring in patients during time, the immunopathological findings associated with disease heterogeneity and, as a consequence, the best therapeutic approach for the disease.

Acknowledgements: We express gratitude to E. Brambilla, E. Clementi, F. Codazzi, L.M.E. Grimaldi, F. Grohovaz, M. Filippi, V. Martinelli, and P. Panina for contributing to the characterization of the cytokine effects on calcium mobilization in multiple sclerosis patients. The work has been in part supported by the Istituto Superiore di Sanità (target project: Multiple Sclerosis) and MURST.

References

1. Cotran RS, Kumar V, Robbins SL (1994) Pathologic basis of disease. WB Saunders, Philadelphia
2. Sacca R, Cuff CA, Ruddle NH (1997) Mediators of inflammation. Curr Opin Immunol 9: 851-857
3. Brouckaert P, Libert C, Everaerdt B et al. (1993) Tumor necrosis factor, its receptors and the connection with interleukin 1 and interleukin 6. Immunobiol 187: 317-329
4. Luster AD (1998) Chemokines–chemotactic cytokines that mediate inflammation. N Engl J Med 338: 436-445

5. Kremlev SG, Chapoval AI, Evans R (1998) Cytokine release by macrophages after interacting with CSF-1 and extracellular matrix proteins: characteristics of a mouse model of inflammatory responses in vitro. Cell Immunol 185: 59-64
6. Abbas AK, Murphy KM, Sher A (1996) Functional diversity of helper T lymphocytes. Nature 383: 787-793
7. Chomarat P, Banchereau J (1997) An update on interleukin-4 and its receptor. Eur Cytok Net 8: 333-344
8. Martin R, McFarland HF, McFarlin DE (1992) Immunological aspects of demyelinating diseases. Annu Rev Immunol 10: 153-187
9. Steinman L (1996) A few autoreactive cells in an autoimmune infiltrate control a vast population of nonspecific cells: a tale of smart bombs and the infantry. Proc Natl Acad Sci USA 93: 2253-2256
10. Lassmann H, Bruck W, Lucchinetti C, Rodriguez M (1997) Remyelination in multiple sclerosis. Mult Scler 3: 133-136
11. Lucchinetti CF, Bruck W, Rodriguez M, Lassmann H (1996) Distinct patterns of multiple sclerosis pathology indicate heterogeneity on pathogenesis. Brain Pathol 6: 259-274
12. Goverman J, Woods A, Larson L et al. (1993) Transgenic mice that express a myelin basic protein-specific T cell receptor develop spontaneous autoimmunity. Cell 72: 551-560
13. Lafaille JJ, Nagashima K, Katsuki M, Tonegawa S (1994) High incidence of spontaneous autoimmune encephalomyelitis in immunodeficient anti-myelin basic protein T cell receptor transgenic mice. Cell 78: 399-408C
14. Esiri MM, Gay D (1997) The immunocytochemistry of multiple sclerosis plaques. In: Raine CS, McFarland HF, Tourtellotte WW (eds) Multiple sclerosis. Clinical and pathogenetic basis. Chapman & Hall, London, pp 173-186
15. Karpus WJ, Ransohoff RM (1998) Chemokine regulation of experimental autoimmune encephalomyelitis: temporal and spatial expression patterns govern disease pathogenesis. J Immunol 161: 2667-2671
16. Battistini L, Borsellino G, Sawicki G et al. (1997) Phenotypic and cytokine analysis of human peripheral blood gamma delta T cells expressing NK cell receptors. J Immunol 159:3723-3730
17. Newcombe J, Li H, Cuzner ML (1994) Low density lipoprotein uptake by macrophages in multiple sclerosis plaques: implications for pathogenesis. Neuropathol Appl Neurobiol 20: 152-162
18. Huterer SJ, Tourtellotte WW, Wherrett JR (1995) Alterations in the activity of phospholipases A2 in post mortem white matter from patients with multiple sclerosis. Neurochem Res 20: 1335-1343
19. Narayana PA, Doyle TJ, Lai D, Wolinsky JS (1998) Serial proton magnetic resonance spectroscopic imaging, contrast-enhanced magnetic resonance imaging, and quantitative lesion volumetry in multiple sclerosis. Ann Neurol 43: 56-71
20. Cuzner ML, Gveric D, Strand C et al. (1996) The expression of tissue-type plasminogen activator, matrix metalloproteases and endogenous inhibitors in the central nervous system in multiple sclerosis: comparison of stages in lesion evolution. J Neuropathol Exp Neurol 55: 1194-1204
21. Li H, Cuzner ML, Newcombe J (1996) Microglia-derived macrophages in early multiple sclerosis plaques. Neuropathol Appl Neurobiol 22: 207-215
22. Selmaj KW, Raine CS, Cannella B, Brosnan CF (1991) Identification of lymphotoxin and tumor necrosis factor in multiple sclerosis lesions. J Clin Invest 87: 949-954
23. Filippi M, Rocca MA, Martino G et al. (1998) Magnetization transfer changes in the normal-appearing white matter precede the appearance of enhancing lesions in patients with multiple sclerosis. Ann Neurol 43: 809-814

24. Soderstrom M, Link H, Sun JB et al. (1994) Autoimmune T cell repertoire in optic neuritis and multiple sclerosis: T cells recognising multiple myelin proteins are accumulated in cerebrospinal fluid. J Neurol Neurosurg Psychiatry 57: 544-551
25. Chou YK, Vainiene M, Whitham R et al. (1989) Response of human T lymphocyte lines to myelin basic protein: association of dominant epitopes with HLA class II restriction molecules. J Neurosci Res 23: 207-216
26. Trotter JL, Hickey WF, van der Veen RC, Sulze L (1991) Peripheral blood mononuclear cells from multiple sclerosis patients recognize myelin proteolipid protein and selected peptides. J Neuroimmunol 33: 55-62
27. Sun J, Link H, Olsson T et al. (1991) T and B cell responses to myelin-oligodendrocyte glycoprotein in multiple sclerosis. J Immunol 146: 1490-1495
28. Olsson T, Zhi WW, Hojeberg B et al. (1990) Autoreactive T lymphocytes in multiple sclerosis determined by antigen-induced secretion of interferon-gamma. J Clin Invest 86:981-985
29. Link H, Sun JB, Wang Z et al. (1992) Virus-reactive and autoreactive T cells are accumulated in cerebrospinal fluid in multiple sclerosis. J Neuroimmunol 38: 63-73
30. Sun JB, Olsson T, Wang WZ et al. (1991) Autoreactive T and B cells responding to myelin proteolipid protein in multiple sclerosis and controls. Eur J Immunol 21:1461-1468
31. Voskuhl RR, Martin R, Bergman C et al. (1993) T helper 1 (Th1) functional phenotype of human myelin basic protein-specific T lymphocytes. Autoimmunity 15: 137-143
32. Oksenberg JR, Panzara MA, Begovich AB et al. (1993) Selection for T-cell receptor V beta-D beta-J beta gene rearrangements with specificity for a myelin basic protein peptide in brain lesions of multiple sclerosis. Nature 362: 68-70
33. Martino G, Consiglio A, Franciotta DM et al. (1997) Tumor necrosis factor α and its receptors (R1 and R2) in relapsing-remitting multiple sclerosis. J Neurol Sci 152:51-61
34. Burns J, Rosenzweig A, Zweiman B, Lisak RP (1983) Isolation of myelin basic protein-reactive T-cell lines from normal human blood. Cell Immunol 81: 435-440
35. Martino G, Grohovaz F, Brambilla E et al. (1998) Proinflammatory cytokines regulate antigen-independent T-cell activation by two separate calcium-signaling pathways in multiple sclerosis patients. Ann Neurol 43: 340-349
36. Mein LE, Hoch RM, Dornmair K et al. (1997) Encephalitogenic potential of myelin basic protein-specific T cells isolated from normal rhesus macaques. Am J Pathol 150: 445-453
37. Boehm U, Klamp T, Groot M, Howard JC (1997) Cellular responses to interferon-gamma. Annu Rev Immunol 15: 749-795
38. Unutmaz D, Pileri P, Abrignani S (1994) Antigen-independent activation of naive and memory resting T cells by a cytokine combination. J Exp Med 180:1159-1164
39. Tough DF, Borrow P, Sprent J (1996) Induction of bystander T cell proliferation by viruses and type I interferon in vivo. Science 272: 1947-1950
40. Martino G, Clementi E, Brambilla E et al. (1994) γ-Interferon activates a previously undescribed Ca^{2+} influx in T lymphocytes from patients with multiple sclerosis. Proc Natl Acad Sci USA 91:4825-4829
41. Dolmetsch RE, Xu K, Lewis RS (1998) Calcium oscillations increase the efficiency and specificity of gene expression. Nature 392: 933-936
42. Trapp BD, Peterson J, Ransohoff RM et al. (1998) Axonal transection in the lesions of multiple sclerosis. N Eng J Med 338: 278-285
43. Vandenbroeck K, Martino G, Marrosu M et al. (1997) Occurrence and clinical relevance of an interleukin-4 gene polymorphism in patients with multiple sclerosis. J Neuroimmunol 76: 189-192

Chapter 19

Magnetic Resonance and Blood-Brain Barrier Dysfunction in Multiple Sclerosis

M. Rovaris[1], C. Tortorella[1], J.C. Sipe[2], M. Filippi[1]

Introduction

Breakdown of the blood-brain barrier (BBB) is an early event in the development of multiple sclerosis (MS) lesions and it is often associated with transvascular inflammatory cell infiltration of the central nervous system (CNS) parenchyma [1]. Because of the frequent association between disruption of the BBB and perivascular lymphocytic infiltration, it has been proposed that BBB damage occurs during the migration of activated T cells into the CNS with the subsequent recognition of antigens and activation of a cascade of cytokine release [2].

Magnetic resonance imaging (MRI) is a sensitive tool for diagnosing and evaluating the dynamics of MS in vivo, including the development and evolution of BBB abnormalities [3-7]. Enhancement on T1-weighted scans after the injection of a standard dose of gadolinium-diethylenetriaminepentacetic acid (Gd-DTPA) is used to evaluate BBB dysfunction [3], although the low pathological specificity of the technique inevitably limits its potential for defining the pathophysiology of the disease [6].

More recently, the introduction of new MR strategies for post-contrast imaging and the application of non-conventional techniques (i.e. magnetization transfer imaging (MTI), magnetic resonance spectroscopy (MRS) and diffusion imaging) have provided further insights into the early stages of MS plaque development, with the potential to increase the pathological specificity of MRI findings [8-11].

The present review outlines the major contributions which may be obtained by conventional and more recent techniques in the study and monitoring of BBB dysfunction in MS.

---

[1] Neuroimaging Research Unit, Department of Neuroscience, Scientific Institute Ospedale San Raffaele, University of Milan, Via Olgettina 60 - 20132 Milan, Italy.
e-mail: m.filippi@hsr.it
[2] Scripps Clinic, Division of Neurology and The Scripps Research Institute, La Jolla, California, USA

Pathological Correlates

Gadolinium-Enhanced MRI

Gd is a rare element of the lanthanide series, with strong paramagnetic properties. When chelated with DTPA, it is widely used as a contrast agent in MRI. Gd administration markedly increases the T1 relaxation time of adjacent mobile water protons, thus producing focal high signal intensity on T1-weighted images in areas where Gd is concentrated in brain tissue [3]. The intensity and size of enhancement depends on the local concentration of Gd, which in turn depends on its intravascular concentration, on the level of BBB permeability and on the size of leakage space [12]. However, the pathophysiological mechanisms underlying Gd leakage are still not completely understood, even if there is some evidence supporting active, energy-dependent, vescicular or microvescicular transport rather than an opening of the tight junctions between adjacent endothelial cells [13-15].

Studies both in animals [15] and in human MS brain biopsy specimens [16, 17] demonstrated that Gd enhancement is consistent with histopathological findings of BBB breakdown. Studies of chronic relapsing experimental allergic encephalomyelitis (CREAE), a model of immune-mediated demyelination developed by Hawkins et al. [15], demonstrated that areas of Gd enhancement seen with MRI corresponded well to areas of BBB breakdown labelled by histochemical markers (i.e. horseradish peroxidase). The same authors [2] showed that the pattern of BBB breakdown in acute and chronic inflammatory demyelination evolves from a diffuse, short-lived disturbance in acute experimental allergic encephalomyelitis (EAE) to a more focal and prolonged breakdown in animals with the chronic relapsing form of the disease. This suggests that prolonged BBB breakdown and inflammation could result in chronic demyelinating lesions [18]. On the other hand, perivascular inflammation appears to be a necessary precondition to the development of enhancement, since non-inflammatory demyelination is unaccompanied by changes in BBB permeability [19, 20].

Recent studies in animals with EAE have shown that Gd enhancement correlates with the number of inflammatory cells within the lesions [21, 22] and mainly represents macrophage activation [23]. Because CREAE and MS present similar morphological and functional changes, it is possible to suggest that enhancement in MS lesions may reflect active inflammation and that immunologically-mediated inflammation is the cause of the initial vascular changes in acute lesions. This hypothesis is supported by evidence that intravenous steroids strongly suppress Gd enhancement, possibly by reducing the activity of a protease (i.e. 92 kDa metalloproteinase) involved in the trafficking of activated T cells across the BBB [24], whereas steroids have no effect on the development of MS lesions on T2-weighted images [25, 26].

Serial MRI scans confirm that enhancement occurs in almost all new lesions in patients with relapsing-remitting (RR) or secondary progressive (SP) MS [27]. Focal areas of increased signal on enhanced images can also be detected before the appearance of lesions on unenhanced T2-weighted scans [6]. Enhancement may reappear in older plaques, with or without a concomitant increase in their size [27], thus suggesting either partial repair of BBB or possible reactivation of BBB abnormalities. The duration of enhancement, as measured by the administration of a standard dose (SD) of Gd, lasts between 4 and 8 weeks for most MS lesions [27-29], even though a shorter enhancing period (less than 1 month) was described in 44% of lesions seen in longitudinal studies with weekly scanning [30, 31].

The heterogeneity of BBB changes is reflected by the different morphological patterns of enhancement, i.e. nodular, patchy or ring-like. It has been suggested that nodular enhancing lesions represent small areas of perivascular inflammation either at the edges of established lesions or in areas of formerly normal appearing white matter (NAWM), whereas ring-enhancing areas are probably areas of acute inflammation at the edge of chronic demyelinated lesions [6, 32]. Studies of the dynamic changes in enhancement patterns over time [6] showed that enhancing lesions change their appearance on scans obtained 15 and 20 minutes after contrast injection compared to scans obtained 2-5 minutes after Gd administration. Most of the lesions changed their pattern from ring to homogeneous enhancement, indicating that the BBB permeability is probaby lower at the centre than in the periphery [6]. Moreover, Kermode et al. [6] reported a gradual increase of the surface area of enhancement over several hours, which could be explained by the extension of oedema beyond the region of active BBB breakdown and the subsequent diffusion of Gd into the extracellular space [6]. Recently, Bruck et al. [33] and van Waesberghe et al. [34] found that ring enhancement is not restricted to reactivation of older MS lesions, but may be the first manifestation of new activity or the evolution of other enhancement patterns, especially in very large plaques.

Some insights about the various pathological substrates of enhancing lesions are provided by the evolution of findings visible on unenhanced T1-weighted images. Most of the enhancing lesions show a corresponding area of hypointensity on pre-contrast T1-weighted images [34], whereas some lesions are isointense to the NAWM. On follow-up unenhanced T1-weighted scans, these lesions can be classified into four categories: (a) persistently isointense lesions, (b) temporarily isointense lesions, (c) persistently hypointense lesions, and (d) temporarily hypointense lesions [34]. Since the degree of hypointensity on T1-weighted scans is mainly related to the extent of extracellular oedema and axonal loss [35], *pattern "a"* may represent purely oedematous or inflammatory lesions, affected by only a minor degree of demyelination. Demyelination associated with mild loss of oligodendrocytes may be the histological phenotype of *pattern "d"*, while persistently hypointense lesions (*pattern "c"*) are probably those in which axonal loss plays a relevant role in determining T1 hypointensity. At a 6-month follow-up [34], 75% of ring-enhancing lesions remain hypointense, thus suggesting that this pattern of enhancement may be a feature of lesions with a more severe central involvement.

Recent strategies in post-contrast imaging have increased the sensitivity of MRI in the detection of enhancing lesions (see below, "clinical correlates") and have provided some insights about the pathological correlates of enhancing lesions. Weekly MRI scanning detected 13% more new enhancing lesions than monthly scanning and demonstrated that almost every new lesion appearing on T2-weighted images shows an initial phase of Gd enhancement [30]. These findings suggest that BBB breakdown is a nearly obligatory event in the early development of new lesions in RR or SP MS [30, 31].

Using a triple dose (TD) of Gd, it has been demonstrated that there are lesions which can be detected only after the administration of TD but not after SD [8, 36, 37], either because of the time course of their BBB leakage – implying that these lesions might be detectable only with TD for part of the inflammatory episode – or because BBB permeability is too restricted for enhancement to be seen on SD. It is known that, soon after the appearance of a new MS lesion, reparative mechanisms are activated [38]. Thus, it is likely that lesions at the beginning or at the end of the inflammatory episode have a less permeable BBB, and that this subtle abnormality can be seen by increasing the intravascular concentration of Gd or by allowing more time for Gd molecules to diffuse [12]. However, in a recent longitudinal MRI study [39], lesions enhancing only after TD had a shorter duration of enhancement, and lesions enhancing after different Gd doses changed their pattern of enhancement on follow-up scans only in the minority of cases (about 15%). In addition, TD lesions have higher MTR values than SD lesions when they start to enhance and the degree of MTR recovery during a three-month follow-up period is higher for the former lesions [8]. This suggests that the extent of BBB opening is correlated to the degree of associated tissue damage and that the magnitude of the two phenomena may be different in different MS enhancing lesions. These findings suggest that enhancing MS lesions form a heterogeneous population and those enhancing only after TD of Gd are characterized by a milder and briefer opening of the BBB.

Magnetization Transfer Imaging

MTI is a recent technique in which image contrast depends on the interactions between protons in a relatively free state and those in a restricted motion state. Adding a magnetization transfer (MT) pulse to enhanced T1-weighted images, the signal of white matter is more suppressed than that of enhancing lesions, since Gd, by shortening proton T1, decreases the effectiveness of the signal suppression induced by an MT pulse [40]. In this way it is possible to obtain a better definition of enhancing lesions, thereby increasing the number of detectable lesions [41, 42].

MTI has been used in preliminary studies aimed at evaluating the pathological characteristics and evolution of enhancing lesions. Lower MT ratio (MTR) values, indicating more pronounced tissue disorganization, were found in: (a) ring-enhancing compared to nodular-enhancing lesions [43-45], (b) lesions enhancing only after the injection of SD Gd compared to those enhancing after TD [8], (c) lesions enhancing on at least two consecutive monthly scans compared to those enhancing on a single scan [46], and (d) T1-weighted hypointense lesions at the time of their initial enhancement compared to isointense areas [34]. Furthermore, two possible evolutions of MTR for new enhancing MS lesions have been described [47]. In some lesions, a moderate decrease of MTR, with a subsequent complete recovery within a few weeks, may reflect early oedema, demyelination and subsequent remyelination [47, 48]. In the other lesions, a marked reduction of MTR with only partial recovery at follow-up may indicate the formation of lesions with severe pathological destruction and subsequent failure of the reparative mechanism [34, 47, 48]. A recent, longitudinal study [34], correlating MT and non-contrast T1-weighted images, described two additional patterns of MTR evolution in enhancing lesions, following an initial marked decrease. The first was rapid restoration of MTR suggesting remyelination, and the second complete failure of restoration of MTR values suggesting concomitant destruction of oligodendrocytes and axons.

Variable changes of MTR may also be detected in the NAWM which is subsequently involved by enhancement [9, 49, 50]. MTR reduction may reflect different but not mutually exclusive pathological substrates. These include (1) increased amounts of products of NAWM degradation [51], (2) increased water content in the hyperplastic astrocytes participating in the demyelinating process [51, 52], and (3) demyelination and remyelination characteristics of the early phase of lesion formation [8, 53-55]. Filippi et al. [8] observed that the greatest reduction of MTR occurs about one month before enhancement. This interval is similar to the time between immune activation and clinical manifestations of the lesions in EAE [56] and it may be the amount of time necessary for immune activation and BBB transmigration of T cells in the early phase of the pathological evolution of MS lesions.

Magnetic Resonance Spectroscopy

MRS allows in vivo identification of local changes in brain chemical composition, which may reflect the pathological evolution of MS lesions [57]. Proton MRS of enhancing lesions [10] has shown that, in some lesions during the first 6-10 weeks following the onset of enhancement, elevated lactate levels can be found and could be due to the concomitant presence of inflammation, local ischaemia and neuronal mitochondrial dysfunction. Spectra from these lesions may also show an elevated choline peak returning to normal over a 4-6 month period and a lipid peak suggesting increased membrane turnover due to continued demyelination [10]. Several recent studies have shown that the lipid signal may persist for up to 4-8 months, with a progressive reduction of its intensity, whereas enhancement always ceased within two months. This suggests that BBB breakdown and demyelination do not always occur simultaneously [10, 57, 58].

Newly-formed MS lesions are also characterized by a reduction of *N*-acetyl aspartate (NAA) peak on MRS, with considerable recovery that begins after 4-6 months [10]. This finding could reflect reversible dysfunction of neuronal mitochondria in acute lesions.

Diffusion-Weighted MRI

Diffusion-weighted MRI allows measurement of the spatial distribution of proton diffusion coefficients within brain tissue [59]. Due to the presence of "restricting barriers" (i.e. cell membranes), the "apparent" diffusion coefficient (ADC) of water in vivo is always lower than its true diffusion coefficient at body temperature; therefore, ADC reflects the structural properties of the cellular compartments for a given tissue.

Diffusion-weighted imaging is still not available on standard MRI scanners because it requires powerful field gradients and generates images which are sensitive to motion artefacts. Some preliminary results have been obtained with echo planar imaging (EPI). Tievsky et al. [11] and Gass et al. [60] found that both acute enhancing and chronic non-enhancing MS lesions have high ADC values, whereas some enhancing lesions show a peripheral rim of restricted diffusion with corresponding low ADC values. The increase in ADC for most of the lesions may reflect expanded extracellular volume, due to oedema (acute phase), demyelination (subacute phase) and axonal loss (chronic phase). A decrease in ADC at the periphery of enhancing lesions may, in turn, reflect accumulation of inflammatory cells. Furthermore, increased ADC was found in NAWM of MS patients in comparison with controls [61]. Diffusion imaging studies with the addition of diffusion tensor information can increase the pathological specificity of these findings [11, 60].

Possible Future Directions

Specialized MRI studies for the diagnosis and management of MS continue to evolve. Currently on the forefront of interest is the possibility of more specific imaging methods that show promise to illuminate the pathophysiology of MS. These methods include specific markers or tracers for various components of the immune system since there is circumstantial evidence that MS is an autoimmune disorder. A recent example is the use of a superparamagnetic iron oxide contrast agent, also known as monocrystalline iron oxide nanoparticles (MION), to label lymphocytes in vitro and in vivo for trafficking studies [62, 63]. This technique has logically been applied in an MRI study of relapsing-remitting EAE in the SJL mouse model using MION intravenously [64]. In this model, MION-enhanced MRI provided unique sensitivity in EAE lesion detection and MRI images were correlated with histopathology. Other possibilities may include antibody-linked paramagnetic molecules that bind to specific cellular components or cytokines produced in the demyelinating lesions.

MRS also has the potential to be developed for specific imaging of spectral shifts designed to track lymphocytes, macrophages, glial cells or compounds thought to be important in the development of MS lesions [65].

Clinical Correlates

Enhanced MRI and MS Clinical Activity

Although enhanced MRI provides limited information about the pathology of MS lesions, this technique is widely used as a marker of disease activity in MS patients [7]. Several studies [4, 5, 66-68] have reported that enhanced brain MRI is 5-10 times more sensitive than clinical monitoring and twice more sensitive than unenhanced MRI in the assessment of disease activity in patients with RR MS, although there is a great degree of both intrapatient and interpatient variability [5]. The contribution of spinal MRI for MS monitoring is not as relevant, since the number of enhancing lesions is about 10 times higher in the brain than in the spinal cord [67]. However, 30%-40% of active lesions in the spinal cord are consistent with ongoing clinical manifestations [67]. MRI activity is different for the different clinical types of MS. Patients with RR MS and SP MS have higher MRI activity, on average about 20 active lesions per year, whereas patients with primary progressive (PP) or benign courses have lower MRI activity, about 3 active and 9 active lesions per year, respectively [28, 66, 69]. The low MRI activity of PP MS patients reflects a less intense degree of inflammation in lesions of these patients, as confirmed by a pathological study [70]. However, recent studies emphasized that the frequency and extent of enhancement is significantly lower in SP MS than in RR MS patients [71, 72], thus suggesting that pathological mechanisms other than BBB disruption with local inflammation/demyelination [1] may play a significant role in determining the development of irreversible tissue damage and, as a possible consequence, progression of clinical disability.

Several studies have found that the number of enhancing lesions increases shortly before and during clinical relapses [4, 73, 74] and correlates with further MRI activity [73, 75]. Koudriavtseva et al. [73] reported in RR MS patients that the presence of enhancing lesions predicts the number of relapses and MRI active lesions in the subsequent six months. In the same study, there was weak correlation between baseline volume of enhancing lesions and change of total lesion volume on unenhanced T2-weighted scans in the subsequent six months. Filippi et al. [72], in a one year follow-up study with monthly MRI scans, found a significant correlation between the number of enhancing lesions and the changes of T2-weighted and MTI lesion load in SP MS but not in RR MS patients. Molyneux et al. [75] noted a correlation between the number of active lesions and changes of T2-weighted lesion load in both RR MS and SP MS patients. Stone et al. [76] found a moderate correlation between the degree of clinical disability and the mean frequency of enhancing lesions over three months in a group of RR MS patients. Losseff et al. [77] showed that in patients with SPMS the number of enhancing lesions detected with monthly MRI scans over six months correlated with clinical progression five years later.

The sensitivity and objectivity of MRI in revealing disease activity makes it an important tool for monitoring the efficacy of treatments that modify the clinical course of MS [7]. The beneficial effect of steroids on BBB disruption leads to a reduction in the duration and severity of MS relapses and has been confirmed by several MRI studies [25, 78, 79] which found a positive, although transient (less than two months, on the average), effect of high-dose intravenous methylprednisolone (IVMP) in reducing the number of enhancing MRI lesions. Several studies [4, 80-83] have shown a high statistical power of both parallel-group and cross-over designed trials monitored with enhanced MRI and, in the recent years, virtually all the large-scale, multicentre phase II and III clinical trials testing the efficacy of several immunomodulating and immunosuppressive agents in MS have been or are using enhanced MRI as a primary (phase II) or secondary (phase III) outcome measure [7, 29]. Results obtained in different trials documented that in RR MS patients interferon (IFN) beta-1a [84-86], IFN-β-1b [83-90] and copolymer-1 [91] significantly reduce both the frequency of relapses [84, 85, 90, 91] and the occurrence of enhancing lesions on serial MRI [84-91]. Therefore, the administration of IFN may temporarily inhibit alteration of the BBB, with a consequent therapeutic benefit. The positive effect of IFN on BBB disruption was confirmed by a recent study [92] in which a concomitant treatment with IFN-β-1a prolonged the decrease in number of enhancing lesions that followed IVMP pulse therapy.

Clinical Applications of New Strategies: Advantages and Limitations

Several strategies have been developed to further increase the sensitivity of enhanced MRI for detecting active lesions in MS, with the ultimate goal to increase the reliability of this technique in monitoring MS disease course [93]. A first strategy tries to maximize the information that can be obtained by conventional scanning by more frequent MRI scanning schedules (i.e. weekly instead of monthly) [30] and an increased delay between Gd injection and image acquisition (20-30 min instead of the conventional 5-7 min) [41, 94]. These strategies only modestly improve (10%-20%) the sensitivity of enhanced MRI, while increasing examination costs, patient discomfort and analysis time. A second strategy aims at increasing the MR signal of enhancing lesions. This method uses a high dose of Gd (0.3 mmol/kg (TD) instead of 0.1 mmol/kg (SD)) [36, 37, 41, 95-97] and acquisition of thin slices [98]. While conflicting results have been reported in PP MS [41, 95], several cross-sectional studies found that TD detects significantly more enhancing lesions and "active" scans than SD in patients with benign, SP and RR MS [96, 36]. In a recent, longitudinal study [37], monthly SD and TD scans were obtained from patients with RR or SP MS for a three-month period. This study showed that serial use of TD-enhanced MRI is safe and confirmed its increased sensitivity in detecting subclinical MS activity. It was also shown that the higher sensitivity of monthly TD MRI may reduce the number of scans needed to show different treatment effects by 30% or more [37, 81]. The last strategy in this category consists of the acquisition of 1 or 3 mm thick slices, instead of the conventional 5 mm [98]. Using 1-mm thick scans, about 25% more enhancing lesions were detected than using 5-mm thick scans [98], but the analysis time was significantly increased, since more than 120 slices were needed to analyze the entire brain. The third strategy increases the likelihood of detecting enhancing lesions by reducing the signal of background tissue with the application of an MT pulse to the conventional enhanced T1-weighted scans [41, 42, 44, 45, 97]. Metha et al. [99] studied 53 enhancing lesions of 10 MS patients with conventional and MT T1-weighted scans before and after contrast injection. MT T1-weighted scans detected 9 lesions not seen on the post-contrast conventional T1-weighted images. One longitudinal [42] and two cross-sectional [41, 97] studies recently compared the sensitivity of MRI obtained after a TD of gadolinium to that of MT T1-weighted scans obtained after SD injection, and found TD to be much more sensitive. The combination of TD, delayed scanning and MT pulse increased detection of the number of enhancing lesions by about 130% [41].

From these observations, it is clear that in using conventional enhanced MRI, a significant amount of ongoing MS activity remains undetected. The contribution of this undetected activity to disease evolution may explain some discrepancies previously found between MRI and other clinical and immunological markers of disease activity [100-103]. Moreover, the increased sensitivity of new enhanced MRI techniques could improve our ability to monitor treatments. This may result in smaller sample sizes and shorter follow-up periods needed to show a treatment effect [37] and in a better understanding of treatment mechanisms of action. For example, using weekly MRI scanning, IFN-β-1b resulted in an immediate reduction of MS activity [86]. However, using monthly SD and TD MRI, Filippi et al. [104] found that the effect of recombinant human IFN-β-1a on enhancing MS lesions varies according to the lesions' intrinsic nature and size, being more significant on small lesions and on those enhancing only after TD of Gd. This suggests that the effect is proportional to the degree of lesion BBB disruption. The combination of serial TD scanning with more frequent scanning schedules might, therefore, be useful to rapidly obtain information about the efficacy of a new experimental drug for MS, although, due to the nature of clinical trials in MS, which are usually multicentre, longitudinal and involving large numbers of patients, newer enhanced MRI strategies can raise difficulties associated with standardization and increased costs.

Conclusions

Conventional enhanced MRI provides some information about the early events of MS lesion formation, but it is a non-specific correlate of increased BBB permeability. Varying patterns of enhancement seem to characterize clinical subgroups of MS patients. The frequency and extent of enhancing lesions are widely used as a reliable marker of ongoing disease activity both in natural history and correlative studies as well as in MRI-monitored MS therapeutic trials.

References

1. McDonald WI, Miller DH, Barnes D (1992) The pathological evolution of multiple sclerosis. Neuropathol Appl Neurobiol 18: 319-334
2. Hawkins CP, Mackenzie F, Tofts P et al. (1991) Patterns of blood brain barrier breakdown in inflammatory demyelination. Brain 114: 801-810
3. Grossman RI, Gonzalez-Scarano F, Atlas SW et al. (1986) Multiple sclerosis: Gadolinium enhancement in MR imaging. Radiology 161: 721-725
4. McFarland HF, Frank JA, Albert PS et al. (1992) Using gadolinium-enhanced magnetic resonance imaging to monitor disease activity in multiple sclerosis. Ann Neurol 32: 758-766
5. Miller DH, Barkhof F, Nauta JJP (1993) Gadolinium enhancement increased the sensitivity of MRI in detecting disease activity in MS. Brain 116: 1077-1094
6. Kermode AG, Tofts P, Thompson AJ et al. (1990) Heterogeneity of blood-barrier changes in multiple sclerosis: an MRI study with gadolinium-DTPA enhancement. Neurology 40: 229-235
7. Filippi M, Miller DH (1996) MRI in the differential diagnosis and monitoring the treatment of multiple sclerosis Curr Opin Neurol 9: 76-186
8. Filippi M, Rocca MA, Rizzo G et al. (1998) Magnetization transfer ratios in multiple sclerosis lesions enhancing after different dose of gadolinium. Neurology 50: 1289-1293
9. Filippi M, Rocca MA, Martino G et al. (1998) Magnetization transfer changes in the NAWM precede the appearance of enhancing lesions in patient with multiple sclerosis. Ann Neurol 43: 809-814
10. Davie CA, Hawkins CP, Barker GJ et al. (1994) Serial proton magnetic spectroscopy in acute multiple sclerosis lesions. Brain 117: 49-54
11. Tievsky AL, Ptak T, Wu O et al. (1997) Evaluation of MS lesions with full tensor diffusion weighted imaging and anisotropy mapping. In: Proceedings of the International Society for Magnetic Resonance in Medicine 1: 666 (abstract)
12. Tofts PS, Kermode AG (1991) Measurement of the blood-brain barrier permeability and leakage space using dynamic MR imaging. I. Fundamental concepts. Magn Reson Med 17: 357-367
13. Brown WJ (1978) The capillaries in acute and subacute multiple sclerosis plaques: a morphometric analysis. Neurology 28: 89-92
14. Bradbury M (1979) The concept of blood brain barrier. John Wiley, Chichester, pp 351-382
15. Hawkins CP, Munro PMG, Mackenzie F et al. (1990) Duration and selectivity of blood brain barrier breakdown in chronic relapsing experimental allergic encephalomyelitis studied by gadolinium-DTPA and protein markers. Brain 113: 365-378
16. Katz D, Taubenberger JK, Raine CS et al. (1990) Gadolinium-enhancing lesions on magnetic resonance imaging: Neuropathological findings. Ann Neurol 28: 243
17. Nesbit GM, Forbes GS, Scheithauer BW et al. (1991) Multiple sclerosis: histopathological and MR and/or CT correlation in 37 cases at biopsy and 3 cases at autopsy. Radiology 180: 467-474
18 Lassman H (1983) Comparative neuropathology of chronic experimental allergic encephalomyelitis and multiple sclerosis. Springer, Berlin Heidelberg New York
19. Lusmden CE (1970) The neuropathology of multiple sclerosis. In: Vinken PJ, Bruyn GW (eds) Handbook of clinical neurology. North-Holland, Amsterdam, pp 217-309
20. Dousset V, Brochet B, Vital A et al. (1995) Lysolecithin-induced demyelination in primates: preliminary in vivo study with MR and magnetization transfer. AJNR Am J Neuroradiol 16: 225-231

21. Seeldrayers PA, Syha J, Morrissey SP et al. (1993) Magnetic resonance imaging investigation of blood-brain barrier damage in adoptive transfer experimental autoimmune encephalomyelitis. J Neuroimmunol 46: 199-206
22. Namer IJ, Steibel J, Piddlesen SJ et al. (1994) Magnetic resonance imaging of antibody-mediated demyelinating experimental allergic encephalomyelitis. J Neuroimmunol 54: 41-50
23. Morrissey SP, Stodal H, Zettl U et al. (1996) In vivo MRI and its histological correlates in acute adoptive transfer experimental allergic encephalomyelitis. Quantification of inflammation and oedema. Brain 119: 239-248
24. Rosemberg GA, Dencoff JE, Correa N et al. (1996) Effects of steroids on CSF matrix metalloproteinase in multiple sclerosis: relation to blood-brain barrier injury. Neurology 46: 1626-1632
25. Barkhof F, Hommes OR, Scheltens P, Valk J (1991) Quantitative MRI changes in gadolinium-DTPA enhancement after high-dose intravenous methylprednisolone in multiple sclerosis. Neurology 41: 1219-1222
26. Burnham JA, Wright RR, Dreisbach J et al. (1991) The effect of high-dose steroids on MRI gadolinium enhancement in acute demyelinating lesions. Neurology 41: 1349-1354
27. Miller DH, Rudge P, Johnson J et al. (1988) Serial gadolinium-enhanced magnetic resonance imaging in multiple sclerosis. Brain 111: 927-939
28. Thompson AJ, Kermode AG, MacManus DG et al. (1990) Patterns of disease activity in multiple sclerosis: clinical and magnetic resonance imaging study. Br Med J 300: 631-634
29. Harris JO, Frank JA, Patronas N et al. (1991) Serial gadolinium-enhanced magnetic resonance imaging scans in patients with early, relapsing-remitting multiple sclerosis: implications for clinical trials and natural history. Ann Neurol 29: 548-555
30. Lai M, Hodgson T, Gawne-Cain M et al. (1996) A preliminary study into the sensitivity of disease activity detection by serial weekly magnetic resonance imaging in multiple sclerosis. J Neurol Neurosurg Psychiatry 60: 339-341
31. Tortorella C, Rocca MA, Codella C et al. (1998) Disease activity in multiple sclerosis studied with weekly triple dose magnetic resonance imaging. Mult Scler 4: 303 (abstract)
32. Prineas JW, Connel F (1978) The fine structure of chronically active multiple sclerosis plaques. Neurology 28(Suppl): 68-75
33. Bruck W, Bitsch A, Kolenda H et al. (1997) Inflammatory central nervous system demyelination: correlation of magnetic resonance imaging findings with lesion pathology. Ann Neurol 42: 783-793
34. van Waesberghe JHTM, van Walderveen MAA, Castelijns JA et al. (1998) Patterns of lesion development in multiple sclerosis: longitudinal observations with T1-weighted spin-echo and magnetisation transfer MR. AJNR Am J Neuroradiol 19: 675-683
35. Barnes D, Munro PMG, Youl BD et al. (1991) The longstanding MS lesion. A quantitative MRI and electron microscopy study. Brain 114: 1271-1280
36. Filippi M, Yousry T, Campi A et al. (1996) Comparison of triple dose versus standard dose gadolinium-DTPA for detection of MRI enhancing lesions in patients with MS. Neurology 46: 379-384
37. Filippi M, Rovaris M, Capra R et al. (1998) A multi-centre longitudinal study comparing the sensitivity of monthly MRI after standard and triple dose gadolinium-DTPA for monitoring disease activity in multiple sclerosis: implications for clinical trials. Brain 121: 2011-2020

38. Prineas JW, Barnard RO, Kwon EE et al. (1993) Multiple sclerosis: remyelination of nascent lesions. Ann Neurol 33: 137-151
39. Rovaris M, Mastronardo G, Gasperini C et al. (1998) MRI evolution of new MS lesions enhancing after different doses of gadolinium. Acta Neurol Scand 98: 90-93
40. Tanttu JI, Sepponem RE, Lipton MJ et al. (1992) Synergistic enhancement of MRI with Gd-DTPA and magnetization transfer. J Comput Assist Tomogr 16: 19-24
41. Silver NC, Good CD, Barker GJ et al. (1997) Sensitivity of contrast enhanced MRI in multiple sclerosis: effects of gadolinium dose, magnetisation transfer contrast and delayed imaging. Brain 120: 1149-1161
42. Gasperini C, Bastianello S, Pozzilli C et al. (1997) A multicentre study comparing the sensitivity of T1-weighted images with and without magnetization transfer after the injection of standard and triple dose of gadolinium in detecting enhancing lesions in MS. J Neurol 244(Suppl 3): S24 (abstract)
43. Campi A, Filippi M, Comi G et al. (1996) Magnetization transfer ratios of enhancing and non-enhancing lesions in multiple sclerosis. Neuroradiology 38: 115-119
44. Petrella JR, Grossman RI, McGowan JC et al. (1996) Multiple sclerosis lesions: relationship between MR enhancement pattern and magnetization transfer effect. AJNR Am J Neuroradiology 17: 1041-1049
45. Hiehle JF, Grossman RI, Ramer NK et al. (1995) Magnetization transfer effect in MR-detected multiple sclerosis lesions: comparison with gadolinium-enhanced spin-echo images and non-enhanced T1-weighted images. AJNR Am J Neuroradiol 16: 69-77
46. Filippi M, Rocca MA, Comi G (1998) Magnetization tranfer ratios of multiple sclerosis lesions with variable durations of enhancement. J Neurol Sci 159: 162-165
47. Dousset V, Brochet B, Gayou A et al. (1995) Magnetization transfer profile of demyelinating CNS lesions. In: Proceedings of the Society for Magnetic Resonance in Medicine 1: 114 (abstract)
48. Alonso J, Rovira A, Cucurella MG et al. (1997) Serial magnetization transfer imaging in multiple sclerosis lesions. In: Proceedings of the International Society for Magnetic Resonance in Medicine 1: 639 (abstract)
49. Goodkin DE, Rooney W, Sloan R et al. (1998) PD, T1, Gadolinium (Gd$^+$) intensities, T2, and MTRs are chronically diffusely abnormal in MS brain and on monthly MRI scans are related to the appearance of new Gd$^+$ lesions in NAWM. Neurology 50 (Suppl 4): 191 (abstract)
50. Pike GB, De Stefano N, Narayana S et al. (1998) A longitudinal study of magnetization transfer in multiple sclerosis. In: Proceedings of the International Society for Magnetic Resonance in Medicine 1: 122 (abstract)
51. Allen IV, McKeown SR (1979) A histological, histochemical and biochemical study of the macroscopically normal white matter in multiple sclerosis. J Neuurol Sci 41: 81-91
52. McKeown SR, Allen IV(1978) The cellular origin of lysosomal enzymes in the plaque in multiple sclerosis: a combined histological and biochemical study. Neuropathol Appl Neurobiol 4: 471-482
53. Lexa FJ, Grossman RI, Rosenquist AC (1994) MR of wallerian degeneration in the feline visual system: characterization by magnetization transfer rate with histopathological correlation. AJNR Am J Neuroradiology 15: 201-212
54 Dousset V, Brochet B, Vital A et al. (1994) MR imaging including diffusion and magnetization transer in chronic relapsing experimental encephalomyelitis: correlation with immunological and pathological data. Proc Soc Magn Reson 1: 483 (abstract)
55. Thorpe JW, Barker GJ, Jones SJ et al. (1995) Quantitative MRI in optic neuritis: correlation with clinical indings and electrophysiology. J Neurol Neurosurg Psychiatry 59: 487-492

56. Cambi F, Lees MB, Williams RM et al. (1983) Chronic experimental encephalomyelitis produced by bovine proteolipid apoprotein: immunological studies in rabbits. Ann Neurol 13: 303-308
57. Grossman RI, Lenkinski RE, Ramer KN et al. (1992) MR proton spectroscopy in multiple sclerosis. AJNR Am J Neuroradiol 13: 1535-1543
58. Narayana PA, Wolinsky JS, Jackson EF, McCarthy M (1992) Proton MR spectroscopy of gadolinium-enhanced multiple sclerosis plaques. J Magn Reson Imaging 2: 263-270
59. Le Bihan D (1988) Separation of diffusion and perfusion in intravoxel incoherent motion (IVIM) MR imaging. Radiology 168: 497-505
60. Gass A, Gaa J, Schreiber W et al. (1997) Echo planar diffusion weighted magnetic resonance imaging in patients with active multiple sclerosis. In: Proceedings of the International Society for Magnetic Resonance in Medicine 1: 658 (abstract)
61. Horsfield MA, Lai M, Webb SL et al. (1996) Apparent diffusion coefficients in benign and secondary progressive multiple sclerosis by nuclear magnetic resonance. Magn Reson Med 36: 393-400
62. Yeh TC, Zhang W, Ildstad ST, Ho C (1995) In vivo dynamic MRI tracking of rat T-cells labelled with superparamagnetic iron-oxide particles. Magn Reson Med 33: 200-208
63. Schoepf U, Marecos EM, Melder EJ et al. (1998) Intracellular magnetic labeling of lymphocytes for in vivo trafficking studies. Biotechniques 24: 642-651
64. Xu S, Jordan EK, Brocke ES et al. (1998) Study of relapsing remitting experimental allergic encephalomyelitis SJL mouse model using MION-46L enhanced in vivo MRI: early histopathological correlation. J Neuroscience Res 52: 549-558
65. Dingley AJ, Veale MF, King NJ, King GF (1994) Two-dimensional 1H NMR studies of membrane changes during the activation of primary T lymphocytes. Immunomethods 4: 127-138
66. Thompson AJ, Miller DH, Youl BD et al. (1992) Serial gadolinium-enhanced MRI in relapsing/remitting multiple sclerosis of varying disease duration. Neurology 42: 60-63
67. Thorpe JW, Kidd D, Moseley IF et al. (1996) Serial gadolinium-enhanced MRI of the brain and spinal cord in early relapsing-remitting multiple sclerosis. Neurology 46: 373-378
68. Barkhof F, Scheltens P, Frequin STMF et al. (1992) Relapsing-remitting multiple sclerosis: Sequential enhanced MR imaging vs clinical findings in determining disease activity. AJR Am J Roentgenol 159: 1041-1047
69. Thompson AJ, Kermode AG, Wicks D et al. (1991) Major differences in the dynamics of primary and secondary progressive multiple sclerosis. Ann Neurol 29: 53-62
70. Revesz T, Kidd D, Thompson AJ et al. (1994) A comparison of the pathology of primary and secondary progressive multiple sclerosis. Brain 117: 759-765
71. Filippi M, Rossi P, Colombo B et al. (1997) Serial contrast-enhanced MR in patients with multiple sclerosis and varying levels of disability. AJNR Am J Neuroradiol 18: 1549-1556
72. Filippi M, Rocca MA, Horsfield MA, Comi G (1998) A one year study of new lesions in multiple sclerosis using monthly gadolinium enhanced MRI: correlations with changes of T2 and magnetization transfer lesion loads. J Neurol Sci 158: 203-208
73. Koudriavtseva T, Thompson AJ, Fiorelli M et al. (1997) Gadolinium enhanced MRI disease activity in relapsing-remitting multiple sclerosis. J Neurol Neurosurg Psychiatry 62: 285-287
74. Smith ME, Stone LA, Albert PS et al. (1993) Clinical worsening in multiple sclerosis is associated with increased frequency and area of gadopentetate dimeglumine-enhancing magnetic resonance imaging lesions. Ann Neurol 33: 480-489

75. Molyneux PD, Filippi M, Barkhof F et al. (1998) Correlations between monthly enhanced MRI lesion rate and changes in T2 lesion volume in multiple sclerosis. Ann Neurol 43: 332-339
76. Stone LA, Smith E, Albert PS et al. (1995) Blood-brain barrier disruption on contrast-enhanced MRI in patients with mild relapsing-remitting multiple sclerosis: relationship to course, gender and age. Neurology 45: 1122-1126
77. Losseff N, Kingsley D, McDonald WI et al. (1996) Clinical and magnetic resonance imaging predictors in primary and secondary progressive MS. Mult Scler 1: 218-222
78. Gasperini C, Pozzilli C, Bastianello S et al. (1997) The influence of clinical relapses and steroid therapy on the development of Gd-enhancing lesions: a longitudinal MRI study in relapsing-remitting patients. Acta Neurol Scand 95: 201-207
79. Miller DH, Thompson AJ, Morrissey SP et al. (1992) High dose steroids in acute relapses of multiple sclerosis: MRI evidence for a possible mechanism of therapeutic effect. J Neurol Neurosurg Psychiatry 55: 450-453
80. Truyen L, Barkhof F, Tas M et al. (1997) Specific power calculations for magnetic resonance imaging (MRI) in monitoring active relapsing-remitting multiple sclerosis (MS): implications for phase II therapeutic trials. Mult Scler 2: 283-290
81. Nauta JJP, Thompson AJ, Barkhof F, Miller DH (1994) Magnetic resonance imaging in monitoring the treatment of multiple sclerosis patients: statistical power of parallel-groups and crossover designs. J Neurol Sci 122: 6-14
82. Sormani MP, Molyneux PD, Gasperini C et al. (1999) Statistical power of MRI-monitored trials in multiple sclerosis: new data and comparison with previous results. J Neurol Neurosurg Psychiatry (*in press*)
83. Tubridy N, Ader HJ, Barkhof F et al. (1998) Exploratory treatement trials in multiple sclerosis using MRI: sample size calculations for relapsing remitting and secondary progressive subgroups using placebo controlled parallel groups. J Neurol Neurosurg Psychiatry 64: 50-55
84. Jacobs DL, Cookfair DL, Rudick RA et al. (1996) Intramuscular interferon beta-1a for disease progression in relapsing multiple sclerosis. Ann Neurol 39: 285-294
85. Pozzilli C, Bastianello S, Koudriavtseva T et al. (1996) Magnetic resonance imaging changes with recombinant human interferon beta-1a: a short term study in relapsing-remitting multiple sclerosis. J Neurol Neurosurg Psychiatry 61: 251-258
86. Simon JH, Jacobs LD, Campion M et al. (1998) Magnetic resonance studies of intramuscular interferon beta-1a for relapsing multiple sclerosis. Ann Neurol 43: 79-87
87. Calabresi P, Stone LA, Bash CN et al. (1997) Interferon beta results in immediate reduction of contrast-enhanced MRI lesions in multiple sclerosis patients followed by weekly MRI. Neurology 48: 1446-1448
88. Paty DW, Li DKB, UBC MS/MRI Study Group, IFNB Multiple Sclerosis Study Group (1993) Interferon beta-1b is effective in relapsing-remitting multiple sclerosis. II. MRI analysis results of a multicenter, randomized, double-blind, placebo-controlled trial. Neurology 43: 662-667
89. Stone LA, Frank JA, Albert PS et al. (1995) The effect of interferon-beta on blood-brain barrier disruptions demonstrated by contrast-enhanced magnetic resonance imaging in relapsing-remitting multiple sclerosis. Ann Neurol 37: 611-619
90. The IFNB Multiple Sclerosis Study Group, the University of British Columbia MS/MRI Analysis Group (1995) Interferon beta-1b in the treatment of multiple sclerosis: final outcome of the randomized controlled trial. Neurology 45: 1277-1285
91. Mancardi GL, Sardanelli F, Parodi RC et al. (1998) Effect of copolymer-1 on serial gadolinium enhanced MRI in relapsing remitting multiple sclerosis. Neurology 50: 1127-1133

92. Gasperini C, Pozzilli C, Bastianello S et al. (1998) Effects of steroids on Gd-enhancing lesions before and during recombinant beta interferon 1a treatment in relapsing-remitting multiple sclerosis. Neurology 50: 403-406
93. Barkhof F, Filippi M, Miller DH et al. (1997) Strategies for optimizing MRI techniques aimed at monitoring disease activity in multiple sclerosis. J Neurol 244: 76-84
94. Filippi M, Yousry T, Rocca MA et al. (1997) Sensitivity of delayed gadolinium-enhanced MRI in multiple sclerosis. Acta Neurol Scand 95: 331-334
95. Filippi M, Campi A, Martinelli V et al. (1995) Comparison of triple dose versus standard dose gadolinium-DTPA for detection of MRI enhancing lesions in patients with primary progressive multiple sclerosis. J Neurol Neurosurg Psychiatry 59: 540-544
96. Filippi M, Capra R, Campi A et al. (1996) Triple dose of gadolinium-DTPA and delayed MRI in patients with benign multiple sclerosis. J Neurol Neurosurg Psychiatry 60: 526-530
97. van Waesberghe JHTM, Castelijns JA, Roser W et al. (1997) Single dose gadolinium with magnetization transfer contrast versus triple dose gadolinium in detecting enhancing multiple sclerosis lesions. AJNR Am J Neuroradiol 18: 1279-1285
98. Filippi M, Yousry T, Horsfield MA et al. (1996) A high-resolution three-dimensional gradient echo sequence improves the detection of disease activity in multiple sclerosis. Ann Neurol 40: 901-907
99. Metha RC, Pike BG, Enzmann DR (1995) Improved detection of enhancing and non-enhancing lesions of multiple sclerosis with magnetization transfer. AJNR Am J Neuroradiol 16: 1771-1778
100. Hartung HP, Reiners K, Archelos JJ et al. (1995) Circulating adhesion molecules and tumor necrosis factor receptor in multiple sclerosis: correlation with magnetic resonance imaging. Ann Neurol 38: 186-193
101. Martino G, Filippi M, Martinelli V et al. (1996) Clinical and radiological correlates of a novel T lymphocyte gamma-interferon-activated Ca^{2+} influx in patients with relapsing-remitting multiple sclerosis. Neurology 46: 1416-1421
102. Rieckmann P, Albrecht M, Kitze B et al. (1995) Tumor necrosis factor-alpha messenger RNA expression in patients with relapsing-remitting multiple sclerosis is associated with disease activity. Ann Neurol 37: 82-88
103. Rieckmann P, Altenhofen B, Riegel A et al. (1997) Soluble adhesion molecules (sVCAM-1 and sICAM-1) in cerebrospinal fluid and serum correlate with MRI activity in multiple sclerosis. Ann Neurol 41: 326-333
104. Filippi M, Rovaris M, Capra R et al. (1998) Serial standard- and triple dose to monitor the effect of interferon β-1a on multiple sclerosis activity. Neurology 50(Suppl 4): A323 (abstract)

Chapter 20

Immunotherapies for Multiple Sclerosis

P. Perini, P. Gallo

Introduction

Multiple sclerosis (MS), a chronic inflammatory demyelinating disease of the central nervous system (CNS), is currently regarded as an organ-specific, T-cell-mediated autoimmune disease. Although the immunopathogenesis of MS is largely hypothetical and several immune-mediated processes may account for myelin damage, it is widely accepted that the autoimmune process that leads activated, antigen-specific CD4$^+$ T cells to migrate into the brain and initiate inflammation starts in the peripheral immune system. If autoreactive T lymphocytes play a pivotal role in initiating the disease, then selective immunotherapeutic strategies have to be designed to delete these cells by apoptotic mechanism(s) or to block their activation by inducing a state of anergy or suppression. It is a common feeling that the rapid progress in biotechnology and immunology will make available, within a few years, many potential treatments for MS. This article is a concise overview of only *some* of the immunotherapies, mainly based on the antigen-specific modulation of the immune response, elaborated in recent years. Most of the immune-selective therapies here described have been designed for and successfully applied to prevent, suppress and/or treat the animal model of inflammatory autoimmune demyelinating disease, the experimental allergic encephalomyelitis (EAE). Some of these therapies have been also applied to human pathology, and will be herewith described with more details. However, many unresolved crucial points have to be considered regarding the application of immunotherapies to MS, including the lack of a specific or unique MS antigen, the relevance of the Th1/Th2 balance in controlling the activity and progression of human autoimmune pathology, the role played by intramolecular and intermolecular epitope/antigen spreading in conditioning the chronic evolution of MS, the role played by exogenous antigens in the activation of autoreactive T cells (i.e. molecular mimicry, superantigen activation) and finally, the implication of several genes - some of them codifying immunological factors - in MS susceptibility. Moreover, any therapeutical strategy for MS should be: based on solid scientific evidence, clinically acceptable (with mild side effects), specific and selective (capable to induce a highly restricted immunosuppression), inducing long-lasting therapeutic benefits and, finally, able to shut off ongoing disease (i.e. prevent relapses). We would like to stress that none of the approaches herewith described fulfill all of these criteria.

Immunotherapies Targeting MHC Molecules

The assembly of the trimolecular complex (TMC), formed by the interaction of major histocompatibility complex (MHC) molecules with antigen peptide and antigen-specific T cell receptor (TCR) confers specificity to the immune response. Accordingly, since all three components can be targeted by highly selective immunological interventions, the TMC is considered the crucial target for antigen-specific (antigen-restricted) immunotherapies.

The MHC molecule that presents autoantigen is the least variable component of the TMC. Monoclonal antibodies (MAb) to MHC molecules [1] or to the MHC-antigen complex [2], MHC-binding "competitor" peptides [3,4], and soluble HLA-DR-myelin complexes [5] have been suggested to interfere with the onset of autoimmune diseases associated with certain MHC alleles. Indeed, the murine model of EAE has been successfully treated or fully prevented with (1) monoclonal antibodies directed to MHC molecules [1], (2) synthetic competitor peptides that bind "specific" MHC sequences and block autoantigen recognition, thus suppressing disease development (with disease inhibition being dependent on the relative MHC binding affinity of the disease-inducing peptide) [6], and (3) autoantibodies to class II MHC molecules generated by vaccination with peptides mimicking hypervariable regions of MHC class II beta chain [7].

As mentioned previously, an interesting approach to inhibiting TMC assembly is to administer specific peptides complexed to the complementary sequence of MHC molecules [8-10]. Actually, administration of complexes composed of soluble I-A^s and peptides of either myelin basic protein (MBP$_{91-103}$) or proteolipid protein (PLP$_{139-151}$) prevented EAE development in SJL/J mice [9].

A phase I clinical trial is currently in progress to test the ability of a soluble complex of HLA-DR2 and MBP$_{84-102}$ (the dominant MBP peptide in humans) (Anergix, Anergen) to tolerize autoreactive T cells. Preliminary data suggest that these complexes have a longer half-life than do soluble antigens (hours vs. minutes), have a good distribution, and are well tolerated by patients. In a previous open trial on 11 patients, a peptide-specific response and a decrease in MBP-reactive cell numbers were observed. Several recombinant HLA-DR2-MBP peptide complexes are also undergoing preclinical testing in order to define the route of administration and the delivery system, to improve both the bioavailability and the plasma half-life of the peptides, and to define more appropriately the specificity of this therapy [5].

Immunotherapies Targeting the TCR

In the murine EAE model, restricted TCR usage by myelin-reactive $CD4^+$ lymphocytes has been demonstrated. Accordingly, it is possible to prevent EAE development and progression by vaccination with autoreactive T cell clones or TCR peptides. This approach induced regulatory/suppressor (anti-clonotypic and anti-activated) T cells [11-14]. Phase I and II clinical trials based on T cell or TCR peptide vaccination have been carried out in MS patients. The TCR may be targeted also with monoclonal antibodies (see the section Monoclonal Antibody-Mediated Immunotherapies).

T Cell Vaccination

The normal human T cell repertoire contains potentially self-reactive T lymphocytes that are normally suppressed by counter-regulatory mechanisms. In autoimmune disease suppressive mechanisms are lost, so that autoaggressive T cells are activated. One possible way to re-establish self-tolerance is by vaccination with in-vitro-inactivated, autoantigen-specific $CD4^+$ T cell clones [15]. The first evidence of efficacious T cell vaccination against EAE was obained about twenty years ago [16, 17].

In a phase I trial, six patients were inoculated with MBP-specific T cells cloned from their own blood. MBP-reactive T cell clones were selectively activated in vitro, then inactivated by irradiation, and finally re-injected into the patients [18]. The treatment was safe and well tolerated, but the patient number did not allow any clinical conclusion. A $CD8^+$ anti-clonotypic T cell response to the inoculated MBP-specific T cell clones and the depletion of MBP-specific cells were observed in the vaccinated MS patients. In the majority of these patients, MBP-reactive T cells remained undetectable in circulation over a period of 1-3 years after vaccination, while they reappeared in some individuals, coinciding with clinical exacerbation. Magnetic resonance imaging (MRI) showed a mean 8% increase in brain lesion size in the vaccinated patients compared with 39.5% increase in the controls [19]. Interestingly, the reappearing MBP-reactive T cells were of different clonal origin from those analyzed before immunization, thus suggesting a shift of the T cell repertoire to other MBP determinants. Analysis of the T cell response to the vaccines disclosed that the anti-clonotypic response elaborated by treated patients was formed by $CD8^+$ MHC class I-restricted cytotoxic (i.e. capable of lysing the immunizing clones in a clonotype-specific manner) lymphocytes that recognized the hypervariable regions of the TCR. Moreover, the response was restricted to the immunizing clones and did not affect the MBP-reactive clones not used for immunization [20].

Major limits of this therapeutic approach are: (1) the human T cell response to a potential encephalitogenic peptide can be extremely complex [21, 22] and multiple TCR are used to recognize an immunodominant epitope [23]; (2) the functional characteristics that T cell clones exhibit in vitro may dramatically differ from those they exert in vivo, as suggested by the "three monkey experiment" performed in Wekerle's laboratories: only one of three MBP-specific T cell clones was encephalitogenic in vivo when reinjected into the autologous monkey [24]; (3) several potential autoantigens, other than MBP, may be involved in MS immunopathogenesis (e.g. myelin oligodendrocyte glycoprotein (MOG) and PLP) [25-28], and their immunogenic potentials need to be further investigated [29], and (4) both the target antigens and the repertoire of encephalitogenic T lymphocytes may vary significantly over time [30-32].

TCR Peptide Vaccination

Regulatory T and B cells are naturally induced to recognize unique determinants, or idiotypes, located preferentially within the hypervariable regions of the TCR sequence, including the complementary-determining regions (CDRs) that interact with the MHC-antigen complex to form the TMC [33, 34]. Regulatory, anti-idiotypic T cells may recognize naturally processed TCR peptides expressed on the T cell surface in association with MHC molecules [35]. EAE studies have been performed to identify the potentially immunogenic TCR idiotypes and to test the hypothesis that TCR sequences could induce anti-idiotypic regulation. Vaccination of rats with TCR Vβ8.2-CDR2 peptide induced anti-idiotypic T cells and antibodies that inhibited the activation of pathogenic T cells, and prevented both development and progression of EAE [36, 37].

Based on preliminary studies which showed (1) the overexpression of TCR Vβ5.2 and Vβ6.1 by MBP-specific T cells from the peripheral blood and the plaques of demyelination [38-40], and (2) the increased frequency of TCR peptide-specific T cells in MS patients treated by intradermal vaccination with CDR2 peptides from Vβ5.2 and Vβ6.1 TCR [41], a double-blind pilot trial was carried out in MS patients [42]. Twenty three HLA-DRB1*1501* patients with chronic progressive MS were treated for 12 months with native Vβ5.2$_{38-58}$ peptide (or with a 1:1 mixture of the native peptide and Y49T-substituted peptide, i.e., threonine instead of tyrosine at position 49), 100 μg weekly for 4 weeks, and then monthly for an additional 10 months, for a total of 14 injections. Patients who responded to vaccine showed a boosted frequency of TCR peptide-specific T cells, reduced frequency of MBP-specific T cells, and remained clinically stable without side effects during one year of therapy. Nonresponders had an increased MBP response and progressed clinically. Interestingly, TCR-reactive T cells were predominantly T helper 2 (Th2)-like, and directly inhibited MBP-specific Th1 responses in vitro, primarily through the release of interleukin-10 (IL-10) and interleukin-4 (IL-4), but not transforming growth factor beta (TGF-β) [42]. Therefore, synthetic peptides may activate an anti-idiotypic regulatory network directed at TCR determinants, and downregulate Th1 cell activation, proliferation and cytokine production in vivo in MS patients. At present, TCR peptide vaccination seems limited by the same problems previously outlined for T cell vaccination.

A novel DNA-based vaccination technique is at present being developed; two different methodologies have been explored. In the first, an antigen-encoding

DNA fragment was incorporated into appropriate vectors: EAE induction was significantly suppressed by immunizing animals with recombinant vaccinia virus expressing a $V\beta_{8.2}$ gene [43]. In the second, the antigen-encoding DNA fragment was used as a naked DNA vaccine: vaccination with DNA encoding a variable region of the TCR was capable of preventing EAE and activating a Th2-like immunological response [44].

Anergy Induction by Specific Neuroantigens, Altered Peptide Ligands Glatiramer Acetate (Cop-1)

A promising strategy for inducing a state of antigen-specific anergy is to administer native or modified myelin antigens in four different forms: (1) free antigen, (2) antigen coupled to autologous cells, (3) antigen-MHC complexes (see section Immunotherapies Targeting MHC Molecules), and (4) "partially agonistic" altered peptide ligands (APL). These strategies have all been successfully applied to prevent, suppress and/or treat EAE.

Intravenous administration of soluble antigens can lead to tolerogenic signals [45], while immunization of mice with MBP/MBP-peptides and incomplete Freund's adjuvant (IFA) can prevent EAE development and ameliorate ongoing paralysis [46-49]. The basic principle of this phenomenon is that antigen recognition in the absence of costimulatory signals, provided by professional antigen-presenting cells (APC), induces a state of anergy instead of a state of activation in antigen-specific T lymphocytes. However, a phase I clinical trial aimed at inducing tolerance to MBP was unsuccessful and some patients became sensitized after repeated MBP injections [50].

Whole spinal cord homogenate, MBP, PLP, and peptides derived from myelin antigens coupled to syngeneic splenocytes have also inhibited EAE [51-55]. However, the critical requirement for ethylenecarbodiimide (ECDI) in the coupling of splenocytes with myelin antigens to induce tolerance was interpreted as the ability of ECDI to inhibit expression of APC-derived costimulatory signals, thus leading to Th1 anergy [56].

Autoimmune responses may be regulated by APL. In SJL/J (I-A^u) mice, two synthetic modified peptides of MBP, N_{1-20} and AcN_{9-20}, which competed with the encephalitogenic peptide Ac_{1-11}, prevented EAE development and suppressed in vitro proliferative responses of T cells to the encephalitogenic peptide [57]. These findings suggest that the mechanism of protection consists in the competition between natural and modified peptides for MHC molecules. In $PLP_{139-151}$-induced EAE, the disease was mediated by Th1 lymphocytes which showed diversity in their TCR usage but which all recognized tryptophan-$_{144}$ and histidine-$_{147}$ as the primary and secondary TCR contact points, respectively [58]. A panel of altered self-peptides generated by a single (Q144) or double substitution (L144/R147) in the primary and secondary TCR contact points of $PLP_{139-151}$ has been tested in experimental models. Immunization with the modified peptides preferentially induced Th2-like (IL-4 and IL-10 secreting) and Th0-like (interferon gamma

(IFN-γ) and IL-10 secreting) cells, and prevented the induction of EAE not only by native PLP molecule, but also by other myelin antigens, such as MBP and MOG [59]. These data demonstrate that single or double amino-acid substitutions in an autoantigenic peptide significantly influence T cell differentiation, and suggest a novel mechansim by which altered self peptides (e.g. a T cell receptor antagonist peptide) inhibit autoimmune disease. In a recently concluded phase I trial, an altered peptide designed around an immunodominant epitope of MBP was safe and well tolerated by MS patients [60].

Glatiramer acetate, also known as copolymer 1 (Cop-1, Copaxone), is the acetate salt of a random polymer of four amino acids (L-alanine, L-lysine, L-glutamic acid, and L-tyrosine at a molar ratio of 6.0:4.7:1.9:1.0) [61]. Cop-1 suppressed both acute monophasic and chronic relapsing EAE in various animal species. More recently, the suppressive effect of Cop-1 was found not to be restricted to MBP-induced EAE, but also to be extended to EAE induced by other myelin antigens, such as PLP [62, 63] or MOG [64]. Although the mechanism of Cop-1 action has not yet been elucidated, competition for the peptide-binding site of the class II MHC antigen seems plausible [65]. This results in the inhibition of antigen-specific T-cell activation by interference with antigen presentation [66, 67], thus including Cop-1 in the group of MHC competitors. However, other possible therapeutic effects include the activation and expansion of antigen-specific regulatory T cells and the immune deviation from a Th1 to a Th2 immune response [2]. However, two initial pilot trials were not definitely persuasive on the clinical efficacy of Cop-1 in relapsing-remitting MS (RR MS) patients [68, 69]. In a recent two-year double-blind trial carried out in the USA on RR MS patients, Cop-1 induced a 29% reduction in relapse rate (1.19 vs. 1.68 in the placebo group) [70], but MRI monitoring showed no significant inhibition of disease activity in patients treated with Cop-1 compared to placebo [71].

In summary, neuroantigen peptide therapy can be used to prevent, suppress, and treat EAE in various animal species [3, 55, 72-80]. Peptide can be administered to mice as non-immunogenic native peptide, as altered peptide ligand, oras peptide coupled with either MHC molecules or syngeneic cells. The mechanisms involved in EAE inhibition can be one or more of the following: competition with the MHC or the TCR binding site, immune deviation from a Th1 to a Th2 response, deletion of specific antigen-reactive T cells, or bystander (non-antigen specific) suppression.

Many variables have to be taken into consideration when examining the possible clinical application of peptide-based therapies to human diseases. For instance, when specific neuroantigens are administered in soluble form, the following factors have to be evaluated: the dose (low dose: suppression; high dose: apoptosis), the form of the antigen used (water soluble or insoluble), the route of administration (subcutaneous, intramuscular or intravenous), the timing of antigen delivery with respect to the natural history and course of the disease (before immunization, before the appearance of clinical signs, during active disease), the use of adjuvants or concomitant immune stimulatory or suppressive agents (cytokines, steroids), whether sensitization might occur with repeated treatment, and finally, the development of neutralizing antibodies to the administered peptide (for a critical review on neuroantigen peptide-based therapy, see [81]).

Oral Tolerance

Proteins that pass through the gastrointestinal tract generate antigen-specific systemic hyporesponsiveness [82]. By this way, the mucosal immune system (gut-associated lymphoid tissue, GALT) may adsorb foreign peptides without becoming sensitized to them. In the EAE model, if mice are fed MBP and then immunized, the immune response to MBP is reduced and the disease is abrogated [83-86]. Interestingly, the mechanism reponsible for the antigen-specific hyporesponsiveness may change depending on the amount of antigen fed: a low dose favors active suppression by inducing regulatory cells, while a high dose favors anergy or apoptosis [87, 88]. Oral tolerance is being tested clinically in some human organ-specific autoimmune diseases, such as rheumatoid arthritis, uveitis, thyroiditis, juvenile diabetes, in neurological diseases such as MS and myasthenia gravis, and in transplantation (for a review see [82]). The primary goal of this therapeutic strategy is the generation of regulatory T lymphocytes that migrate from gut to brain and release anti-inflammatory cytokines locally. Indeed, experimental findings suggest that oral administration of antigen preferentially induces Th2-like T cells in the mucosal-associated lymphoid tissue. These cells, which produce IL-4, IL-10 and large amounts of TGF-β and are triggered in an antigen-specific fashion, migrate to the target organ where they mediate antigen-nonspecific bystander suppression [89]. Therefore, attractive advantages of this therapy are that it is not necessary to know precisely the target antigens and that a mixture of antigens from the target organ (e.g. all myelin) can be administered to obtain the same final result, i.e. the delivery of anti-inflammatory cytokines to the selected organ.

However, while a preliminary trial in RR MS patients suggested a clinical effect on the number of relapses [90], a phase II/III pivotal multicenter clinical trial failed to show any benefit over placebo. New trials in MS are being planned, including the use of recombinant human proteins, combination therapy (e.g. oral antigen plus Th2-inducer cytokines such as IL-4 and IL-10) [91], and synergistic enhancers of oral tolerance (in animals, the combination of oral IFN-τ and oral MBP prevented development of EAE using doses at which each individual therapy was ineffective [92].

Monoclonal Antibody-Based Immunotherapies

Monoclonal antibodies (MAb) are potentially useful immunosuppressive agents, and can be applied to re-establish tolerance to self-antigens. In autoimmunity, MAb against functional molecules on the T-cell surface may be a tool to manipulate the immune system by reinducing tolerance in the T cells which are driving the disease, while leaving the rest of the immune system relatively intact [93]. Indeed, short treatments with MAb to CD4 and CD8 molecules guided the immune system of experimental animals to accept organ grafts and to arrest autoimmune diseases [94-96]. Two strategies were designed in order to achieve therapeutic tolerance using MAb:

1. to use MAb to achieve *central (thymic) tolerance*. This appproach requires a two-stage attack: first, MAb is used to ablate or inactivate peripheral T cells; then a permanent source of antigen that can access the thymus is introduced, inducing tolerance in any T cells that develop later;
2. to establish a form of *peripheral tolerance* that would be long-lasting, even after MAb therapy is stopped [97].

However, some disavantages in applying murine MAb for treating human disease were that they needed to be given systemically, had a short $t_{1/2}$, were potentially immunogenic, and induced massive depletion of the target cells. The development of humanized (chimeric) MAb [98, 99] has much improved their use in human autoimmune diseases.

The use of anti-CD4 depleting MAb suppressed both the onset and the progression of EAE in several experimental settings [100-105]. EAE prevention or reversal was also obtained with anti-Vβ8 [12, 106, 107], anti-TCR [108], and anti-Ia molecules [1, 109, 110]. Anti-CD4 depleting MAbs preferentially eliminated resting, naive T cells, but spared memory or recently activated T cells [111]. When a nondepleting anti-CD4 MAb was used, a modification of the cytokine repertoire (namely, an increase in IL-4 and IL-13 production and a decrease in IFN-γ production) was observed in T cells that respond to the encephalitogenic peptide, thus suggesting that one major mechanism behind MAb-induced immunosuppression was a Th1 to Th2 immune deviation [112].

Two anti-CD4 MAb have been tested in MS patients. A murine anti-human CD4 MAb, B-F5 IgG1 [113] was tested in 21 patients. Neither clinical improvement nor deterioration were observed during the 10-day course of treatment. Expanded disability status scale (EDSS) improvement (>1 point) was observed in six patients one month post-treatment, associated with a decrease in $CD3^+$ and $CD4^+$ cell numbers. Ten patients developed xenogenic antibodies. The most interesting observation of the trial was that five patients with the relapsing-remitting form of the disease were relapse-free at the sixth month post-therapy. The results of a phase I and II (randomized, double-blind, placebo-controlled) clinical trial with a chimeric anti-CD4 Mab, cM-T412 [114-116], showed a substantial and sustained reduction in the number of circulating $CD4^+$ lymphocytes, but not reduced MS activity, as measured by monthly gadolinium-enhanced MRI exams. The statistically significant reduction in the number of clinical relapses observed over a period of 18 months was interpreted as the expression of physician unblinding [116].

A MAb directed to the CD52 molecule on T cells (called CAMPATH-1H, genetically engineered to be identical to human IgG1 except in the antigen-binding complementary-determining regions (CDRs) [98, 117] was tested in patients with rheumatoid arthritis and found to deplete all circulating lymphocytes for many weeks and to depress CD4$^+$ cell numbers beyond three months. No major complications were observed, and significant clinical benefit was seen in 7 of 8 patients [118, 119]. Recently, CAMPATH-1H has been applied to treat RR MS patients [120]. A prolonged T cell depletion which abrogated the formation of new clinical and radiological lesions was observed in 29 patients. However, pre-existing symptoms were transiently re-experienced and disability from lesions acquired prior to therapy continued to progress. Interestingly, T cell clones generated after treatment showed less proliferation and IFN-γ secretion in response to mitogens in vitro. However, a major complication was the development of autoimmune thyroid disease suggesting that the regulation of antibody-mediated immunity was dysregulated by the therapy [120].

Therapy with MAb to B7-1 (an essential costimulatory molecule expressed on activated APC, but also on T and B cells) was demonstrated to suppress clinical relapses, to improve CNS pathology, and to block epitope spreading in relapsing-remitting EAE [121]. Moreover, evidence was obtained that anti-B7-1 induced the development of Th2 cell clones, capable of suppressing EAE induction and progression when injected in immunized mice [58]. Similar results were obtained with a CTLA4-Ig fusion protein [122], thus indicating that blocking the costimulatory pathway causes Th1 to Th2 immune deviation [123]. More recently, MAb to anti-EPR-1 (effector cell protease receptor-1) induced significant immune suppression by inhibiting protease-dependent mechanisms of lymphocyte costimulation [124]. At present, no clinical test is ongoing to evaluate the possible application of MAb directed to costimulatory molecules to human disease, MS included.

A strategy to induce tolerance to exogenous and endogenous antigens with MAb is to administrate a nondepleting MAb (e.g. anti-CD4) before antigen challenge. Treatment with a nondepleting anti-CD4 MAb (H129.19, rat IgG2a) induced immunological tolerance to viral antigens in adult mice, very likely through a mechanism of clonal deletion of the mature T lymphocyte [125, 126]. A two day treatment with anti-CD4 before MBP priming, at a dosage that saturated CD4 receptor but did not deplete CD4$^+$ cells, fully prevented EAE development in PL/J mice. This treatment appears to induce specific MBP unresponsiveness, sparing the potential reactivity of the rest of the CD4$^+$ cells to unrelated antigens [105].

Although the use of MAb to T cell surface molecules alone or in combination with peptides/antigens might be the basis for future clinical trials, it is generally felt that more safety data are needed and that the rationale for these treatments should more precisely be defined.

Adhesion molecules (AM) are cell surface proteins that play a pivotal role in the transendothelial migration of immune system cells. Based on their structure, AM can be classified in selectins, integrins, cadherins, and members of the immunoglobulin superfamily [127, 128]. Augmented expression of AM on the endothelial cell surface at the blood-brain barrier has been described in EAE and MS brain specimens [129, 130], and increased serum and cerebrospinal fluid AM levels have been detected in MS patients with active progressive disease [131]. In Lewis rats, adoptive transfer EAE was prevented by a single intraperitoneal injection of an anti-α4β1 integrin MAb [132], while MAb against ICAM-1 suppressed

MBP-induced EAE but not adoptive transfer EAE [133]. In guinea pigs, an anti-α4 integrin MAb prevented and reversed actively induced EAE [134]. Based on these studies, an anti-α4-integrin humanized monoclonal antibody (Antegren) was designed for MS treatment. Preclinical studies, using clinical parameters, brain morphology and MRI as endpoints, have demonstrated that Antegren is protective in the guinea pig model of EAE [135]. Phase I and II clinical trial in MS patients are in progress.

Therapies Targeting the Cytokine Network

As previously mentioned, an extensive and dynamic network formed by cytokines, soluble cytokine receptors, cytokine receptor antagonists and natural autoantibodies to cytokines regulates immune and inflammatory responses. Organ-specific autoimmune diseases are thought to be primarily mediated by IFN-γ-producing Th1 cells. However, while the existence of functionally distinct T cell subsets, namely Th1 and Th2 [136, 137], producing different sets of cytokines has been established in both experimental animals and in humans, the existence of clear-cut Th1- and Th2-related human pathologies is still a matter of debate. Actually, most immune responses do not reflect an absolute Th1 or Th2 pattern, and involve both Th1 and Th2 cytokines [138]. Th1 lymphocytes produce IL-2, IFN-γ, and tumor necrosis factor alpha (TNF-α), activate cell-mediated immune responses, and induce delayed-type hypersensitivity. Th2 cells produce IL-4, IL-5, IL-6, IL-10, and IL-13, activate B cells, play a role in antibody-mediated immune responses, and down-regulate cell-mediated immunity (IL-4, IL-10 and IL-13 have downregulatory effects on Th1 cells) [139]. It is now generally accepted that development of Th1 or Th2 cells reflects two opposite, differentiation pathways available to uncommitted, pluripotent T cells during the period of activation.

Th1-derived cytokines have been described in EAE acute inflammatory lesions [140, 141]. Moreover, myelin-specific autoreactive $CD4^+$ lymphocytes displayed a Th1 phenotype, and the adoptive transfer of these cells induced EAE in syngeneic mice [142-145]. Conversely, the recovery of acute and relapsing EAE was associated with the presence of a Th2-like response in the CNS, and regulatory T cells capable to suppress EAE were found to express a Th2 functional profile [58, 146]. These finding suggest that the induction and activation of antigen-specific Th2 lymphocytes may prevent and suppress autoimmune demyelinating diseases [147]. Indeed, systemic administration of anti-inflammatory cytokines, such as type I interferons (IFN-α and IFN-β), IL-4 or TGF-β had therapeutic potential in EAE [148-156].

Cytokine-induced "immune deviation" as antigen-specific therapy for inflammatory autoimmune demyelination involves a selective deviation of harmful Th1 responses towards an anti-inflammatory Th2 phenotype [157]. When SJL mice were inoculated with highly pathogenic, MBP-specific Th1 cell lines together with exogenous IL-4, clinical signs of EAE were suppressed and MBP-reactive Th2-like $CD4^+$T-cells – which produced large amounts of IL-4 – were isolated from the spleen of the recipients [158]. Interestingly, IL-4 induction was associated with upregulation instead of downregulation of Th1 cytokines, namely IFN-γ and IL-2. Moreover, MBP-specific Th2 cells were induced only when IL-4 was administered with MBP-activated T cells, thus suggesting that IL-4 expression requires coactivation with a TCR-mediated signal. Consequently, administration of an anti-inflammatory cytokine (e.g. IL-4) during exacerbation of an autoimmune disease should selectively guide activated, autoantigen-specific T cells to produce the same cytokine (IL-4). Although induction of autoreactive Th2 lymphocytes by cytokines associated with selected autoantigens seems a promising approach for treatment of autoimmune demyelination, further experiments are required to characterize the mechanisms operative during immune deviation. The role played by Th2 cells in the immunopathogenesis of EAE has recently been reconsidered [159]: Th2 cells caused EAE in immunodeficient hosts rather than protect them from the disease, and disease induction by Th1 cells was not reduced in any recipient by coadministration of previously activated Th2 cells. These apparently contradictory findings suggest that Th2 populations generated under different conditions or from animals of different genetic backgrounds also differ in some properties, although they produce the same cytokine pattern.

The treatment of RR MS patients with IFN-β (IFN-β1a and IFN-β1b) [160-162] is currently thought to induce a sort of immune deviation in the cytokine network. However, experimental findings gave contradictory results. On one hand, IFN-β favored the production of a Th2-like pattern of cytokines, especially IL-10 [163, 164], and suppressed IFN-γ production by T cells [165]. On the other hand, IFN-β inhibited the differentiation of Th2 from naive cells, and promoted the differentiation in vitro of Th1-type T cells characteristic of cutaneous inflammatory responses [166].

Along with other cytokines and adhesion molecules, TNF-α and TNF-β (also called lymphotoxin, LT) are present in MS plaques [167, 168]. In vitro studies have demonstrated that TNF-α had a demyelinating effect [169, 170], and that peripheral blood mononuclear cells from MS patients with active disease expressed higher levels of TNF-α mRNA [171, 172]. Moreover, TNF-α injection aggravated EAE [173]. Finally, in other human autoimmune diseases (i.e. Crohn's disease and rheumatoid arthritis) intravenous treatment with the humanized monoclonal anti-TNF-α antibody cA2 resulted in a significant improvement lasting 2-4 months after a single infusion [174-177]. However, in a open-label phase I safety trial, two rapidly progressive MS patients treated with cA2 MAb showed no change in clinical status, while transient increases in gadolinium-enhancing lesions, cerebrospinal fluid cells and IgG index were observed after each treatment [178]. Recently, TNF-α was suggested to act as a potent anti-inflammatory cytokine in autoimmune-mediated demyelination: upon immunization with MOG, mice lacking TNF-α develop severe neurological impairment with high mortality and extensive inflammation and demyelination. Moreover, inactivation of the TNF-α gene converted MOG-resistant mice to a state of high susceptibility. Furthermore, treatment with TNF-α drammatically reduced disease severity in TNF-α -/- mice. These findings suggest that TNF-α may be not essential for the induction and expression of inflammatory and demyelinating lesions, and that it may limit extent and duration of severe CNS pathology [179].

The contradictory findings reported above point out the potential danger of therapies based on immune deviation of Th1 cells toward Th2 cells when applied to autoimmune diseases of unknown pathogenesis [180-182]. It is also clear that the concept of network must be kept firmly in mind when a cytokine-based therapy is proposed for chronic inflammatory diseases. The systemic administration of cytokines implicates remarkable side effects, the necessity of high doses in order to have sufficient cytokine concentration in the target organ, and the appearance of regulatory mechanisms after repeated injections (e.g. autoantibodies, soluble receptors, receptor antagonists). For these reasons, local (i.e. at the target tissue level) delivery of suppressive cytokines may represent a therapeutic strategy for the treatment of autoimmune diseases.

Gene Therapy for Local Delivery of Suppressor Cytokines

The development of viral vectors for gene transfer in the nervous system (see the outstanding review of Hermens and Verhaagen [183]) has constituted an important step forward in designing cytokine/growth factor-based immunotherapies. Cytokine gene therapy for organ-specific autoimmune diseases using viral vectors can be approached by introducing heterologous genes of suppressive cytokines in the autoreactive T cells or in the target organ. Antigen-specific T lymphocytes transduced with viral vectors can deliver regulatory cytokines in a site-specific manner, thus limiting systemic side effects. Encephalitogenic MBP-specific T lymphocytes which had been retrovirally modified to express IL-4 delayed and reduced the severity of EAE when adoptively transferred to MBP-immunized mice [184]. In this experimental setting, disease amelioration was the direct expression of intracerebral, rather than systemic, delivery of IL-4. Recently, non-replicative herpetic vectors were used to introduce the IL-4 gene in Biozzi mice immunized with MOG_{40-55} [185, 186]. A significant amelioration of both clinical and pathological CNS features was observed when herpetic vectors containing IL-4 were injected into the cisterna magna before and after the appearance of EAE signs. The protective effect was not due to systemic IL-4 action because no anti-MOG Th1 to Th2 immune deviation was observed in treated animals. These preliminary results indicate that herpetic vector-mediated intracerebral release of anti-inflammatory cytokines may represent a useful strategy for the treatment of CNS autoimmune diseases, especially when the putative autoantigen is unknown.

Conclusions

Based on stimulating results obtained in EAE, the animal model of autoimmune demyelination, an impressive number of promising immunoselective therapies are being developed for the treatment of MS. Some of these therapies are at an early stage of development, while others that have been applied successfully in EAE may lead to a simplistic way of thinking about MS etiology, pathogenesis and treatment. Since the target antigens are not known and the restricted TCR usage often found in EAE is generally not found in MS, antigen-specific therapies are still far from a rationale application in MS. At present, "semiselective" therapies targeting the cytokine network seem to have a realistic prospect for clinical application. However, MS is a heterogeneous disease, not only from the clinical point of view but also from the pathological one, and many immunological mechanisms may account for myelin damage. Therefore, only a better understanding of MS immunopathology will provide the basis for more rational and effective therapeutic interventions.

References

1. Steinman L, Rosenbaum JT, Sriram S, McDevitt HO (1981) In vivo effects of antibodies to immune response gene products: prevention of experimental allergic encephalomyelitis. Proc Natl Acad Sci USA 78: 7111-7118
2. Aharoni R, Teitelbaum D, Sela M, Arnon R (1997) Copolymer 1 induces T cells of the T helper type 2 that crossreact with myelin basic protein and suppress experimental autoimmune encephalomyelitis. Proc Natl Acad Sci USA 94: 10821-10826
3. Lamont AG, Sette A, Fujinami R et al. (1990) Inhibition of experimental autoimmune encephalomyelitis induction in SJL/J mice by using a peptide with high affinity for IAs molecules. J Immunol 145: 1687-1693
4. Gautam AM (1995) Self and non-self peptides treat autoimmune encephalomyelitis: T cell anergy or competition for major histocompatibility complex class II binding? Eur J Immunol 25: 2059-2063
5. Spack EG (1998) Therapeutic and diagnostic application of soluble HLA-DR2: myelin peptide complexes in multiple sclerosis. Presented at the IBC's 6th International Symposium on Multiple Sclerosis. Continued Breakthroughs and Clinical Trials. Washington DC, 10-11 December 1998
6. Wauben MHM, Joosten I, Schlief A et al. (1994) Inhibition of experimental autoimmune encephalomyelitis by MHC class II binding competitor peptides depends on the relative MHC binding affinity of the disease-inducing peptide. J Immunol 152: 4211-4120
7. Bright JJ, Topham DJ, Nag B et al. (1996) Vaccination with peptides from MHC class II beta chain hypervariable region causes allele-specific suppression of EAE. J Neuroimmunol 67: 119-124
8. Nicolle MW, Nag B, Sharma SD et al. (1994) Specific tolerance to an acetylcholine receptor epitope induced in vitro in myasthenia gravis CD4+ lymphocytes by soluble major histocompatibility complex class II-peptide complexes. J Clin Invest 93: 1361-1369
9. Sharma SD, Nag B, Su XM et al. (1991) Antigen-specific therapy of experimental autoimmune encephalomyelitis by soluble class II major histocompatibility complex-peptide complexes. Proc Natl Acad Sci USA 88: 11465-11469
10. Spack EG, McCutcheon M, Corbelletta N et al. (1995) Induction of tolerance in experimental autoimmune myasthenia gravis with solubilized MHC class II: acetylcholine receptor peptide complexes. J Autoimmun 8: 787-807

11. Vandenbark AA, Offner H, Reshef T et al. (1985) Specificity of T cell lines for peptides of myelin basic protein. J Immunol 135: 229-235
12. Acha-Orbea H, Mitchell DJ, Timmerman L et al. (1988) Limited heterogeneity of T-cell receptors from lymphocytes mediating auto-immune encephalomyelitis allows specific immune intervention. Cell 54: 263-273
13. Lider O, Reshef T, Beraud E et al. (1988) Anti-idiotypic network induced by T cell vaccination against experimental autoimmune encephalomyelitis. Science 239: 181-185
14. Lohse AW, Mor F, Karin N, Cohen IR (1989) Control of experimental autoimmune encephalomyelitis by T cells responding to activated T cells. Science 244: 820-824
15. Zhang J, Raus J (1993) T cell vaccination in autoimmune diseases. From laboratory to clinic. Hum Immunol 38: 87-96
16. Ben-Nun A, Cohen IR (1981) Vaccination against autoimmune encephalomyelitis (EAE): attenuated autoimmune T lymphocytes confer resistance to induction of active EAE but not to EAE mediated by the intact T lymphocyte line. Eur J Immunol 11: 949-952
17. Ben-Nun A, Wekerle H, Cohen IR (1981) Vaccination against autoimmune encephalomyelitis with T-lymphocyte line cells reactive against myelin basic protein. Nature 11: 949-952
18. Zhang J, Medaer R, Stinissen P et al. (1993) MHC restricted clonotypic depletion of human myelin basic protein-reactive T cells by T cell vaccination. Science 261: 1451-1454
19. Medaer R, Stinissen P, Truyen L et al. (1995) Depletion of myelin basic protein autoreactive T cells by T-cell vaccination: pilot trial in multiple sclerosis. Lancet 346: 807-808
20. Zhang J, Vandevyver C, Stinissen P, Raus J (1995) In vivo clonotypic regulation of human myelin basic protein-reactive T cells by T cell vaccination. J Immunol 155: 5868-5877
21. Hafler DA, Saadeh MG, Kuchroo VK et al. (1996) TCR usage in human and experimental demyelinating disease. Immunol Today 17: 152-159
22. Martin D, McFarland HF, McFarlin DE (1992) Immunological aspects of demyelinanting disease. Annu Rev Immunol 10: 153-187
23. Holhlfeld R, Mainl E, Weber F et al. (1995) The role of autoimmune T lymphocytes in the pathogenesis of multiple sclerosis. Neurology 25: 531-538
24. Meinl E, Hoch RM, Dormair K et al. (1997) Encephalitogenic potential of myelin basic protein specific T cells isolated from normal rhesus macaques. Am J Pathol 159: 445-453
25. Sun JB, Link H, Olsson T et al. (1991) T and B cell responses to myelin-oligodendrocyte glycoprotein in multiple sclerosis. J Immunol 146: 1490-1495
26. Sun JB, Olsson T, Wang WZ et al. (1991) Autoreactive T and B cells responding to myelin proteolipid protein in multiple sclerosis and controls. Eur J Immunol 21: 1461-1468
27. Kerlero de Rosbo N, Milo R, Lees MB et al. (1993) Reactivity of myelin antigens in multiple sclerosis. Peripheral blood lymphocytes respond predominantly to myelin oligodendrocyte glycoprotein. J Clin Invest 92: 2602-2608
28. Zhang J, Burger D, Saruhan G et al. (1993) The T lymphocyte response against myelin-associated glycoprotein and myelin basic protein in patients with multiple sclerosis. Neurology 43: 403-407
29. Anderton SM, Wraith DC (1998) Hierarchy in the ability of T cell epitopes to induce peripheral tolerance to antigens from myelin. Eur J Immunol 28: 1251-1261
30. Sercarz EE, Lehmann PV, Ametani A et al. (1993) Dominance and crypticity of T cell antigenic determinants. Annu Rev Immunol 11: 729-766

31. Miller SD, McRae BL, Vanderlugt CL et al. (1995) Evolution of the T cell repertoire during the course of experimental immune-mediated demyelinating disease. Immunol Rev 144: 225-244
32. Yu M, Johnson JM, Tuohy VK (1996) A predictable sequential determinant spreading cascade invariably accompanies progression of experimental autoimmune encephalomyelitis: a basis for peptide-specific therapy after onset of clinical disease. J Exp Med 183: 1777-1788
33. Davis MM, Bjorkman PJ (1998) T-cell antigen receptor genes and T-cell recognition. Nature 334: 395-402
34. Williams WV, Weiner DB, Wadsworth S, Greene MI (1988) The antigen-major histocompatibility complex-T-cell receptor interaction: A structural analysis. Immunol Rev 7: 339-344
35. Vandenbark AA, Hashim GA, Offner H (1996) T cell receptor peptides in treatment of autoimmune disease: rationale and potential. J Neurosci Res 43: 391-402
36. Vandenbark AA, Hashim GA, Offner H (1989) Immunization with a synthetic T-cell receptor V-region peptide protects against experimental autoimmune encephalomyelitis. Nature 341: 541-544
37. Offner H, Hashim GA, Vandenbark AA (1991) T cell receptor peptide therapy triggers autoregulation of experimental encephalomyelitis. Science 251: 430-432
38. Kotzin BL, Karuturi S, Chou YK et al. (1991) Preferential T-cell receptor β-chain variable gene use in myelin basic protein-reactive T-cell clones from patients with multiple sclerosis. Proc Natl Acad Sci USA 88: 9161-9165
39. Oksenberg JR (1993) Selection of T-cell receptor Vβ-Dβ-Jβ gene rearrangements with specificity for a myelin basic protein peptide in brain lesions of multiple sclerosis. Nature 362: 68-70
40. Shimonkevitz R, Murray R, Kotzin B (1995) Characterization of T-cell receptor Vβ usage in the brain of a subject with multiple sclerosis. Ann N Y Acad Sci 756: 305-306
41. Bourdette DN, Whitham RH, Chou YK et al. (1994) Immunity to TCR peptides in multiple sclerosis. I. Successful immunization of patients with synthetic Vβ5.2 and Vβ6.1 CDR2 peptides. J Immunol 152: 2510-2519
42. Vandenbark AA, Chou YK, Whitham R et al. (1996) Treatment of multiple sclerosis with T-cell receptor peptides: results of a double-blind pilot trial. Nat Med 2: 1109-1115
43. Chunduru SK, Sutherland RM, Stewart GA et al. (1996) Exploitation of the Vβ8.2 T cell receptor in protection against experimental autoimmune encephalomyelitis using a live vaccinia virus vector. J Immunol 156: 4940-4945
44. Waisman A, Ruiz PJ, Hirschberg DL et al. (1996) Suppressive vaccination with DNA encoding a variable region gene of the T cell receptor prevents autoimmune encephalomyelitis and activates Th2 immunity. Nat Med 2: 857-859
45. Swierkosz JE, Swanborg RH (1997) Immunoregulation of experimental allergic encephalomyelitis: condition for induction of suppresor cells and analysis of mechansim. J Immunol 119: 1501-1506
46. Alvord EC Jr, Shaw CM, Hruby S, Kies MW (1965) Encephalitogen-induced inhibition of experimental allergic encephalomyelitis: prevention, suppression and therapy. Ann N Y Acad Sci 122: 333-345
47. MacPherson CF, Yo SL (1973) Studies on brain antigens. VI. Prevention of experimental allergic encephalomyelitis by a water-soluble spinal cord protein, 1-scp. J Immunol 110: 1371-1375
48. MacPherson CF, Armstrong H, Tan O (1977) Prevention of experimental allergic encephalitis in guinea pigs with spinal cord protein: optimum pretreatment schedules and reappraisal of plausible mechanisms. Immunology 33: 161-166

49. Marušic S, Tonegawa S (1997) Tolerance induction and autoimmune encephalomyelitits amelioration after adminiistration of myelin basic protein-derived peptide. J Exp Med 186: 507-515
50. Alvord EC Jr, Shaw CM, Hruby S, Kies MW (1979) Has myelin basic protein received a fair trial in the treatment of multiple sclerosis? Ann Neurol 6: 461-468
51. Kennedy MK, Dal Canto MC, Trotter JL, Miller SD (1988) Specific immune regulation of chronic-relapsing experimental autoimmune encephalomyelitis in mice. J Immunol 141: 2986-2993
52. Tan L-J, Kennedy MK, Dal Canto MC, Miller SD (1991) Successful treatment of paralytic relapses in adoptive experimental autoimmune encephalomyelitis via neuroantigen-specific tolerance. J Immunol 147: 1797-1802
53. Vandenbark AA, Vainiene M, Ariail K et al. (1996) Prevention and treatment of relapsing autoimmune encephalomyelitis with myelin peptide-coupled splenocytes. J Neurosci Res 45: 430-438
54. McKenna RM, Carter BG, Paterson JA, Sehon AH (1983) The suppression of experimental allergic encephalomyelitis in Lewis rats by treatment with myelin basic protein-cell conjugates. Cell Immunol 81: 391-402
55. Malokty MK, Pope L, Miller SD (1994) Epitope and fuctional specificity of peripheral tolerance induction in experimental autoimmune encephalomyelitis in adult Lewis rats. J Immunol 153: 841-851
56. Jenkins MK, Schwartz RH (1987) Antigen presentation by chemically modified splenocytes induces antigen-specific T cell unresponsiveness in vitro and in vivo. J Exp Med 165: 301-319
57. Sakai K, Zamvil SS, Mitchell DJ et al. (1989) Prevention of experimental encephalomyelitis with peptides that block interaction of T cells with major histocompatibility complex proteins. Proc Natl Acad Sci USA 86: 9470-9474
58. Kuchroo VK, Das MP, Brown JA et al. (1995) B7-1 and B7-2 costimulatory molecules activate differentially the Th1/Th2 developmental pathways: application to autoimmune disease therapy. Cell 80: 707-718
59. Nicholson LB, Murtaza A, Hafler BP et al. (1997) A T cell receptor antagonist peptide induces T cells that mediate bystander suppression and prevent autoimmune encephalomyelitis induced with multiple myelin antigens. Proc Natl Acad Sci USA 4: 9279-9284
60. Gaur A (1998) Development of altered peptide ligand-based therapy for multiple sclerosis. Presented at the IBC's 6th International Symposium on Multiple Sclerosis. Continued Breakthroughs and Clinical Trials. Washington DC, 10-11 December 1998
61. Arnon R (1996) The development of Cop-1 (Copaxone), an innovative drug for the treatment of multiple sclerosis. Immunol Lett 50: 1-15
62. Keith AR, Arnon R, Tietelbaum D et al. (1979) The effect of Cop-1, a synthetic polypeptide on chronic-relapsing EAE in guinea pigs. J Neurol Sci 42: 267-273
63. Tietelbaum D, Fridkis-Hareli M, Arnon R, Sela M (1996) Copolymer 1 inhibits chronic relapsing experimental allergic encephalomyelitis induced by proteolipid protein (PLP) peptides in mice and interferes with PLP-specific T cell responses. J Neuroimmunol 64: 209-217
64. Ben-Nun A, Mendel I, Bakimer R et al. (1996) The autoimmune reactivity to myelin oligodendrocyte glycoprotein (MOG) in multiple sclerosis is potentially pathogenic: effect of copolymer 1 on MOG-induced disease. J Neurol 243(Suppl 1): S14-S22
65. Fridkis-Hareli M, Teitelbaum D, Gurevich E et al. (1994) Direct binding of myelin basic protein and synthetic copolymer 1 to class II major histocompatibility complex molecules on living antigen presenting cells - specificity and promiscuity. Proc Natl Acad Sci USA 91: 4872-4876

66. Racke MK, Martin R et al. (1992) Copolymer-1-induced inhibition of antigen-specific T cell activation: interference with antigen presentation. J Neuroimmunol 37: 75-84
67. Tietelbaum D, Milo R et al. (1992) Synthetic copolymer-1 inhibits human T-cell line specific for myelin basic protein. Proc Natl Acad Sci USA 89: 137-141
68. Bornstein MB, Miller A, Slagle S et al. (1988) A pilot trial of Cop-1 in exacerbating-remitting multiple sclerosis. N Engl J Med 317: 408-414
69. Bornstein MB, Miller A, Slagle S et al. (1991) A placebo-controlled, double-blind, randomized, two-center, pilot trial of Cop-1 in chronic progressive multiple sclerosis. Neurology 41: 533-539
70. Johnson KP, Brooks BR, Cohen JA et al. (1995) Copolymer I reduces relapse rate and improves disability in relapsing-remitting multiple sclerosis. Results of a phase III multicentre, double-blind, placebo-controlled trial. Neurology 45: 1268-1276
71. Cohen JA, Grossman RI, Udupa JK et al. (1995) Assessment of the efficacy of copolymer-1 in the treatment of multiple sclerosis by quantitative MRI. Neurology 45 (Suppl 4): A418 (abstract)
72. Critchfield JM, Racke MK, Zuñiga-Pflücker JC (1994) T cell deletion in high antigen dose therapy of autoimmune encephalomyelitis. Science 263: 1139-1143
73. Gaur A, Wiers B, Liu A et al. (1992) Amelioration of autoimmune encephalomyelitis by myelin basic protein synthetic peptide-induced anergy. Science 258: 1491-1494
74. Gaur A, Boehme SE, Chalmers D et al. (1997) Amelioration of relapsing experimental autoimmune encephalomyelitis with altered myelin basic protein peptides involves different cellular mechanisms. J Neuroimmunol 74: 149-158
75. Karin N, Mitchell DJ, Broke S et al. (1994) Reversal of experimental autoimmune encephalomyelitis by a soluble peptide variant of a myelin basic protein epitope: T cell receptor antagonism and reduction of interferon gamma and tumor necrosis factor alpha production. J Exp Med 180: 2227-2237
76. Miller SD, Wetzig RP, Claman HN (1979) The induction of cell mediated immunity and tolerance with protein antigens coupled with syngeneic lymphoid cells. J Exp Med 149: 758-773
77. Miller SD, Tan LJ, Pope L et al. (1992) Antigen-specific tolerance as a therapy for autoimmune encephalomyelitis. Int Rev Immunol 9: 203-222
78. Pope L, Paterson PY, Miller SD (1992) Antigen-specific inhibition of the adoptive transfer of experimental autoimmune encephalomyelitis in Lewis rats. J Neuroimmunol 37: 177-190
79. Racke MK, Crichfield JM, Quigley L et al. (1996) Intravenous antigen administration as a therapy for autoimmune demyelinating disease. Ann Neurol 39: 46-56
80. Willenborg DO, Staten EA, Witting GF (1978) Experimental allergic encephalomyelitis: modulation by intraventricular injection of myelin basic protein. Exp Neurol 61: 527-536
81. Willenborg DO, Staykova MA (1988) Approaches to the treatment of central nervous system autoimmune disease using specific neuroantigen. Immunol Cell Biol 76: 91-103
82. Weiner HL, Friedman A, Miller A et al. (1994) Oral tolerance: immunologic mechanisms and treatment of animal and human organ-specific autoimmune diseases by oral administration of autoantigens. Annu Rev Immunol 12: 809-837
83. Higgings P, Weiner HL (1988) Suppression of experimental autoimmune encephalomyelitis by oral administration of myelin basic protein and its fragments. J Immunol 140: 440-445
84. Bitar DM, Whitacre CC (1998) Suppression of experimental autoimmune encephalomyelitis by the oral administration of myelin basic protein. Cell Immunol 112: 364-370

85. Brod SA, Al-Sabbgh A, Sobel RA et al. (1992) Suppression of experimental autoimmune encephalomyelitis by oral administration of myelin antigens. IV. Suppression of chronic relapsing disease in the Lewis rat and strain 13 guinea pig. Ann Neurol 29: 615-622
86. Miller A, Lider O, Al-Sabbagh A, Weiner HL (1992) Suppression of experimental autoimmune encephalomyelitis by oral administration of myelin basic protein. V. Hierarchy of suppression by myelin basic protein from different species. J Neuroimmunol 39: 243-250
87. Friedman A, Weiner HL (1994) Induction of anergy or active suppression following oral tolerance is determined by frequency of feeding and antigen dosage. Proc Natl Acad Sci USA 91: 6688-6692
88. Chen Y, Inobe J et al. (1995) Peripheral deletion of antigen-reactive T cells in oral tolerance. Nature 376: 177-180
89. Weiner HL (1997) Oral tolerance: immune mechanisms and treatment of autoimmune diseases. Immunol Today 18: 335-341
90. Weiner HL, Mackin GA, Matsui M et al. (1993) Double-blind pilot trial of oral tolerization with myelin antigens in multiple sclerosis. Science 259: 1321-1324
91. Slavin A, Maron R, Komagata Y, Weiner HL (1998) Oral administration of IL-4 and IL-10 enhances the induction of low dose oral tolerance. J Neuroimmunol 90(1): 84 (abstract)
92. Soos JM, Johnson HM, Weiner HL, Zamvil SS (1998) Interferon tau induction of Th2 cytokines and synergy with oral MBP for treatment of experimental allergic encephalomyelitis. J Neuroimmunol 90(1): 84 (abstract)
93. Cobbold SP, Qin S, Leong LYW et al. (1992) Reprogramming the immune system for peripheral tolerance with CD4 and CD8 monoclonal antibodies. Immunol Rev 129: 165-201
94. Qin S, Cobbold S, Benjamin R, Waldmann H (1989) Induction of classical transplantation tolerance in the adult. J Exp Med 169: 779-794
95. Qin S, Wise M, Cobbold S et al. (1990) Induction of tolerance in peripheral T-cells with monoclonal antibodies. Eur J Immunol 20: 2737-2745
96. Shizuro JA, Alters SE, Fathman CG (1992) Anti-CD4 monoclonal antibodies in therapy: creation of nonclassical tolerance in the adult. Immunol Rev 129: 105-130
97. Waldmann H, Cobbold S (1993) The use of monoclonal antibodies to achieve immunological tolerance. Immunol Today 1: 247-251
98. Hale G, Clark MR, Marcus R et al. (1988) Remission induction in non-Hodgkin lymphoma with reshaped human monoclonal antibody CAMPATH-1H. Lancet II: 1394-1396
99. Pulito VL, Roberts VA, Adair JR et al. (1996) Humanization and molecular modeling of the anti-CD4 monoclonal antibody, OKT4A. J Immunol 156: 2840-2850
100. Brostoff SW, Mason DW (1984) Experimental allergic encephalomyelitis: successful treatment in vivo with a monoclonal antibody that recognized T helper cells. J Immunol 133: 1938-1942
101. Waldor MK, Sriram S, Hardy R et al. (1985) Reversal of experimental allergic encephalomyelitis with monoclonal antibody to a T cell subset marker. Science 227: 415-417
102. Sedgwick JD, Mason DW (1986) The mechanism of inhibition of experimental allergic encephalomyelitis in the rat by monoclonal antibody against CD4. J Neuroimmunol 13: 217-232
103. Sriram S, Roberts CA (1986) Treatment of established chronic relapsing experimental allergic encephalomyelitis with anti-L3T4 antibodies. J Immunol 136: 4464-4469

104. O'Neill JK, Baker D, Davison AN et al. (1993) Control of immunomediated disease of the central nervous system with monoclonal (CD4-specific) antibodies. J Neuroimmunol 45: 1-14
105. Biasi G, Facchinetti A, Monastra G et al. (1997) Protection from experimental autoimmune encephalomyelitis (EAE): non-depleting anti-CD4 mAb treatment induces peripheral T-cell tolerance to MBP in PL/J mice. J Neuroimmunol 73: 117-123
106. Urban JL, Kumar V, Kono DH et al. (1988) Restricted use of TCR Vβ genes in murine auto-immune encephalomyelitis raises possibilities for antibody therapy. Cell 54: 577-592
107. Zaller DM, Osman G, Kanagawa O, Hood L (1990) Prevention and treatment of murine experimental allergic encephalomyelitis with T-cell receptor Vβ-specific antibodies. J Exp Med 8: 579-621
108. Owhashi M, Heber-Katz E (1988) Protection from experimental allergic encephalomyelitis conferred by a monoclonal antibody directed against a shared idiotype on rat T-cell receptor specific for myelin basic protein. J Exp Med 168: 2153-2164
109. Sriram S, Steinman L (1983) Anti-IA antibody suppresses active encephalomyelitis: treatment model for disease linked to IR genes. J Exp Med 158: 1362-1367
110. Sriram S, Topham DJ, Carroll L (1987) Haplotype specific suppression of experimental allergic encephalomyelitis with anti-IA antibodies. J Immunol 139: 1485-1489
111. Chace JH, Cowdery JS, Field EH (1994) Effect of anti-CD4 on CD4 subsets. I. Anti-CD4 preferentially deletes resting, naive CD4 cells and spares activated CD4 cells. J Immunol 152: 405-413
112. Stumbles P, Mason D (1995) Activation of CD4$^+$ T cells in the presence of a nondepleting monoclonal antibody to CD4 induces a Th2-type response in vitro. J Exp Med 182: 5-13
113. Racadot E, Rumbach L, Bataillard M et al. (1993) Treatment of multiple sclerosis with anti-CD4 monoclonal antibody. A preliminary report on B-F5 in 21 patients. J Autoimmun 6: 771-786
114. Lindsey JW, Hodgkinson S, Metha R et al. (1994) Repeated treatment with chimeric anti-CD4 antibody in multiple sclerosis. Ann Neurol 36: 183-189
115. Llewellyn-Smith N, Lai M, Miller DH et al. (1997) Effects of anti-CD4 antibody treatment on lymphocyte subsets and stimulated tumor necrosis factor alpha production: a study of 29 multiple sclerosis patients entered into a clinical trial of cM-T412. Neurology 48: 810-816
116. van Oosten BW, Lai M, Hodgkinson S et al. (1997) Treatment of multiple sclerosis with the monoclonal anti-CD4 antibody Cm-T412: results of a randomized, double-blind, placebo-controlled, MR-monitored phase II trial. Neurology 47: 1531-1534
117. Reichmann L, Clark M, Waldmann H, Winter G (1988) Reshaping human antibodies for therapy. Nature 332: 323-326
118. Isaacs JD, Watts RA, Hazleman B et al. (1992) Humanized monoclonal antibody therapy for rheumatoid arthritis with an antiglobulin response. Lancet 340: 748-752
119. Lookwood CM, Thiru S, Isaacs JD et al. (1993) Long-term remission of intractable systemic vasculitis with monoclonal antibody therapy. Lancet 341: 1620-1622
120. Compston DAS, Coles AJ, Miller DH, Waldmann H (1998) CAMPATH-1H exposes three mechanisms underlying the natural history of multiple sclerosis in the individual patient. J Neuroimmunol 90(1): 96
121. Miller SD, Vanderlugt C, Lenschow DJ et al. (1995) Blockade of CD28/B7-1 interaction prevents epitope spreading and clinical relapses in murine EAE. Immunity 3: 739-745

122. Perrin PJ, Scott D, Quigley L et al. (1995) Role of B7: CD28/CTLA-4 in the induction of chronic relapsing experimental allergic encephalomyelitis. J Immunol 154: 1481-1490
123. Khoury SJ, Akalin SJ, Cannon C et al. (1995) CD28-B7 costimulatory blockade by CTLA4-Ig prevents actively induced experimental autoimmune encephalomyelitis and inhibits Th1 but spares Th2 cytokines in the central nervous system. J Immunol 155: 4521-4524
124. Duchosal MA, Rothermel AL, McConahey et al. (1996) In vivo immunosuppression by targeting a novel protease receptor. Nature 380: 352-356
125. Biasi G, Facchinetti A, Panozzo M et al. (1991) Moloney murine leukemia virus tolerance in anti-CD4 monoclonal antibody-treated adult mice. J Immunol 147: 2284-2289
126. Facchinetti A, Panozzo M, Pertile P et al. (1992) In vivo and in vitro death of mature T cells induced by separate signals to CD4 and $\alpha\beta$TCR. Immunobiology 185: 380-389
127. Hogg N, Berlin C (1995) Structure and function of adhesion receptors in leukocyte trafficking. Immunol Today 16: 327-330
128. Imhof BA, Dunon D (1995) Leukocyte migration and adhesion. Adv Immunol 58: 345-416
129. Raine CS (1994) Multiple sclerosis: immune system adhesion molecule expression in the central nervous system. J Neuropathol Exp Neurol 53: 328-337
130. Raine CS, Lee SC, Scheinberg LC et al. (1990) Adhesion molecules on endothelial cells in the central nervous system: an emerging area in the neuroimmunology of multiple sclerosis. Clin Immunol Immunopathol 57: 173-187
131. Hartung H-P, Archelos JJ, Zielasek J et al. (1995) Circulating adhesion molecules and inflammatory mediators in demyelination: a review. Neurology 45(Suppl 6): S22-S32
132. Yednock TA, Cannon C, Fritz LC et al. (1992) Prevention of experimental autoimmune encephalomyelitis by antibodies against $\alpha 4\beta 1$ integrin. Nature 356: 63-66
133. Archelos JJ, Jung S, Mäurer M et al. (1993) Inhibition of experimental autoimmune encephalomyelitis by an antibody to the intercellular adhesion molecule ICAM-1. Ann Neurol 34: 145-154
134. Kent SJ, Karlik SJ, Cannon C et al. (1995) A monoclonal antibody to $\alpha 4$ integrin suppresses and reverses active experimental allergic encephalomyelitis. J Neuroimmunol 58: 1-10
135. Walicke PA (1998) Update on Antegren (Natalizumab) injection. Presented at the IBC's 6th International Symposium on Multiple Sclerosis. Continued Breakthroughs and Clinical Trials. Washington DC, 10-11 December 1998
136. Mosmann TR, Coffman RL (1989) TH1 and TH2 cells. Different patterns of lymphokine secretion lead to different functional properties. Annu Rev Immunol 7: 145-173
137. Romagnani S (1994) Lymphokine production by human T cells in disease states. Annu Rev Immunol 12: 227-257
138. Allen JE, Maizels RM (1997) Th1-Th2: reliable paradigm or dangerous dogma. Immunol Today 18: 387-392
139. Romagnani S (1997) The Th1/Th2 paradigm. Immunol Today 18: 263-266
140. Merrill JE, Kono DH, Clayton J et al. (1992) Inflammatory leukocytes and cytokines in the peptide-induced disease of experimental allergic encephalomyelitis in SJL and B10.PL mice. Proc Natl Acad Sci USA 89: 574-578
141. Liblau RS, Singer SM, McDevitt HO (1995) Th1 and Th2 CD4$^+$ T cells in the pathogenesis of organ-specific autoimmune diseases. Immunol Today 16: 34-38
142. Zamvil SS, Steinman L (1990) The T lymphocyte in experimental allergic encephalomyelitis. Annu Rev Immunol 8: 579-621

143. Van der Veen RC, Stohlman SA (1993) Encephalitogenic Th1 cells are inhibited by Th2 cells with related peptide specificity: relative roles of interleukin-4 (IL-4) and IL-10. J Neuroimmunol 48: 213-222
144. Miller SD, Karpus WJ (1994) The immunopathogenesis and regulation of T-cell mediated demyelinating diseases. Immunol Today 15: 356-361
145. Nicholson LB, Greer JM, Sobel RA et al. (1995) An altered peptide ligand mediates immune deviation and prevents autoimmune encephalomyelitis. Immunity 3: 397-405
146. Chen Y, Kuchroo VK, Inobe J et al. (1994) Regulatory T cell clones induced by oral tolerance: suppression of autoimmune encephalomyelitis. Science 265: 1237-1240
147. Olsson T (1995) Cytokine-producing cells in experimental autoimmune encephalomyelitis and multiple sclerosis. Neurology 45(Suppl 6): S11-S15
148. Abreu SL (1995) Interferon in experimental autoimmune encephalomyelitis (EAE): Effects of exogeneous interferon on the antigen-enhanced adoptive transfer of EAE. Int Arch Allergy Appl Immunol 76: 302-307
149. Brod SA, Scott M, Burns DK, Phillips JT(1995) Modification of acute experimental autoimmune encephalomyelitis in the Lewis rat by oral administration of type I interferons. J Interfer Cytokine Res 15: 115-122
150. Brod SA, Khan M (1996) Oral administration of interferon-α is superior to subcutaneous administration of interferon-α in the suppression of chronic relapsing experimental autoimmune encephalomyelitis. J Autoimmun 9: 11-20
151. Inobe JI, Chen Y, Weiner HL (1996) In vivo administration of IL-4 induces TGF-β-producing cells and protects animals from experimental autoimmune encephalomyelitis. Ann N Y Acad Sci 778: 390-392
152. Kuruvilla AP, Shah R, Hochwald GM et al. (1991) Protective effect of transforming growth factor beta 1 on experimental autoimmune diseases in mice. Proc Natl Acad Sci USA 88: 2918-2921
153. Johns LD, Flanders KC, Ranges GE, Sriram S (1991) Successful treatment of experimental allergic encephalomyelitis with transforming growth factor β1. J Immunol 147: 1792-1796
154. Racke MK, Dhib-Jalbut S, Cannella B et al. (1991) Prevention and treatment of chronic relapsing experimental allergic encephalomyelitis by transforming growth factor-β1. J Immunol 146: 3012-3017
155. Stevens DB, Gould KE, Swanborg RH (1994) Transforming growth factor β1 inhibits tumor necrosis factor-α/lymphotoxin production and adoptive transfer of disease by effector cells of autoimmune encephalomyelitis. J Neuroimmunol 51: 77-83
156. Yu M, Nishiyama A, Trapp BD, Tuohy VK (1996) Interferon-β inhibits progression of relapsing-remitting experimental autoimmune encephalomyelitis. J Neuroimmunol 64: 91-100
157. Röcken M, Racke M, Shevach EM (1996) IL-4-induced immune deviation as antigen-specific therapy for inflammatory autoimmune disease. Immunol Today 17: 225-231
158. Racke MK, Bonomo A, Scott DE et al. (1994) Cytokine-induced immune deviation as a therapy for inflammatory autoimmune disease. J Exp Med 180: 1961-1966
159. Lafaille JJ, Vandet Keere F, Hsu AL et al. (1997) Myelin basic protein-specific T helper α(Th2) cells cause experimental autoimmune encephalomyelitis in immunodeficient hosts rather than protect them from the disease. J Exp Med 186: 307-312
160. IFNB Multiple Sclerosis Study Group (1993) Interferon beta-1b is effective in relapsing-remitting multiple sclerosis. I. Clinical results of a multicenter, randomized, double-blind, placebo-controlled trial. Neurology 43: 655-661
161. IFNB Multiple Sclerosis Study Group (1995) Interferon beta-1b in the treatment of multiple sclerosis. Final outcome of the randomized controlled trial. Neurology 45: 1277-1285

162. Jacobs LD, Cookfair DL, Rudick RA et al. (1996) Intramuscular interferon beta-1a for disease progression in relapsing multiple sclerosis. Ann Neurol 39: 285-294
163. Rep MHG, Hintzen RQ, Polman CH, van Lier RAW (1996) Recombinant interferon-blocks proliferation but enhances interleukin-10 secretion by activated human T cells. J Neuroimmunol 67: 111-118
164. Rudick RA, Ransohoff RM, Peppler R et al. (1996) Interferon beta induces interleukin-10 expression: relevance to multiple sclerosis. Ann Neurol 40: 618-627
165. Noronha A, Toscas A, Jensen MA (1993) Interferon β decreases T cell activation and interferon γ production in multiple sclerosis. J Neuroimmunol 46: 145-154
166. McRae BL, Picker LJ, van Seventer GA (1997) Human recombinant interferon-β influences T helper subset differentiation by regulating cytokine secretion pattern and expression of homing receptors. Eur J Immunol 27: 2650-2656
167. Cannella B, Raine CS (1995) The adhesion molecule and cytokine profile of multiple sclerosis lesions. Ann Neurol 37: 424-435
168. Selmaj K, Raine CS, Cannella B, Brosnan CF (1991) Identification of lymphotoxin and tumor necrosis factor in multiple sclerosis lesions. J Clin Invest 87: 949-954
169. Robbins DS, Shirazi Y, Drysdale BE et al. (1987) Production of cytotoxic factor for oligodendrocytes by stimulated astrocytes. J Immunol 139: 2593-2600
170. Selmaj K, Raine CS (1988) Tumor necrosis factor mediates myelin and oligodendrocyte damage in vitro. Ann Neurol 23: 339-346
171. Rieckmann P, Albrecht M, Kitze B et al. (1994) Cytokine mRNA levels in mononuclear blood cells from patients with multiple sclerosis. Neurology 44: 1523-1526
172. Rieckmann P, Albrecht M, Kitze B et al. (1995) Tumor necrosis factor-alpha messenger RNA expression in patients with relapsing-remitting multiple sclerosis is associated with disease activity. Ann Neurol 37: 82-88
173. Kuroda Y, Shimamoto Y (1991) Human tumor necrosis factor-alpha augments experimental allergic encephalomyelitis in rats. J Neuroimmunol 34: 159-164
174. Derkx B, Taminiau J, Radema S et al. (1994) Tumor necrosis factor antibody treatment in Crohn's disease. Lancet 342: 173-174
175. Van Dullemen HM, Van Deventer SJH, Hommes DW et al. (1995) Treatment of Crohn's disease with anti-tumor necrosis factor chimeric monoclonal antibody (cA2). Gastroenterology 109: 129-135
176. Elliott MJ, Maini RN, Feldmann M et al. (1994) Randomised double-blind comparison of chimeric monoclonal antibody to tumor necrosis factor alpha (cA2) versus placebo in rheumatoid arthritis. Lancet 344: 1105-1110
177. Elliott MJ, Maini RN, Feldmann M et al. (1994) Repeated therapy with monoclonal antibody to tumor necrosis factor alpha (cA2) in patients with rheumatoid arthritis. Lancet 334: 1125-1127
178. van Oosten BW, Barkhof F, Truyen L et al. (1996) Increased MRI activity and immune activation in two multiple sclerosis patients treated with the monoclonal anti-tumor necrosis factor antibody cA2. Neurology 47: 1531-1534
179. Liu J, Marino MW, Wong G et al. (1998) TNF is a potent anti-inflammatory cytokine in autoimmune-mediated demyelination. Nat Med 4: 78-83
180. Khoruts A, Miller SD, Jenkins MK (1995) Neuroantigen-specific Th2 cells are inefficient suppressors of experimental autoimmune encephalomyelitis induced by effector Th1 cells. J Immunol 155: 5011-5017
181. McFarland HF (1996) Complexities in the treatment of autoimmune disease. Science 274: 2054-2057
182. Genain CP, Abel K, Belmar N et al. (1996) Late complication of immune deviation therapy in a nonhuman primate. Science 274: 2054-2057

183. Hermens WTJMC, Verhaagen J (1998) Viral vectors, tools for gene transfer in the nervous system. Progr Neurobiol 55: 399-432
184. Shaw MK, Lorens JB, Dhawan A et al. (1997) Local delivery of interleukin 4 by retrovirus-transduced T lymphocytes ameliorates experimental autoimmune encephalomyelitis. J Exp Med 185: 1711-1714
185. Martino G, Furlan R, Galbiati F, et al. (1998) A gene therapy approach to treat demyelinating disease using nonreplicative herpetic vectors engineered to produce cytokines. Mult Scler 4: 222-227
186. Furlan R, Poliani PL, Galbiati F, et al. (1998) Central nervous system delivery of interleukin-4 by a nonreplicative herpes simplex type 1 viral vector ameliorates autoimmune demyelination. Hum Gene Ther 9: 2605-2617

Chapter 21

Animal Models of Demyelination of the Central Nervous System

A. Uccelli

Experimental Autoimmune Encephalomyelitis

Experimental autoimmune encephalomyelitis (EAE) is possibly the best animal model for studying autoimmune diseases and in particular demyelinating diseases of the central nervous system (CNS) such as multiple sclerosis (MS). Since the classic studies of Rivers et al. [1] in monkeys immunized with CNS homogenate, EAE has been an invaluable tool for dissecting mechanisms of the immune response against self-antigens within the CNS, as well for testing new therapies for the treatment of autoimmune diseases. An autoimmune response leading to EAE in susceptible species can be obtained by active immunization with CNS proteins or by passive transfer of T lymphocytes reactive against myelin antigens to syngeneic recipients. The role of T lymphocytes in EAE was first demonstrated by Paterson who succeeded in transferring disease by means of T cells from immunized animals [2]. Since then, many researchers have attempted to characterize the role of T cells in EAE. Over the years it became clear that activated CD4+ T cells mediate EAE upon recognition of the target antigen bound to class II molecules of the major histocompatibility complex (MHC) [3]. Encephalitogenic T cells can be retrieved from the blood of immunized as well as naive animals, supporting the concept that autoaggressive lymphocytes are part of the natural immune repertoire [4, 5].

Virtually all mammalian species can be susceptible to EAE as long as they are properly immunized; several species and strains have been utilized including mice, rats and guinea pigs (Table 1). The clinical, pathological and immunological picture of autoimmune models of demyelination depends upon the mode of sensitization, the nature of the immunogen, and the genetic background of each species and strain.

Modes of sensitization include the route of immunization, primarily subcutaneously, and the use of immunogens emulsified with an equal volume of complete Freund's adjuvant containing *Mycobacterium tuberculosis* to create an antigen depot. Boosts with *Bordetella pertussis* are often used to help open the blood-brain barrier (BBB).

Whole myelin homogenate as well as distinct myelin proteins, including myelin

---

Neuroimmunology Unit, Department of Neurological Sciences and Neurorehabilitation, University of Genoa, Italy. e-mail: labunit@cisi.unige.it

Table 1. Models of experimental autoimmune encephalomyelitis

Species	MHC	Immunogen	Encephalitogenic epitope (amino acid sequence)	TCR	Course	Reference
Mice	**Inbred**					
Biozzi/ABH	dq1	MH	–	–	RR	Baker et al. [27]
		MOG	1-22/43-47/134-148	–	CR	Amor et al. [28]
SJL	H-2^S	MH	–	–	AM	Brown [15]
		MBP	89-101	Vβ17	AM	Zamvil [3]
		PLP	139-151	–	RR	Tuohy [95]
		PLP	40-70/100-119/178-209	–	RR	Greer [30]
		MOG	92-106	–	CR	Amor [28]
PL/J	H-2^U	MH	–	–	AM	Zamvil [3]
		MBP	Ac1-9	Vβ8.2/Vα2	CR	Zamvil [3]
		MBP	35-47	–	CR	Zamvil [3]
		MOG	35-55	–	CR	Kerlero de Rosbo [97]
C57BL/6	H-2^B	MOG	35-55	Vβ8	CP	Mendel [96]
C3H.SW	H-2^K	MOG	35-55	Vβ8	CP	Mendel [96]
B.10.PL	H-2^U	MH	Ac1-9	Vβ8.2/Vα2	AM	Zamvil [3]
		MBP	1-37	–	CR	Zamvil [3]
Rats	**Inbred**					
Lewis	RTI.B^1	MH	–	–	AM	Hoffman et al. [98]
		MBP	68-88	Vβ8.2/Vα2	AM	Burns et al. [18]
		MOG	35-55	–	RR	Linington et al. [8]
DA	av1	MH	–	–	CR	Lorentzen et al. [29]
Guinea Pig	**Inbred**					
13	–	MH	–	–	AM	Freud [99]
Primates	**Outbred**					
Macaque	–	MH	Heterogeneous	–	AM	Rivers [1]
Marmoset	–	MH	Heterogeneous	Diverse	CR	Massacesi et al. [78]

MHC, major histocompatibility complex; *MH*, myelin homogenate; *MOG*, myelin oligodendrocyte glycoprotein; *MBP*, myelin basic protein; *PLP*, proteolipid protein; *RR*, relapsing-remitting; *CR*, chronic relapsing; *AM*, acute monophasic; *CP*, chronic progressing; –, Not determined.

basic protein (MBP), myelin oligodendrocyte glycoprotein (MOG), and proteolipid protein (PLP), have been used to induce EAE in different species. The role of MBP, an abundant hydrophilic myelin protein, was first characterized by Ben-Nun and colleagues who successfully induced EAE by transferring MBP-specific T cell lines to naive Lewis rats [6]. Passive transfer studies also elucidated the role of other myelin antigens, including PLP [7] and MOG [8]. MOG, a minor glycoprotein exposed on the surface of the myelin sheath, is the target of both humoral and cellular immune responses. The importance of the humoral response to MOG was first demonstrated by Linington et al. who showed the presence of sharp demyelination following the injection of anti-MOG auto-antibodies into animals with T-cell-mediated EAE [9]. Other encephalitogenic proteins include a lipid-bound form of MBP [10] as well as non-myelin auto-antigens such as the astrocyte-derived calcium binding protein S100β [11].

Depending upon the species, the antigen and the mode of sensitization, the course of EAE can be monophasic acute, chronic relapsing, or even primarily progressive, mimicking human MS. The classic picture of acute EAE is characterized by perivascular inflammation mainly represented by CD4+ and CD8+ T lymphocytes and macrophages within the cerebral white matter. Nevertheless, manipulation of the above-mentioned factors can lead to a wide spectrum of neuropathological patterns including demyelination, remyelination, gliosis, loss of axons and, in certain species, also necrosis [12].

EAE in Inbred Species

Olitsky and Yager were the first to establish EAE in mice [13]. Since then, thousands of scientists have used inbred rodents as the most suitable species for EAE studies. EAE has been successfully induced in guinea pigs [14] as well as in several strains of rats and mice. Some mice strains such as SJL/J develop a relapsing-remitting disease following active immunization with spinal cord homogenate [15]. Pathological investigations show the presence of mononuclear cell infiltrates within the CNS and demyelination. A similar disease can also be obtained by the adoptive transfer of MBP-specific T lymphocytes [16]. In this H-2^S strain, a few epitopes of MBP concentrated within the sequence of amino acids 89-101 are recognized by encephalitogenic T cells which preferentially use the T cell receptor (TCR) Vβ17 segment [17]. A more restricted response to encephalitogenic determinants of MBP has been reported for EAE-susceptible H-2^U mice such as PL/J and B.10.PL. Both encephalitogenic and non-encephalitogenic T cells recognize the N-terminal peptide Ac1-9 utilizing the same TCR Vβ8.2/Vα2 (or Vα4) gene combination [3]. Interestingly, the same TCR gene segments are characteristic of the T cell response to MBP in Lewis rats [18]. In Lewis rats, EAE induced either by active immunization or by passive transfer of T cells is an acute monophasic disease mostly characterized by inflammation [19]. In this species, EAE is mediated by CD4+ encephalitogenic T lymphocytes that are specific for the peptide MBP 68-88 in the context of the class II MHC molecule RT1.B^1 [20] and that rapidly home to and persist at the site of inflammation [21]. The dominant use of a restricted TCR repertoire has been successfully exploited in designing immunospecific therapies by means of anti-Vβ monoclonal antibodies [22] and active vaccination with epitopes of the encephalitogenic TCR molecule [23]. Nevertheless, the presence of an epitope dominance, as well as limited TCR usage within the T cell response to MBP, seems to be confined to the early phases of EAE and remains controversial in humans. Later stages of EAE are characterized by a more diverse recognition of previously cryptic determinants inside the MBP molecule (intra-molecular spreading) and within other CNS antigens (inter-molecular spreading) [24]. The appearance of a diverse T cell repertoire is further confirmed by the presence, in the spinal cord of rats with EAE, of a heterogeneous Vβ population during the recovery phase of disease [25]. This may result from apoptosis of encephalitogenic T cell clones [26] followed by secondary recruitment of activated cells from the recirculating T cell pool specific for minor antigens.

Recently, a chronic relapsing form of EAE has been successfully induced in Biozzi (AB/H) mice by immunization with spinal cord homogenate [27] or MOG [28]. The full spectrum of pathological lesions seen in MS, including sharp demyelination and remyelination, is seen in this species. Another useful model, recently described in DA rats, is characterized by a chronic relapsing course, inflammation and demyelination [29].

In contrast to the rather restricted response to MBP, a diverse recognition of determinants within other myelin antigens such as PLP [30] and MOG [28] seems to occur in most rodents. However as a general rule for all inbred species, strains of different haplotype appear to react with different epitopes of myelin antigens.

The presence of encephalitogenic T cells reacting against self-antigens in naive animals is of remarkable conceptual importance and supports the fact that autoaggressive T cells escaping thymic deletion are maintained within the normal circulating T lymphocyte pool [4, 5]. A possible explanation for self-reactive T cells escaping negative selection has been recently clarified in a MBP-/- transgenic model in which endogenous MBP inactivated high avidity clones reactive against the immunodominant epitope and made that determinant appear cryptic [31].

Thus, other factors are also necessary to cause autoimmune disease. The genetic background is a major factor conferring susceptibility to EAE; a number of murine loci have already been identified [32]. Both MHC and non-MHC genes have been reported to control the development and severity of EAE [33]. The role of environmental factors has been elegantly elucidated by Goverman and colleagues [34] who created transgenic mice expressing an MBP-specific TCR; these mice spontaneously developed EAE only when challenged with microbial stimuli. In a similar model, the complete cohort of anti-MBP TCR transgenic mice, deficient for mature T and B cells, developed spontaneous EAE, suggesting that other cells may have a protective role counteracting encephalitogenic cells [35]. A protective or regulatory role has been claimed for almost all cells involved in the immune response including CD4$^+$ [36], CD8$^+$ [37], and CD4$^-$CD8$^-$ [38] T cells, macrophages [39], B cells [40], γδ T cells [41] and NK cells [42].

The environment exerts a major effect on EAE, leading to the activation of potentially autoaggressive T cells which consequently home to the brain and induce disease [43]. Activation state is the necessary prerequisite for T cells to migrate through the BBB irrespective of their antigen specificity. Activation of myelin-specific T lymphocytes in the peripheral compartment can occur through a mechanism of molecular mimicry [44] and by stimulation with microbial superantigens [45]. Other theoretical possibilities such as activation of T cells carrying two sets of receptors, one specific for a foreign protein and another for a self-antigen, have never been demonstrated to play a role in autoimmunity [46]. Migration through the BBB involves adhesion molecules on both T cells (LFA-1 and VLA-4) and endothelial cells (ICAM-1 and VCAM-1) [47]. Following migration through the BBB, neuroantigen-specific CD4$^+$ T cells are reactivated in situ by fragments of myelin antigens presented in the framework of class II MHC molecules on the surface of local antigen-presenting cells including macrophages, microglia and, although less efficient, astrocytes [48]. These events are associated with the release of proinflammatory cytokines leading to the upregulation of MHC molecules on a variety of resident, antigen-presenting cells [49]. The kinetics of cytokines along the EAE clinical course suggests the role of T helper 1 (Th1) cytokines such as tumor necrosis factor (TNF)-α, TNF-β, interleukin (IL)-12, and interferon (IFN)-γ before and at the peak of disease. The recovery phase correlates with Th2 cytokines such as transforming growth factor (TGF)-β, IL-10 and possibly IL-4 [50]. The onset of overt inflammation also maintains endothelial activation and leads to a second wave of inflammatory recruitment, including T cells and macrophages that damage tissue by means of TNF-α [51], oxygen and nitrogen intermediates, perforin and complement [52] and demyelinating antibodies [9].

Thus far, no consistent evidence differentiates between activated MBP-specific T lymphocytes with no encephalitogenic capabilities and their pathogenic counterparts, in spite of identical growing and stimulation conditions, sharing of epitope specificity, and MHC restriction. Some studies suggest that differences in encephalitogenicity correlate with a predominant Th1 cytokine profile [53], their brain homing capacity [54] and the ability to mediate a delayed-type hypersensitivity (DTH) response [55]. Moreover, encephalitogenic T cells, despite their CD4+ phenotype, were cytotoxic for cells (e.g. astrocytes) presenting myelin antigens in an MHC-restricted manner [56]. On the other hand, non-encephalitogenic T cells may initiate autoimmune regulatory mechanisms through the production of IL-3 [57]. At least in Lewis rats, encephalitogenicity of MBP-specific T cells may correlate with the cytokine profile which depends on the MHC haplotype of the strain [58].

A rather simplistic picture of the EAE cytokine network suggests that TNF-α, TNF-β, IFN-γ and IL-12 (proinflammatory cytokines) have a disease-promoting role while TGF-β, IL-10 and possibly IL-4 (anti-inflammatory cytokines) protect from disease. Although a detailed analysis of the current literature on this topic is beyond the scope of this chapter, an enormous amount of data states that the real picture is much more complicated. Several factors influence the cytokine profile of effector and regulatory cells in EAE and, therefore, the final outcome of the immune response within the target organ. These include age of the animal [59], nature of antigen-presenting cells [60, 61], local cytokine micro-environment [62], selective engagement with costimulatory molecules [63] interaction with altered forms of the immunizing antigen [64], and the route of immunization [65].

Based on the hypothesis that Th1 cytokines play a promoting effect on autoimmunity while Th2 cytokines may have a protective role, immune deviation toward a Th2 profile has been exploited for successfully treating EAE by administration of anti-inflammatory cytokines [66], altered peptide ligands [64], monoclonal antibodies (MAb) affecting B7/CD28 interactions [63] or anti-inflammatory cytokines [67, 68], and by induction of oral tolerance [65]. Despite the success of most experimental treatments targeting Th1 cytokines, the Th1 versus Th2 dichotomy underscores the complexity of interactions that lead to reciprocal cross-regulation of Th1/Th2 responses. This has been dramatically elucidated by the severe aggravation of EAE in primates due to an enhanced Th2 response occurring after discontinuation of treatment for immune deviation [69]. Moreover, conflicting results arise from the utilization of genetically manipulated mice either lacking or over-expressing cytokines, as is the case in a number of TNF-α studies which demonstrated a different clinical phenotype depending on the experimental conditions. For example, over-expression of TNF-α in transgenic mice leads to spontaneous inflammation and demyelination within the CNS [70], while TNF-α deficient knock-out mice can still develop EAE, thus challenging the role of this cytokine in EAE pathogenesis [71]. Surprisingly, in a recent study, MOG-immunized, TNF-deficient mice developed severe EAE but were remarkably ameliorated by the administration of TNF-α, possibly supporting a protective role for this cytokine [72]. Targeting cytokine genes has helped to elucidate the role of other cytokines such as INF-γ [73], IL-4 and IL-10 [74], nitric oxide [75], and also of Fas/Fas-ligand and perforin pathways [76]. Nevertheless, it must be kept in mind that genetic manipulation results in experimental conditions that only partially represent the in vivo situation and that likely underscore the redundancy of the cytokine system.

Overall, the deep knowledge of the immunogenetics of inbred species and the possibility of successfully manipulating these animals, together with their accessibility and moderate costs, make these species the first choice for studies on autoimmune diseases of the CNS.

EAE in Outbred Species

EAE in non-human primates represented the first experimental model for demyelinating diseases of the CNS [1]. The recent advances in primate housing and handling techniques, knowledge of primate anatomy, immunology and genetics, and the compatibility with most human reagents and diagnostic techniques have sparked wide interest in EAE in these species. A unique advantage of monkeys arises from their outbred condition that closely resembles the human status. Moreover, the transfer of immunocompetent cells in outbred primates is allowed by the possibility of crossing the trans-species barrier among closely related species [77] and by the natural bone marrow chimerism in some others [5]. Therefore, in primates it is possible to elucidate the role of pathogenic cells by means of passive transfer experiments in a polymorphic setting. Recently, EAE has been induced in the common marmoset *Callithrix jacchus*, a unique primate species whose offspring develop in utero as genetically distinct twins or triplets sharing bone marrow-derived elements through a common placental circulation [78]. It has been recently demonstrated that *C. jacchus* TCR genes are extensively conserved [79] and that class II MHC region genes, despite a relatively low polymorphism, encode the evolutionary equivalents of the HLA-DR and -DQ molecules [80]. A fully demyelinating form of EAE has been induced by active immunization with whole myelin [78], MOG or MBP followed by administration of MOG-specific antibodies [81]. Passive transfer experiments have demonstrated that encephalitogenic MBP-specific T cells are part of the normal marmoset repertoire [5]. As in humans, MBP-reactive T cells recognize different determinants by means of a diverse TCR repertoire (A. Uccelli, unpublished results). EAE induced with whole myelin or MOG is characterized pathologically by perivascular inflammation with conspicuous primary demyelination whose topography correlates with magnetic resonance imaging (MRI) abnormalities [82]. On the contrary, immunization with MBP or passive transfer of MBP-reactive T cells leads to mild inflammation and no demyelination ([5] and G.L. Mancardi and A. Uccelli, unpublished results). A complex role for cytokines, possibly released by activated T lymphocytes and macrophages during the immune response, is suggested by the high expression of CD40 and CD40-ligands in marmoset active lesions [83]. The role of Th1 cytokines has been demonstrated by the prevention of disease following treatment with the cAMP-specific type IV phosphodiesterase inhibitor Rolipram [84]. On the other hand, the ambiguous role of Th2 cytokines was highlighted by the enhancement of EAE occurring after discontinuation of a MOG-based tolerization treatment due to an enhanced proliferative and antibody response to the antigen. Hence, it is likely that Th2-like T cells play a different role, protecting or favoring autoimmunity, under different conditions [69].

EAE has also been induced in macaques by immunization with myelin or MBP emulsified in complete adjuvant [85]. The EAE course is primarily hyperacute or acute, often with a lethal outcome, and is characterized by intense inflammation associated with hemorrhages and necrosis resembling acute disseminated encephalomyelitis [86]. The association of EAE-susceptibility with a class II MHC allele [87], the presence of myelin-reactive T cells correlating with the course of EAE [86], the beneficial effect of anti-CD4 antibodies on EAE outcome [88], and the possibility of inducing a mild form of EAE by adoptive transfer of MBP-specific T cells from unprimed animals [89] all provide strong evidence that T cells play a central role in this model.

At the moment, the major advantage of a non-human primate EAE model for human MS resides in the molecular and functional organization of the primate immune system, leading to the possibility of evaluating the safety and efficacy of biological molecules as therapy for MS.

Virus-Induced Demyelinating Diseases

CNS demyelination spontaneously occurs following infection with neurotropic viruses such as Theiler's virus [90]. Theiler's murine encephalomyelitis virus (TMEV), a natural mouse pathogen, is a picornavirus that induces a chronic demyelinating disease with a clinical course and histopathology similar to that of chronic-progressive MS. Viral persistence within the CNS is required for the immune system to mount a cellular and humoral response leading to demyelination [91]. The host response may trigger both protective and pathogenic immune responses which are the result of a balance between persistent viral infection and immune injury mediated by CD4+ or CD8+ T cells and antibodies [92]. In susceptible animals the lack of virus-specific cytotoxicity has been postulated to lead to demyelination. On the other hand, resistant strains clear the infection following acute encephalomyelitis, possibly due to the ability to generate an effective class I-restricted T-cell response [93]. As in EAE, epitope spreading to endogenous myelin determinants has been shown to play a key role in the chronic-progressive course of disease. Demyelination in TMEV-infected mice is initiated by a mononuclear inflammatory response mediated by virus-specific CD4$^+$ T cells targeting viruses, which chronically persists in the CNS. Following myelin destruction, activation of CD4$^+$ T cells specific for multiple myelin epitopes occurs, leading to disease progression [94]. Other models of virus-mediated demyelinating disease of the CNS are obtained by infections with the mouse hepatitis virus strain JHM (MHV-JHM) and corona virus.

Although almost the complete spectrum of MS-like lesions can be observed in virus models of demyelination, the mechanisms underlying the pathogenesis of immune response within the CNS are extremely complex and depend on the mutual interaction between virus and host, thus making it difficult to dissect the role of each factor in the pathogenesis of demyelinating diseases of the CNS.

References

1. Rivers TM, Sprunt DH, Berry GP (1933) Observations on the attempts to produce acute disseminated encephalomyelitis in monkeys. J Exp Med 58: 39-53
2. Paterson PY (1960) Transfer of allergic encephalomyelitis in rats by means of lymph node cells. J Exp Med 111:119-135
3. Zamvil SS, Steinman L (1990) The T lymphocyte in experimental allergic encephalomyelitis. Annu Rev Immunol 8: 579-621
4. Schlusener HJ, Wekerle H (1985) Autoaggressive T-lymphocyte lines recognizing the encephalitogenic portion of myelin basic protein: in vitro selection from unprimed rat T-lymphocyte populations. J Immunol 135: 3128-3133
5. Genain CP, Lee-Parritz D, Nguyen MH et al. (1994) In healthy primate, circulating autoreactive T cells mediate autoimmune disease. J Clin Invest 94: 1339-1345
6. Ben-Nun A, Wekerle H, Cohen IR (1981) The rapid isolation of clonable antigen specific T lymphocyte lines capable of mediating autoimmune encephalomyelitis. Eur J Immunol 11: 195-199
7. Yamamura T, Namikawa T, Endoh M et al. (1986) Passive transfer of experimental allergic encephalomyelitis induced by proteolipid apoprotein. J Neurol Sci 76: 269-275
8. Linington C, Berger T, Perry L et al. (1993) T cells specific for the myelin oligodendrocyte glycoprotein (MOG) mediate an unusual autoimmune inflammatory response in the central nervous system. Eur J Immunol 23: 1364-1372
9. Linington C, Bradl M, Lassmann H et al. (1988) Augmentation of demyelination in rat acute allergic encephalomyelitis by circulating mouse monoclonal antibodies directed against a myelin/oligodendrocyte glycoprotein. Am J Pathol 130: 443-454
10. Massacesi L, Vergelli M, Zehetbauer B et al. (1993) Induction of experimental autoimmune encephalomyelitis in rats and immune response to myelin basic protein in lipid bound form. J Neurol Sci 119: 91-98
11. Kojima K, Berger T, Lassmann H et al. (1994) Experimental autoimmune panencephalitis and uveoretinitis transferred to the Lewis rat by T lymphocytes specific for the S100β molecule, a calcium binding protein of astroglia. J Exp Med 180: 817-829

12. Raine CS (1997) The lesion in multiple sclerosis and chronic relapsing experimental allergic encephalomyelitis: A structural comparison. In: Raine CS, McFarland HF, Tourtellotte WW (eds) Multiple sclerosis: clinical and pathogenetic basis. Chapman and Hall, London, pp 242-286
13. Olitsky PK, Yager RH (1949) Experimental disseminated encephalomyelitis in mice. J Exp Med 90: 213-223
14. Lisak RP, Zweiman B, Kies MW, Driscoll B (1975) Experimental allergic encephalomyelitis in resistant and susceptible guinea pigs: in vivo and in vitro correlates. J Immunol 114: 546-549
15. Brown AM, McFarlin DE (1981) Relapsing experimental allergic encephalomyelitis in the SJL/J mouse. Lab Invest 45: 278-284
16. Mokhtarian F, McFarlin DE, Raine CS (1984) Adoptive transfer of myelin basic protein-sensitized T-cells produces chronic relapsing demyelinating disease in mice. Nature 309: 356-358
17. Sakai K, Sinha AA, Mitchell DJ et al. (1988) Involvement of distinct murine T-cell receptors in the autoimmune encephalitogenic response to nested epitopes of myelin basic protein. Proc Natl Acad Sci USA 85: 8608-8612
18. Burns FR, Li RX, Shen L et al. (1989) Both rat and mouse T-cell receptors specific for the encephalitogenic determinant of myelin basic protein use similar Vα and Vβ chain genes even though the major histocompatibility complex and encephalitogenic determinants being recognized are different. J Exp Med 169: 27-39
19. Lassmann H, Brunner C, Bradl M, Linington C (1988) Experimental allergic encephalomyelitis: the balance between encephalitogenic T lymphocytes and demyelinating antibodies determines size and structure of demyelinated lesions. Acta Neuropathol 75: 566-576
20. Vandenbark AA, Offner H, Reshef T et al. (1985) Specificity of T lymphocyte lines for peptides of myelin basic protein. J Immunol 139: 229-233
21. Kim G, Tanuma N, Kojima T et al. (1998) CDR3 size spectratyping and sequencing of spectratype-derived TCR of spinal cord T cells in autoimmune encephalomyelitis. J Immunol 160: 509-513
22. Zaller DM, Osman G, Kanagawa O, Hood L (1990) Prevention and treatment of murine experimental allergic encephalomyelitis with T-cell receptor Vβ-specific antibodies. J Exp Med 171: 1943-1955
23. Howell MD, Winters ST, Olee T et al. (1989) Vaccination against experimental encephalomyelitis with T cell receptor peptides. Science 246: 668-670
24. Lehmann PV, Forsthuber T, Miler A, Sercarz EE (1992) Spreading of T cell autoimmunity to cryptic determinants of an autoantigen. Nature 358: 155-157
25. Offner H, Buenafe AC, Vainiene M et al. (1993) Where, when and how to detect biased expression of disease relevant Vβ genes in rats with experimental autoimmune encephalomyelitis. J Immunol 151: 506-517
26. Bauer J, Wekerle H, Lassmann H (1995) Apoptosis in brain-specific autoimmune disease. Curr Opin Immunol 7: 839-843
27. Baker D, O'Neill JK, Gschmeissner SE et al. (1990) Induction of chronic relapsing experimental allergic encephalomyelitis in Biozzi mice. J Neuroimmunol 28: 261-270
28. Amor S, Groome N, Linington C et al. (1994) Identification of epitopes of myelin oligodendrocyte glycoprotein for the induction of experimental allergic encephalomyelitis in SJL and Biozzi AB/H mice. J Immunol 153: 4349-4356
29. Lorentzen JC, Issazadeh S, Storch M et al. (1995) Protracted relapsing and demyelinating experimental autoimmune encephalomyelitis in DA rats immunized with syngeneic spinal cord and incomplete Freund's adjuvant. J Neuroimmunol 63: 193-205

30. Greer JM, Sobel RA, Sette A et al. (1996) Immunogenic and encephalitogenic epitope clusters of myelin proteolipid protein. J Immunol 156: 371-379
31. Targoni OS, Lehmann PV (1998) Endogenous myelin basic protein inactivates the high avidity T cell repertoire. J Exp Med 187: 2055-2063
32. Sundvall M, Jirholt J, Yang HT et al. (1995) Identification of murine loci associated with susceptibility to chronic experimental autoimmune encephalomyelitis. Nat Genet 10: 313-317
33. Weissert R, Wallstrom E, Storch MK et al. (1998) MHC haplotype-dependent regulation of MOG-induced EAE in rats. J Clin Invest 102: 1265-1273
34. Goverman J, Woods A, Larson L et al. (1993) Transgenic mice that express a myelin basic protein-specific T cell receptor develop spontaneous autoimmunity. Cell 72: 551-560
35. Lafaille J, Nagashima K, Katsuki M, Tonegawa S (1994) High incidence of spontaneous autoimmune encephalomyelitis in immunodeficient anti-myelin basic protein T cell receptor transgenic mice. Cell 78: 399-408
36. Kumar V, Sercarz EE (1993) The involvement of T cell receptor peptide-specific regulatory CD4+ T cells in recovery from antigen induced autoimmune disease. J Exp Med 178: 909-916
37. Jiang H, Zhang SI, Pernis B (1992) Role of CD8+ T cells in murine experimental allergic encephalomyelitis. Science 256: 1213-1215
38. Kozovska MF, Yamamura T, Tabira T (1996) T-T cell interaction between CD4-CD8- regulatory T cells and T cell clones presenting TCR peptide. Its implications for TCR vaccination against experimental autoimmune encephalomyelitis. J Immunol 157: 1781-1790
39. Huitinga I, Van Rooijen N, De Groot CJA et al. (1990) Suppression of experimental allergic encephalomyelitis in Lewis rats after elimination of macrophages. J Exp Med 172: 1025-1033
40. Wolf SD, Dittel BN, Hardardottir F, Janaway CA (1996) Experimental autoimmune encephalomyelitis induction in genetically B-cell deficient mice. J Exp Med 184: 2271-2278
41. Kobayashi Y, Kaway K, Ito K et al. (1997) Aggravation of murine experimental allergic encephalomyelitis by administration of T-cell receptor γδ-specific antibodies. J Neuroimmunol 73: 169-174
42. Zhang BN, Yamamura T, Kondo T et al. (1997) Regulation of experimental autoimmune encephalomyelitis by natural killer (NK) cells. J Exp Med 186: 1677-1687
43. Naparstek Y, Ben-Nun A, Holoshitz J et al. (1983) T lymphocyte line producing or vaccinating against autoimmune encephalomyelitis (EAE). Functional activation induces peanut agglutinin receptors and accumulation in the brain and thymus of line cells. Eur J Immunol 13: 418-423
44. Ufret-Vincenty RL, Quigley L, Tresser N et al. (1998) In vivo survival of viral antigen-specific T cells that induce experimental autoimmune encephalomyelitis. J Exp Med 188: 1725-1738
45. Brocke S, Gaur A, Piercy C et al. (1993) Induction of relapsing paralysis in experimental autoimmune encephalomyelitis by bacterial superantigen. Nature 365: 642-644
46. Padovan E, Giachino C, Cella M et al. (1995) Normal T lymphocytes can express two different T cell receptor β chains: implications for the mechanism of allelic exclusion. J Exp Med 181: 1587-1591
47. Cannella B, Cross AH, Raine CS (1990) Upregulation and coexpression of adhesion molecules correlates with experimental autoimmune demyelination in the central nervous system. J Exp Med 172: 1521-1524

48. Myers KJ, Dougherty JP, Ron Y (1993) In vivo antigen presentation by both brain parenchymal cells and hematopoietically derived cells during the induction of experimental autoimmune encephalomyelitis. J Immunol 15: 2252-2260
49. Traugott U, McFarlin DE, Raine CS (1986) Immunopathology of the lesion in chronic relapsing experimental autoimmune encephalomyelitis in the mouse. Cell Immunol 99: 394-410
50. Olsson T (1994) Role of cytokines in multiple sclerosis and experimental autoimmune encephalomyelitis. Eur J Neurol 1: 7-19
51. Selmaj KW, Raine CS (1988) Tumor necrosis factor mediates myelin and oligodendrocyte damage in vitro. Ann Neurol 23: 339-346
52. Compston DA, Scolding NJ (1991) Immune-mediated oligodendrocyte injury. Ann N Y Acad Sci 633: 196-204
53. Conboy YM, DeKruyff R, Tate KM et al. (1997) Novel genetic regulation of T helper 1 (Th1)/Th2 cytokine production and encephalitogenicity in inbred mouse strains. J Exp Med 185: 439-451
54. Baron JL, Madri JA, Ruddle NH et al. (1993) Surface expression of $\alpha 4$ integrin by CD4 T cells is required for their entry into brain parenchyma. J Exp Med 177: 57-68
55. Beraud E, Balzano C, Zamora AJ et al. (1993) Pathogenic and non-pathogenic T lymphocytes specific for the encephalitogenic epitope of myelin basic protein: functional characteristics and vaccination properties. J Neuroimmunol 47: 41-54
56. Sun D, Wekerle H (1986) Ia-restricted encephalitogenic T lymphocytes mediating EAE lyse autoantigen-presenting astrocytes. Nature 320: 70-72
57. Jeong MC, Itzikson A, Uccelli A et al. (1998) Differential display analysis of encephalitogenic mRNA. Int Immunol (*in press*)
58. Mustafa M, Vingsbo C, Olsson T et al. (1993) The major histocompatibility complex influences myelin basic protein 68-86-induced T-cell cytokine profile and experimental autoimmune encephalomyelitis. Eur J Immunol 23: 3089-3095
59. Forsthuber T, Yip HC, Lehmann PV (1996) Induction of TH1 and TH2 immunity in neonatal mice. Science 271: 1728-1730
60. Krakowski ML, Owens T (1997) The central nervous system environment controls effector CD4+ T cell cytokine profile in experimental allergic encephalomyelitis. Eur J Immunol 27: 2840-2847
61. Aloisi F, Ria F, Penna G, Adorini L (1998) Microglia are more efficient than astrocytes in antigen processing and in Th1 but not Th2 cell activation. J Immunol 160: 4671-4680
62. Falcone M, Bloom BR (1997) A T helper cell 2 (Th2) immune response against non-self antigens modifies the cytokine profile of autoimmune T cells and protects against experimental allergic encephalomyelitis. J Exp Med 185: 901-907
63. Kuchroo VK, Das MP, Brown JA et al. (1995) B7-1 and B7-2 costimulatory molecules activate differentially the Th1/Th2 developmental pathways: application to autoimmune disease therapy. Cell 80: 707-718
64. Nicholson LB, Greer JM, Sobel RA et al. (1995) An altered peptide ligand mediates immune deviation and prevents autoimmune encephalomyelitis. Immunity 3: 397-405
65. Weiner HL (1997) Oral tolerance: immune mechanisms and treatment of autoimmune diseases. Immunol Today 18: 335-343
66. Racke MK, Bonomo A, Scott DE et al. (1994) Cytokine-induced immune deviation as a therapy for inflammatory autoimmune disease. J Exp Med 180: 1961-1966
67. Ruddle NH, Bergman CM, McGrath KM et al. (1990) An antibody to lymphotoxin and tumor necrosis factor prevents transfer of experimental allergic encephalomyelitis. J Exp Med 172: 1193-1200

68. Leonard JP, Waldburger KE, Goldman SJ (1995) Prevention of experimental autoimmune encephalomyelitis by antibodies against interleukin 12. J Exp Med 181: 381-386
69. Genain CP, Abel K, Belmar N et al. (1996) Late complications of immune deviation therapy in a nonhuman primate. Science 274: 2054-2057
70. Taupin V, Renno T, Bourbonniere L et al. (1997) Increased severity of experimental autoimmune encephalomyelitis, chronic macrophage/microglial reactivity, and demyelination in transgenic mice producing tumor necrosis factor-alpha in the central nervous system. Eur J Immunol 27: 905-913
71. Frei K, Eugster HP, Bopst M et al. (1997) Tumor necrosis factor alpha and lymphotoxin alpha are not required for induction of acute experimental autoimmune encephalomyelitis. J Exp Med 185: 2177-2182
72. Liu J, Marino MW, Wong G et al. (1998) TNF is a potent anti-inflammatory cytokine in autoimmune-mediated demyelination. Nat Med 4: 78-83
73. Ferber IA, Brocke S, Taylor-Edwards C et al. (1996) Mice with a disrupted IFN-gamma gene are susceptible to the induction of experimental autoimmune encephalomyelitis (EAE). J Immunol 156: 5-7
74. Bettelli E, Das MP, Howard ED et al. (1998) IL-10 is critical in the regulation of autoimmune encephalomyelitis as demonstrated by studies of IL-10- and IL-4-deficient and transgenic mice. J Immunol 161: 3299-3306
75. Sahrbacher UC, Lechner F, Eugster HP et al. (1998) Mice with an inactivation of the inducible nitric oxide synthase gene are susceptible to experimental autoimmune encephalomyelitis. Eur J Immunol 28: 1332-1338
76. Malipiero U, Frei K, Spanaus KS et al. (1997) Myelin oligodendrocyte glycoprotein-induced autoimmune encephalomyelitis is chronic/relapsing in perforin knockout mice, but monophasic in Fas- and Fas ligand-deficient lpr and gld mice. Eur J Immunol 27: 3151-3160
77. Bontrop RE, Otting N, Slierendregt BL, Lanchbury JS (1995) Evolution of major histocompatibility complex polymorphisms and T-cell receptor diversity in primates. Immunol Rev 143: 43-62
78. Massacesi L, Genain CP, Lee-Parritz D et al. (1995) Chronic relapsing experimental autoimmune encephalomyelitis in new world primates. Ann Neurol 37: 519-530
79. Uccelli A, Oksenberg JR, Jeong M et al. (1997) Characterization of the TCRB chain repertoire in the New World monkey *Callithrix jacchus*. J Immunol 158: 1201-1207
80. Antunes SG, de Groot NG, Brok H et al. (1998) The common marmoset: a new world primate species with limited MHC class II variability. Proc Natl Acad Sci USA 95: 11745-11750
81. Genain CP, Nguyen MH, Letvin NL et al. (1995) Antibody facilitation of multiple sclerosis-like lesions in a non-human primate. J Clin Invest 96: 2966-2974
82. 't Hart BA, Bauer J, Muller HJ et al. (1998) Histopathological characterization of magnetic resonance imaging-detectable brain white matter lesions in a primate model of multiple sclerosis: a correlative study in the experimental autoimmune encephalomyelitis model in common marmosets (*Callithrix jacchus*). Am J Pathol 153: 649-663
83. Laman JD, van Meurs M, Schellekens MM et al. (1998) Expression of accessory molecules and cytokines in acute EAE in marmoset monkeys (*Callithrix jacchus*). J Neuroimmunol 86: 30-45
84. Genain CP, Roberts T, Davis R et al. (1995) Prevention of autoimmune demyelination in non-human primates by a cAMP-specific phosphodiesterase inhibitor. Proc Natl Acad Sci USA 92: 3601-3605

85. Shaw CM, Alvord EC, Hruby S (1988) Chronic remitting-relapsing experimental allergic encephalomyelitis induced in monkeys with homologous myelin basic protein. Ann Neurol 24: 738-748
86. Massacesi L, Joshi N, Lee-Parritz D et al. (1992) Experimental allergic encephalomyelitis in cynomolgus monkeys. Quantitation of T cell responses in peripheral blood. J Clin Invest 90: 399-404
87. Slierendregt BL, Hall M, 't Hart B et al. (1995) Identification of an MHC-DPB1 allele involved in susceptibility to experimental autoimmune encephalomyelitis in rhesus macaques. Int Immunol 7: 1671-1679
88. Van Lambalgen R, Jonker M (1987) Experimental allergic encephalomyelitis in the rhesus monkey. Treatment of EAE with anti-T lymphocyte subset monoclonal antibodies. Clin Exp Immunol 67: 305-312
89. Meinl E, Hoch RM, Dornmair K et al. (1997) Differential encephalitogenic potential of myelin basic protein-specific T cell isolated from normal rhesus macaques. Am J Pathol 150: 445-453
90. Theiler M (1934) Spontaneous encephalomyelitis of mice: a new virus disease. Science 80: 122
91. Rodriguez M, Pavelko KD, Njenga MK et al. (1996) The balance between persistent virus infection and immune cells determines demyelination. J Immunol 157: 5699-5709
92. Lin X, Pease LR, Murray PD, Rodriguez M (1998) Theiler's virus infection of genetically susceptible mice induces central nervous system-infiltrating CTLs with no apparent viral or major myelin antigenic specificity. J Immunol 160: 5661-5668
93. Lin X, Pease LR, Rodriguez M (1997) Differential generation of class I H-2D- versus H-2K-restricted cytotoxicity against a demyelinating virus following central nervous system infection. Eur J Immunol 27: 963-970
94. Miller SD, Vanderlugt CL, Begolka WS et al. (1997) Persistent infection with Theiler's virus leads to CNS autoimmunity via epitope spreading. Nat Med 3: 1133-1136
95. Tuohy VK, Lu Z, Sobel RA, Laursen RA, Lees MB (1989) Identification of an encephalitogenic determinant of myelin proteolipid protein for SJL mice. J Immunol 142: 1523-1527
96. Mendel I, Kerlero de Rosbo N, Ben-Nun A (1995) A myelin oligodendrocyte glycoprotein peptide induces typical chronic experimental autoimmune encephalomyelitis in H-2b mice: fine specificity and T cell receptor V beta expression of encephalitogenic T cells. Eur J Immunol 215: 1951-1959
97. Kerlero de Rosbo N, Mendel I, Ben-Nun A (1995) Chronic relapsing experimental autoimmune encephalomyelitis with a delayed onset and an atypical clinical course, induced in PL/J mice by myelin oligodendrocyte glycoprotein (MOG)–derived peptide: preliminary analysis of MOG T cell epitopes. Eur J Immunol 25: 985-993
98. Hoffman PM, Gaston DD, Spitler LE (1973) Comparison of experimental allergic encephalomyelitis induced with spinal cord, basic protein, and synthetic encephalitogenetic peptide. Clin Immunol Immunopathol 1: 364-371
99. Freund, Lipton MM, Morrison LR (1950) Arch Patol 50: 108

Chapter 22

MHC and Non-MHC Genetics of Experimental Autoimmune Encephalomyelitis

T. Olsson, I. Dahlman, E. Wallström

Introduction

Our present understanding of the ethiopathogenesis of multiple sclerosis (MS) and its experimental models, e.g. various forms of experimental autoimmune encephalomyelitis (EAE), includes three cornerstones: (1) autoimmunity to components of the central nervous system (CNS), (2) genetic predisposition, and (3) environmental influences.

Details in steps leading to autoaggressive T and B cell-mediated damage to the CNS have been, and are continuously dissected, mainly in EAE. However, genetically regulated bottlenecks which allow disease in one individual, but not in another are still largely unknown. Exact definition of these may allow development of rational therapy. The reasons to study MS/EAE genetics are thus not to develop prenatal genetic diagnosis or any other genetic preventive measures, but instead to find proper targets for therapy. In this chapter the genetics of MS is first briefly summarized, followed by a more extensive description of technical and biological issues related to the dissection of the genetic regulation of EAE.

Genetic Basis for the Susceptibility to MS

There are several observations supporting a genetic basis for MS. First, twin studies have shown an approximate concordance rate between monozygotic twins of 30%, while dizygotic twins are concordant in 2%-4%, similar to siblings [1]. This should be compared to the prevalence in the general population of 0.1%-0.2%. Besides documenting a genetic basis for MS, the figures also demonstrate the need for environmental events to trigger the disease. Second, one study in Canada on children adopted before one year of age have shown influence of genetic factors alone on the familial aggregation of MS [2]. This argues against unique environmental agents causing MS, but does not rule out common infectious agents or factors such as diet and climate. Adding evidence against unique environmental factors, maternal and paternal half-sibs display similar age-adjusted MS rates,

---

Neuroimmunology Unit, Center for Molecular Medicine, Department of Medicine, Karolinska Institutet, 171 76 Stockholm, Sweden. e-mail: tomas.olsson@cmm.ki.se

also arguing against intrauterine and perinatal factors, as well as genomic imprinting and mitochondrial genes [3].

The drops in concordance rates from monozygotic to dizygotic twins show that more than one gene is involved. Thus, the disease is polygenic and the number of genes has been theoretically estimated at 5-10 [4]. The HLA complex is so far the only well established genome region known to influence MS. This has mainly been documented in studies of sporadic cases compared to healthy controls [5], but recently also in whole genome scans of family materials [6–8]. However, estimations of the relative risk provided by the HLA complex varies between 5%-60% [5, 9, 10]. In any case, yet undefined non-HLA genes must also be of importance.

Genetic influences of the HLA Complex on MS

Since approximately 30 years ago, it has been known that certain haplotypes of this gene region predispose for MS. In particular individuals carrying HLA-DR2 (DRB1*1501-DRB5*0101-DQA1*0102-DQB1*0602) are at higher risk [11]. HLA-DR3 appears to provide some risk increase [11], while in certain populations DR4 may be important [12]. Notably, no single HLA type seems to exclude MS.

The HLA region codes for numerous immunoregulatory genes. Since the region mostly is inherited "in blocks", e.g. there are few recombinations in the region, it is difficult to pinpoint particular genes in humans responsible for the increase in risk. However, the class II HLA region, harboring the class II MHC genes, shows the strongest association [13]. The class II molecules present protein fragments to T cells. A common speculation for the risk increase of certain HLA types is that allelic differences in the capability to present autoantigenic peptides leads to higher numbers or more pathogenic autoreactive T cells. This could take place either in the thymus, with differences in the clonal deletion of autoreactive cells, or in the periphery with differences in activation of such cells. The presumed genetic influence on MS by molecules in the trimolecular complex, i.e. the class II molecules, also forms the background to the studies aiming at defining so-called immunodominant immune responses to epitopes of candidate myelin autoantigens like proteolipid protein (PLP), myelin basic protein (MBP) and myelin oligodendrocyte glycoprotein (MOG) [14]. Mapping of dominant encephalitogenic myelin epitopes, which depend on the MHC haplotype, has allowed antigen-specific therapy in EAE. Thus, there is hope that definition of such dominant responses to a restricted number of epitopes in humans may allow selective immunotherapy. Furthermore, considering the genetic influence of the class II HLA region, it is natural that many of the new, more or less immunoselective therapies tried in phase I trials aim at interfering with the trimolecular complex (MHC-antigenic peptide-T cell receptor). However, it is still not excluded that neighboring genes are instrumental, or that the DR or DQ molecules are the most important. For example, recent evidence suggests that certain class I HLA alleles can either increase or decrease the risk for MS [15]. Finally, the mechanisms for the MHC influence are still unclear and experimentation on this may to a large extent only be done in proper experimental models of MS, such as certain forms of EAE.

Non-HLA Gene Influences on MS

Approaches to detect non-HLA genes regulating MS susceptibility have included: (1) association studies of candidate genes of sporadic MS cases compared to healthy controls, and (2) whole genome scans of families studied with linkage analysis. The status of these two forms of analysis will be presented in the following discussion.

Association studies of candidate genes have repeatedly shown an influence of certain HLA haplotypes on the risk to develop MS, but as a rule for a long series of other candidate genes, initial claims of influence have not been reproduced in independent materials (reviewed in [9, 11]). There are several reasons for these difficulties. The common assumption that MS is due to an immunological dysregulation has focused attention on genes important for inflammation/immunity. However, these may be numerous (thousands), and it will be difficult to make educated guesses on which are important. A second obstacle may relate to ethnicity. Genes disposing for MS may well differ in populations with different ethnic origins. The different HLA haplotype association in Sardinia [12] may be one example of this. Third, the control material must be accurate, e.g. population-based, ethnically comparable and from the same geographic region as the MS material. This has not always been the case in association studies. Despite these problems, association studies with larger materials and new gene technologies may prove valuable, especially if candidate genes or regions can be defined in a rational way, for example in animal models.

A series of whole genome scans has recently been published [6-8, 16, 17]. The methodology used has mainly been mapping of anonymous microsattelite markers to identify alleles shared by affected sibling-pairs. Combined experience from these studies are as follows:
1. markers within or close to the HLA complex have shown some degree of linkage in all scans;
2. no non-HLA region has shown unequivocal linkage to disease;
3. regions with suggestive linkage largely vary between studies. However,
4. certain regions start to emerge as "hot spots", such as the 17q22-q24 region [16];
5. stratification for the risk increasing HLA haplotype (HLA-DR2) reveals certain loci that depend on this locus and others that are independent [17];
6. a gene, apart from the HLA complex, with major impact on disease is highly unlikely;
7. linkage analysis of family materials will require hundreds of families to positionally clone genes, since each gene has only a modest or low impact on MS.

There are several reasons for these partly frustrating results:
1. genetic heterogeneity. Different genes may predispose for the same phenotype, e.g. MS, and these may vary between families and ethnic origins. In this case, collection of additional families may "dilute out" effects of important genes;
2. the disease disposing genes may contribute with small increases in the risk to developing MS. In this case it may not be possible to collect large enough family to detect a minor genetic effect;
3. furthermore, positional cloning is as yet not achieved in human polygenic diseases and it will be difficult to experimentally evaluate a gene of putative relevance.

Many groups now are on their way, or are considering to, study large population-based samples by association analysis, focusing on "hot spots" defined in linkage analysis. This is in addition to the technique named transmission dysequilibrium test (TDT), in which the allele frequency in parents of the affected individual is used as control. The development of the human genome project and the availability of dense new genetic markers may enable performing whole genome scans using association analysis.

An alternative and additional strategy is to study gene influences in animal models and by synteny comparisons, define candidate genes to be tested by association in human materials. Even if not the same as in MS, genes defined in animals may unravel pathogenetic pathways of relevance also in human disease.

Considerations on EAE Models Suitable for Genetic Analysis

The term "model" implies that the experimental disease in focus for study not is identical to the human disease. Instead, the experimental models offer possibilities to study disease-causing mechanisms that are common between species. To reveal pathways of importance for human disease, the models should aim at mimicking MS as closely as possible.

A first point to consider is whether to use spontaneously occurring disease or disease induced by experimental manipulation. In organ-specific inflammatory disease afflicting the Langerhans islets, i.e. diabetes, rodent models exist which can be argued to develop spontaneously such as in nonobese diabetic (NOD) mice or Bio-Breeding (BB) rats. However, also in these, there are strong environmental influences on disease incidence. There is so far no description of spontaneously occurring MS-like disease in non-manipulated animals. Instead, all models either involve active immunization with myelin antigens and adjuvant, transfer of disease with myelin autoreactive immunocompetent cells, or viral infection. Also in MS, it can be argued that the disease is not spontaneous, but instead induced, although the triggering agent is unknown. This is clear in consideration of the 30% concordance rate between monozygotic twins. Thus, we think that the need for active induction of EAE in rodents may be an advantage rather than a drawback.

A second important point of relevance is the disease course. Perhaps the most characteristic feature of MS is chronicity. Therefore gene influences allowing such a disease course are important to find, and the EAE model used should be chronic. Commonly employed EAE models in mice or rats have monophasic disease courses and are therfore less attractive for genetic analysis. However, increasing numbers of chronic and/or relapsing EAE models in both mice and rats have emerged.

In mice, a variety of immunization protocols induce chronic disease in B10 RIII [18], SJL/J [19] or Biozzi strains [20]. Most mouse models require quite intense adjuvants to develope disease. In addition to the mycobacteria and mineral oil in Freund's complete adjuvant (FCA), mice are commonly injected with pertussis toxin. The amount and types of adjuvants used may well affect the sets of gene influences detected. In any case, identified genome regions with influences on disease may have relevance for human disease. There are now defined genome regions that regulate adjuvant-induced inflammation leading to arthritis [21–23].

The most commonly used strain of rat, LEW, usually develops disease with a monophasic course when immunized with either whole spinal cord homogenate or MBP. However, there are also rat EAE models displaying chronicity. The DA or LEW AV1 rat strains, immunized either with whole spinal cord homogenate or MOG in adjuvant develop chronic relapsing disease, and are therefore attractive for genetic analysis [24–26]. Of importance for the genetic analysis is that there are other rat strains which are resistant or at least much less susceptible with the same immunization protocols (see the section "Non-MHC Gene Influences on EAE").

A third matter of interest is the mimicry of lesional pathology. This seems to be affected mainly by the myelin autoantigen used for immunization. The list of CNS antigens that are able to induce EAE is steadily increasing. Commonly used myelin antigens include MBP, PLP and MOG. Use of MBP or PLP tends to induce EAE with considerable inflammation in the CNS, but with little or no demyelination, a hallmark of human MS [27]. MOG is so far unique among myelin antigens in its ability to induce a combined autoimmune pathogenic T and B cell response, leading to demyelination. The surface exposure of MOG on the outer lamellae of the myelin sheath allows binding of demyelinating autoantibodies [28]. The lesional pathology of MOG-induced EAE in rats closely resembles that seen in MS [25, 26]. This model is therefore attractive for genetic analysis.

There are thus options to use both mouse and rat EAE models for genetic studies. Mouse models have advantages by requiring less space, and follow-up studies using gene deletions and transgenic expression of candidate genes can be performed. In addition, more genes have been mapped in mice and there is better access to immunological reagents of importance for the phenotypic analysis. Apparent advantages with rat EAE models are a more MS-like disease, less intense protocols for disease induction and, in our experience, a more stable and reproducible phenotype, e.g. disease outcome, of putative importance in later positional cloning attempts. Mapping studies in both species are necessary, since the more species and strains that are examined, the more likely it is to find a maximum number of disease regulatory genes. This chapter mainly adresses rat EAE.

Technical Approaches for Genetic Studies in Rodents

In the following we discuss existing strains, strategies and techniques for genetic analysis focusing on the rat species.

Use and Availability of MHC/Non-MHC Congenes in Rats

Since EAE is regulated by genes within and outside the MHC complex, it is almost invaluable to have access to congenic inbred strains where either the non-MHC background genome is kept constant and the MHC haplotypes are varied, or vice versa. Thus, when studying non-MHC background gene influences, it is desirable to keep the MHC haplotype constant since the genetic influence from this region on a particular EAE form in most cases already is known. In this way, the number of animals neccessary to detect other gene regions with influence will be drastically reduced. When studying regulatory influences due to polymorphisms of genes in the MHC complex, it is instead of interest to keep the background genome constant. In certain instances, there are also rat strains with recombinations within the MHC complex, allowing a rough mapping of influences within this region. A few examples of such rat strains illustrating these matters are given in Table 1. The method to obtain a congenic strain by selective backcrossing is discussed in the following section.

Compared to mice, the availability of inbred congenic and recombinant congenic strains in rats is much more limited. This has several reasons, one being that rats are larger than mice and therefore more costly to breed and propagate. Another reason is the lack of programs for embryo freezing. The congenic rats that are available can be bought from commercial suppliers such as Charles

Table 1. Examples of inbred rat strains with differing rat MHC (RT1) haplotypes or non-MHC background genomes. Donor strains of congenic gene fragment are shown in parentheses

Type of congene	Strain	Haplotype	Class I	Class II	Class III
Constant MHC on varied background	DA	av1	a	a	av1
	ACI	av1	a	a	av1
	PVG-RT1a (DA)	av1	a	a	av1
Varied MHC on constant background	LEW.1AV1(DA)	av1	a	a	av1
	LEW.1A(AVN)	a	a	a	a
	LEW.N(BN)	n	n	n	n
	LEW.W(WP)	u	u	u	u
	LEW	l	l	l	l
Intra-MHC recombinant strains	LEW.1AR1	r2	a	u	u
	LEW.1AR2	r3	a	a	u
	LEW.1WR1	r4	u	u	a
	LEW.1WR2	r6	u	a	a

River Laboratories (www.criver.com) or obtained through cooperation with university institutions such as Zentrale Tierlaboratorium, Medizinische Hochschule Hannover (Prof. Hans Hedrich; www.mh-hannover.de/institut/tierlabor/hometier.htm). So far, mainly MHC congenic and intra-MHC recombinant congenic rats have been available, most often on the LEW or the PVG non-MHC background [29].

With the current interest in mapping non-MHC genes in complex experimental autoimmune disorders such as EAE and autoimmune arthritis, it can be expected that congenic strains for non-MHC genome regions may also become available.

When planning experiments with congenic strains acquired from commercial suppliers or other research groups it is advisable to consider that in congenic strains, there may be remaining gene fragments from the congenic donor strain. The congenic strain may also have changed genetically due to genetic drift during the propagation of the rats after the congenization. To test this possibility experimentally, an F2 intercross between the congenic strain and the donor strain can be investigated for the phenotype of interest and correlated to the presence or absence of the congenic fragment [29]. If genetic drift is suspected, the congenic strain may be crossed with the founder strain again, in the same manner as described for the generation of a new congenic strains (discussed in the following section). Experiments in F2 intercrosses may also detect effects of remaining gene fragments at other loci in the congenic strain, derived from the donor. It is therefore always advisable to substantiate phenotypic findings in congenic strains.

Principles and Techniques for Linkage Analysis in Mouse and Rat Crosses

Genetic Mapping in Controlled Crosses. Mapping susceptibility genes for complex disorders in crosses between inbred rodent strains has the advantage that genetic heterogeneity can be avoided and the environmental influence reduced. Susceptibility gene mapping can be separated into (1) identification and (2) finemapping of susceptibility loci.

Strategies for Identification of Susceptibility Loci. Three different breeding schemes can be used to identify susceptibility loci in experimental animals: F2 intercross, backcross and recombinant inbred strains (Fig. 1) (reviewed in [30]). In all cases, the investigator initially identifies two inbred strains whose phenotype differs with regard to the trait of interest, i.e. susceptibility or resistance to EAE. These strains are bred to produce heterozygous F1 animals, all of whom have an identical genotype and a phenotype that in most cases is intermediate between the two parental strains. When an F2 intercross is used for genetic mapping, the heterozygous rats are then brother-sister mated to produce a large F2 population. The F2 individuals will carry different combinations of the two parental genomes and as a consequence the phenotype will differ between these animals due to differences in the genetic background. Genes controlling a phenotype can be identified in this F2 population by linkage analysis, using markers that are evenly distributed across the genome. If a susceptibility locus exists close to a marker, then the phenotype of the animals in the three genotype classes (homozygous strain A, heterozygous AB, homozygous strain B) should differ from one another. Statistical analysis is used to precisely map such loci and to rule out effects of chance alone.

As an alternative to the F2 intercross, a backcross can be used for genetic mapping. The drawback with a backcross is that dominantly and recessively acting genes cannot be detected in the same cross. Therefore, F2 intercrosses are more common in linkage analysis of polygenic diseases.

Recombinant inbred strains are a set of strains carrying different combinations of two parental genomes, which are produced by brother-sister meeting of an F2 intercross for at least 20 additional generations. Once the recombinant inbred strains have been genotyped, they can be used to map susceptibility loci for any phenotype which segregates between the strains. The limiting factor for the use of recombinant inbred strains for genetic mapping is the availability of a set of strains which differs with regard to the studied phenotype.

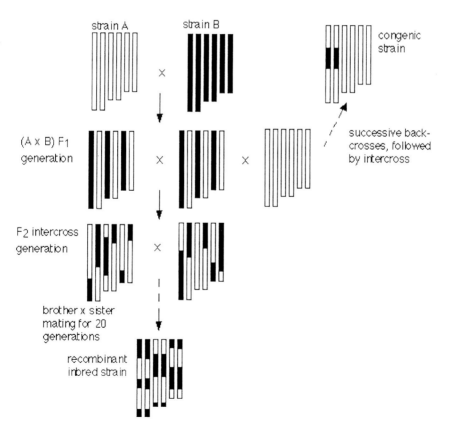

Fig. 1. Common experimental crosses used in genetic mapping. An individual is here represented by three pairs of homologous chromosomes (*thick lines*). See text for details about the crosses. *White*, strain A genome; *black*, strain B genome

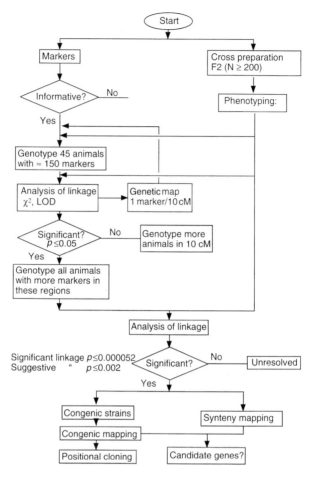

Fig. 2. Flowchart illustrating the steps in a genome-wide search for susceptibility loci for a complex disease

In practice, a whole genome search for susceptibility loci for a complex disease requires at least 200 F2 animals (Fig. 2). Initially, these rats are phenotyped. This implies assessment of disease outcome, preferably as quantitative as possible. Besides determining whether the animals are susceptible or resistant to the disease of interest, this may also include measuring phenotypes which are thought to be causally related to the disease outcome. In EAE this may be features such as degree of CNS inflammation or degree of myelin antigen-specific autoimmunity. For a complex trait in particular, it is valuable to measure more "basic" (less complex) phenotypes which may be involved in the pathogenesis of the disease. Such phenotypes probably have a simpler genetic regulation than the clinical disease outcome which will represent the net outcome of a series of genes when the environmental trigger is kept constant.

As the next step, a subset of the F2 animals, i. e. the most susceptible and resistant, are selected for a genome scan. These animals are genotyped with approximately one marker every 10-20 cM. The genetic distance between adjacent markers is then determined by the use of a linkage program such as MAPMAKER [31]. The rational for selecting the extreme animals for this initial genome scan is that for a complex trait, for any disease regulatory gene the extreme individuals are more likely to have the susceptible or resistant genotype, respectively. Next, all genotype and phenotype data is analyzed to identify regions linked to disease susceptibility. If the phenotype is quantitative, computer programs can be used in this linkage analysis, whereas for qualitative traits chi-square analysis is more commonly applied for statistical analysis.

In genome regions showing support for linkage to a phenotype in the initial genome scan, additional animals are genotyped with more markers located close to the region of interest. The linkage analysis is then repeated.

Due to the multiple independent tests in a genome scan, a very low nominal p-value is required for genome-wide significance [32]. If such loci are identified, their further characterization starts with breeding of congenic strains (see the section "Congenization and Fine-Mapping of Susceptibility Loci"). In contrast, if at this stage no region shows significant linkage to disease susceptibility, this means that the influence of individual loci on the disease is too small to be detected with the available number of animals. Thus, in this case the number of animals in the study has to be enlarged.

Genetic mapping requires markers covering the genome. The markers currently employed are microsatellite markers, which are short nucleotide repeats, 1-6 basepairs long. They are highly polymorphic within the population, i.e. different individuals have different numbers of repeats. This length variation can easily be assessed by polymerase chain reaction (PCR) using microsatellite-specific primers, followed by electrophoresis. Microsatellites are common in the genome and the mouse and rat genetic maps contain many thousands of microsatellite markers [33].

Congenization and Fine-Mapping of Susceptibility Loci. A susceptibility locus for a complex disease identified by linkage analysis usually extends over several cM. The first step to be able to study the influence of the locus on the disease in isolation and to fine-map the region is to breed a congenic strain, e.g. move this genome region from a donor to a recipient strain. This is done by selective backcrossing of the F1 animals onto one of the parental strains for ten generations followed by an intercross [34]. With each crossing, random genomic recombinations occur. In each generation, the animals are genotyped within the region to check that the donor allele is preserved, and such animals are used in the further backcrossing. In addition, with each generation more of the donor genome unlinked to disease susceptibility is lost (Fig. 1). So-called speed congenic strains can be obtained by genotyping the background genome and selecting the animal which has lost most of the alleles from donor strain for further breeding. Hereby, the number of backcrosses can be reduced [35, 36].

After having established a congenic strain in this way, a next step is to repeat extensive phenotypic characterization of the congenic strain in comparison with the parental strain. Ideally, these studies may give clues to critical steps in the pathogenesis of the disease and reveal possible candidate genes in the region.

If a discriminatory phenotype remains between the parental and congenic strain, the congenic region which may extend over 20 cM can be narrowed down by further breeding. Large crosses between the congenic and parental strain are established to obtain a number of strains with recombinations within the congenic region. By determining the smallest region segregating with disease susceptibility, the region harboring the susceptibility gene can be reduced to approximately 1 cM (Fig. 3), which in the rat and mouse usually represents more than 1 Mb. This is not precise enough to clone the gene. Therefore, a physical map for the region, consisting of overlapping YAC and BAC clones must be obtained. The disease gene can be identified by sequencing the BAC and/or YAC clones and by further functional analyses.

Comparative Mapping. Comparative mapping implies definition of genome regions containing conserved gene orders in different species. In genetic mapping studies using experimental animals, it serves two purposes: (i) to determine whether the syntenic or identical human genome region predisposes to the disease and (ii) to identify candidate genes within the region.

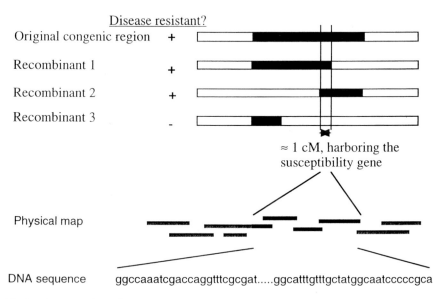

Fig. 3. Fine mapping of a genome region linked to disease susceptibility. Recombinant strains 1 and 2 retain the phenotype of the original congenic strain, therefore the disease gene must be located within the congenic region shared by these strains. As a first step towards cloning the disease gene, a physical map for this region is constructed consisting of a set of overlapping YAC or BAC clones

The similarity between the human and mouse or rat genomes is best illustrated by imagining that the human genome has been cut into approximately 100 pieces which have randomly been put together again. Within these regions the gene order is usually reasonably well conserved. In addition, a number of individual genes have been translocated. Comparative maps between the human, mouse and rat can be found on the Mouse Genome Informatics homepage [37].

Current Knowledge of the MHC and Non-MHC Gene Regulation of EAE

In the following we briefly discuss these matters with focus on the rat species and our own findings in the field.

MHC Gene influences on EAE

The striking influence of the HLA complex on human organ-specific inflammatory disease represents a strong motif to understand the mechanisms for this influence in experimental models, by which new avenues for therapeutic intervention may be opened.

Studies on both pathogenetic mechanisms affected by genes in the MHC complex and the exact positioning of these are complicated by a number of special features of this genome region. The MHC complex is a cluster of many genes, of which many regulate various aspects of immune responses and inflammation. Products of some of these genes are functionally linked. Thus if a MHC haplotype effect on disease is found, there are numerous possible candidates in the region which theoretically could influence an inflammatory disease such as MS or EAE. In addition, the genes within the region (not only class I and II genes) are often highly polymorphic between individuals, probably due to evolutionary pressure; this complicates guesses on candidate genes which may affect disease. There are remarkably few recombinations within this region and close-by genes are in strong linkage dysequilibrium, complicating genetic studies in humans and also the possibilities to raise rodents with recombinations in the region.

Strategies to study MHC gene influences on EAE have included studies in: (1) MHC congenic rodents, revealing influences from any of the genes within the complex of a given haplotype, (2) studies of intra-MHC recombinant strains, roughly delineating gene influences within the complex, and (3) immunobiological/functional studies giving hints for mechanisms of influence. Transgenic expression of a particular candidate gene in a "knock-in" fashion would provide formal proof of a disease regulatory gene, as has been done for a class II gene in collagen-induced arthritis in mice [38], but has so far not been done in EAE.

Historically, the MHC influence on rodent EAE has been documented using MHC congenic strains, keeping background genes constant and varying the MHC haplotype in both rats [39] and mice [40, 41]. MHC influences have mostly been been described as all-or-none phenomena in EAE. This has been the case especially when using myelin peptides for immunization. Absence or presence of disease can then easily be interpreted as absence or presence of binding of the particular peptide to class II MHC molecules. However, MHC haplotype influences are apparent also when using whole spinal cord homogenate or full-length myelin proteins for immunization. Furthermore, the HLA influence is relative in human MS, i.e. no particular HLA type seems to exclude disease.

We studied MHC haplotype influences on MOG-induced EAE in rats on a constant LEW non-MHC background genome and found that MHC did not exert an all-or-none effect on disease susceptibility. Instead, it determined the degree of disease susceptibility, recruitment of MOG-specific immunocompetent cells and CNS pathology in a hierarchal and allele-specific manner. By varying the intensity in the immunization protocol, it also was apparent that a minute autoantigenic challenge was neccesary for induction of disease in the most susceptible MHC haplotypes. Thus, genes in the MHC complex can regulate the ease by which an environmental factor triggers disease [26].

It is well accepted that different MHC haplotypes increase the risk for different organ-specific inflammatory diseases, such as DR2 for MS and DR4 for rheumatoid arthritis. A common speculation for this phenomenon is that the disease-relevant target autoantigens are preferentially and differently handled by the antigen-presenting class I or II molecules within the MHC. In rat EAE, it became apparent that the MHC haplotype not only dictates the afflicted organ, but in addition the disease relevant myelin autoantigen within this organ. Thus, the RT1. N MHC haplotype is resistant to MBP-induced EAE, but highly permissive to MOG-induced EAE. The opposite is the case for the RT1.L haplotype, and the RT1.AV1 haplotype is myelin antigen promiscuous, being permissive for EAE induction by MBP, PLP and MOG [26]. This raises questions if different human HLA types predisposes for MS driven by different myelin autoantigens and if promiscuity of particular MHC haplotypes in target organ antigen reactivity, for example allowing intermolecular antigen spreading, could be a reason for predisposition to autoimmune disease.

Use of intra-MHC recombinant strains between the susceptible RT1. A and resistant RT1. U haplotype (Table 1) mapped major influences to the class II region. It will be of great interest to functionally and genetically dissect this MHC class II influence on a very MS-like disease. Although nearby genes not are excluded, a natural speculation for this influence is that polymorphisms in the B or D molecules may decisively influence MOG peptide binding, in turn regulating encephalitogenecity of recruited MOG-specific autoreactive T cells. It will be important to study details in such a regulation; for example, what are the functional characteristics in form of cytokines, chemokines and adhesion molecules of class II MHC regulated autoreactive T cells that mediate disease, as compared to those where the class II MHC does allow appearance of MOG-reactive T cells without disease-inducing capability?

Interestingly, and perhaps expectedly, the intra-MHC recombinant strains revealed additional allele-specific regulatory influences from the class I and III regions in MOG-induced EAE [26]. An allele-specific disease-promoting influence came from the class I region, perhaps implicating MHC class I-CD8⁺ driven CNS autoaggressiveness. If so, also this arm of immunity can be considered in attempts to design selective immunotherapy. The class III region influences are currently under detailed study in our laboratory.

The mechanisms by which the MHC regulates EAE are thus largely unclear. At a phenomenological level, we have observed that the MHC haplotype regulates the functional differentiation of immunocompetent cells in form of the spectrum of cytokines produced in whole spinal cord-induced EAE. The RT1.AV1 MHC haplotype on different non-MHC backgrounds is characterized by high production of the proinflammatory cytokine interferon-γ (IFN-γ) in the CNS and absence of putative disease downmodulatory cytokines such as interleukin-10 (IL-10) and transforming growth factor beta (TGF-β). These rats had relapsing disease, while other MHC haplotypes with monophasic disease, displayed these immuno-downmodulatory cytokines in the CNS [42]. The MHC thus seems to regulate disease course, perhaps by regulating cytokine production.

The complexity in interpretation of MHC influences is somewhat reduced when using short myelin antigen peptides as immunogens. If anticipating that any ensuing disease is MHC class II-CD4⁺ mediated, theoretically, the outcome of such a peptide immunization could depend on the following events: presence or absence of peptide binding by the B and/or D class II MHC molecules. Absence of any peptide binding will prohibit ensuing immune responses, while presence of binding either may have led to (a) thymic deletion of peptide-specific T cells, since myelin antigens are present in the thymus, if not mimicking peptides may have been active, or (b) peripheral activation of peptide autoreactive T cells that have been positively selected in the thymus. If so, the ensuing immune response may be functionally harmless, suppressive, or autoaggressive, i.e. encephalitogenic. To some extent there is experimental evidence for both of these theoretical possibilities in rat EAE.

We have analyzed the MHC restriction patterns using the two MBP peptides that have proven encephalitogenic in different rat strains, i. e. MBP 63-88 and 89-101. In summary, the experiments with MBP 63-88 demonstrate that certain MHC class II alleles permit a Th1 biased autoreactive response and EAE, while other alleles permit a combined Th1/Th2 autoimmune response and no EAE. In addition, the MHC class I region can exert protective influences, probably by recruiting peptide-specific CD8⁺ cells producing TGF-β [43]. These influences are peptide-specific since a different set of MHC haplotypes display these forms of regulatory responses using the 89-101 peptide [44]. This in turn strongly suggest that polymorphisms in the MHC class I and II molecules and no other genes are responsible for these effects, since these are the structures able to discriminate between peptides.

There is some relation between class II MHC molecule binding of peptides and ensuing immune response. Using overlapping peptides of MBP, we have observed that disease-inducing stretches of MBP in general bind well to isolated B or D molecules, while peptides that have induced Th2 responses bind with intermediate or low affinity [45]. This conforms to observations in other model systems where peptide binding affinity seems to regulate the Th1/Th2 differentiation [46, 47].

Still, despite 30 years of knowledge on MHC influences on autoimmune disease, we do not understand in detail how this gene region sometime permits a harmless immune response and in other situations an autoaggressive organ attack.

Non-MHC Gene Influences on EAE

As in human MS, there is overwhelming evidence demonstrating EAE-regulating influences from genes outside the MHC complex in mice and rats. In most cases this has been proven by using animals with the same MHC but differing non-MHC background genomes, in turn displaying drastically different susceptibility to EAE.

With the invention of the mapping techniques described above, these regulating influences has become approachable. As expected with this complex disease, several genome regions have appeared in different species and strains, and EAE is thus a polygenic disease as is human MS. The genome regions defined represent quite long stretches of DNA with numerous genes. Positional cloning work has started in many laboratories, but so far no single gene has been delineated. Furthermore, with few exceptions little is known on the regulatory influence of these regions on basic putative immunological mechanisms important for disease outcome. Such studies can preferably be made in animals congenic for the regulatory fragment. The delay in data on this is probably due to the time-consuming procedure to raise congenic strains and/or loss of the phenotypic expression of the influencing gene once a congenic strain has been made.

In the following we shortly summarize findings first in mice, then in rats, and possible synteny with human disease.

Microsatellite maps were first available for mice, and genome regions influencing mouse EAE were consequently described. The loci have received their names in order of their descriptions. Since the MHC complex already was known to influence disease, this region has been termed EAE1.

The first reported gene mapping in mice used B10 RIII mice susceptible to chronic EAE after immunisation with MBP 89-101 peptide. These were crossed with MHC identical resistant RIII mice [48]. In this cross EAE2 on mouse chromosome 15 and EAE3 on mouse chromosome 3 displayed significant linkage to disease. Additional regions were found in crosses of other mouse strains with differing immunization protocols [49, 50]. Recently, by using a very large cohort of F2 mice (SJL/JxB10. S/DvTe) several more regions were defined-up to EAE10 [51]. Interestingly, different loci regulated various features of disease such as onset, severity and duration in different manners.

In rats, the MHC non-MHC contribution to disease has been dissected in a DA × BN F2 cross [42] and in DA × E3 recombinant inbred rats [52] after immunization with whole spinal cord homogenate. In both cases there were non-MHC regulatory influences. In the DA × E3 experiments, the non-MHC genes controlled antigen-induced expression of TGF-β, in turn associated with resistance to EAE.

A large DA × BN F2 cross revealed a major impact of the MHC complex in which the n haplotype was protective. However, in addition and in epistatic interaction with the MHC, a major locus controlling EAE was found on rat chromosome 9. This locus was also associated with increased expression of the proinflammatory cytokine IFN-γ (53).

Use of rat strains with the MHC RT1.AV1 haplotype but differing background genomes, i. e. DA, LEW AV1, ACI and PVG AV1 showed that the two first strains were MOG EAE susceptible and the two last resistant [26]. Interestingly, the background genome regulated the functional differentiation of autoreactive T cells. Whereas the proliferative response to MOG ex vivo was similar between the strains, only the disease susceptible strains displayed cells producing significant amounts of the proinflammatory cytokine IFN-γ.

In an F2 cross between MOG EAE susceptible DA rats and less susceptible PVG AV1 rats, we examined the influence of previously defined arthritis-regulating loci from the DA rat. Interestingly, the two regions CIA3 and OIA2 on rat chromosome 4 regulated MOG EAE phenotypes. CIA3 had a drastic influence by strongly regulating survival and also the anti-MOG antibody isotype pattern, consistent with a disease-promoting Th1 bias of the immune response. OIA2 did not affect measured clinical phenotypes, but was associated with an antibody isotype pattern consistent with a Th2 bias [54].

In an F2 cross between MOG EAE susceptible DA rats and resistant ACI rats, we have recently found one more seemingly organ-specific disease-regulating genome region on rat chromosome 18 (I. Dahlman et al., manuscript in preparation).

Some of the regions described in mice and rats coincide with regard to regulation of different inflammatory diseases. This may either be due to identical polymorphic genes regulating the different diseases or to clustering of polymorphic genes with close functions. The MHC complex can be argued to be such a case. A similar reasoning can be applied to synteny comparisons done with a series of human inflammatory diseases [55], where regions found in family studies of different diseases and experimental diseases seem to cluster. A first example of such synteny comparison suggesting relevance of a rodent susceptibility locus being relevant for human disease was provided by EAE2 in Finnish MS material [56]. We have preliminary evidence that 3 out of 3 studied susceptibility regions defined in rat EAE are relevant for MS.

There is an interesting time ahead, when the gene regions regulating experimental disease can be narrowed down, examined for function and hopefully positional cloning is achieved. This will require both patience and intense efforts.

Acknowledgements: The authors' work cited in the chapter have been supported by grants from the Swedish medical research council, the EU Biomed 2 program (contract No BMH4-97-2027), the Swedish Society for Neurologically Disabled, and Petrus and Augusta Hedlunds Foundation.

References

1. Ebers GC, Bulman D, Sadovnick A et al. (1986) A population-based twin study in multiple sclerosis. N Eng J Med 315: 1638-1642
2. Ebers GC, Sadovnick AD, Risch NJ and the Canadian Collaborative Study Group (1995) A genetic basis for familial aggregation in multiple sclerosis. Nature 377: 150-151
3. Sadovnick AD, Ebers GC, Dyment DA, Risch NJ and the Canadian Collaborative Study Group (1996) Evidence for genetic basis of multiple sclerosis. The Lancet 347: 1728-1730
4. Risch N (1990) Linkage strategies for genetically complex traits. I. Multilocus models. Am J Hum Gen 46: 222-228
5. Olerup O, Hillert J (1991) HLA class II associated genetic susceptibility in multiple sclerosis: A critical evaluation. Tissue Antigens 38: 1-15
6. Ebers GC, Kukay K, Bulman DE et al. (1996) A full genome search in multiple sclerosis. Nat Genet 13: 472-476
7. Sawcer S, Jones HB, Feakes R et al. (1996) A genome screen in multiple sclerosis reveals susceptibility loci on chromosome 6p21 and 17q22. Nat Genet 13: 464-468
8. The Multiple Sclerosis Genetics Group (1996) A complete genomic screen for multiple sclerosis underscores a role for the major histocompatibility complex. Nat Genet 13: 469-471
9. Ebers GC, Sadovnick AD (1994) The role of genetic factors in multiple sclerosis susceptibility. J Neuroimmunol 54: 1-17
10. Haines JL, Terwedow HA, Burgess K et al. (1998) Linkage of the MHC to familial multiple sclerosis suggests genetic heterogeneity. Human Mole Genet 7: 1229-1234
11. Hillert J (1996) Genetics of multiple sclerosis. In: Cook S (ed) Handbook of multiple sclerosis. Marcel Dekker, New York, pp 19-51
12. Marrosu MG, Murru MR, Costa G et al. (1998) DRB1-DQA1-DQB1 loci and multiple sclerosis predisposition in the Sardinian population. Hum Mole Genet 7: 1235-1237
13. Hillert J, Olerup O (1993) Multiple sclerosis is associated with genes within or close to the HLA-DR-DQ subregion on a normal DR15, DQ6DW 2 haplotype. Neurology 43: 163 (abstract)
14. Wallström E, Khademi M, Andersson M et al. (1998) Increased reactivity to myelin oligodendrocyte glycoprotein peptides and epitope mapping in DR2(15)+ multiple sclerosis. Eur J Immunol 28: 3329-3335
15. Fogdell-Hahn A, Ligers A, Grönning M et al. (1998) Multiple sclerosis: a modifying influence of HLA class I genes in an HLA class II associated disease. Eur J Immunogenet 25(Suppl 1): 62
16. Kuokkanen S, Gschend M, Rioux JD et al. (1997) Genomewide scan of multiple sclerosis in Finnish multiplex families. Am J Hum Genet 61: 1379-1387
17. Chataway J, Feakes R, Corraddu F et al. (1998) The genetics of multiple sclerosis: principles, background and updated results of the United Kingdom systematic genome screen. Brain 121: 1869-1887
18. Janson C, Olsson T, Höjeberg B, Holmdahl R (1991) Chronic experimental allergic encephalomyelitis induced by the 89-110 myelin basic protein peptide in mice with I-A^r. Eur J Immunol 21: 693-699
19. Brown A, McFarlin DE, Raine CS (1982) Chronologic neuropathology of relapsing experimental allergic encephalomyelitis in the mouse. Lab Invest 46: 171-185
20. Baker D, O'Neill JK, Gschmeissner SE et al. (1990) Induction of chronic relapsing allergic encephalomyelitis in Biozzi mice. J Neuroimmunol 28: 261-270

21. Lorentzen JC, Glaser A, Jacobsson L et al. (1998) Identification of rat susceptibility loci for adjuvant-oil induced arthritis. Proc Natl Acad Sci USA 95: 6383-6387
22. Vingsbo C, Sahlstrand P, Brun JG et al. (1996) Pristane-induced arthritis in rats: a new model for rheumatoid arhtritis with a chronic disease course influenced by both major histocompatibility complex and non-major histocompatibility complex genes. Am J Pathol 149: 1675-1683
23. Vingsbo-Lundberg C, Nordquist N, Olofsson P et al. (1998) Genetic control of arthritis onset, severity and chronicity in a model for rheumatoid arthritis in rats. Nat Genet 20: 401-404
24. Lorentzen J, Issazadeh S, Storch M et al. (1995) Protracted, relapsing and demyelinating experimental autoimmune encephalomyelitis in DA rats, immunized wtih syngeneic spinal cord and incomplete Freund's adjuvant. J Neuroimmunol 63: 193-205
25. Storch MK, Stefferl A, Brehm U et al. (1998) Autoimmunity to myelin oligodendrocyte glycoprotein mimics the spectrum of multiple sclerosis pathology. Brain Pathology 8: 681-694
26. Weissert R, Wallström E, Storch M et al. (1998) MHC haplotype dependent regulation of clinical profile and lesional pathology of myelin oligodendrocyte glycoprotein induced experimental autoimmune encephalomyelitis. J Clin Invest 102: 1265-1273
27. Berger T, Weerth S, Kojima K et al. (1997) Experimental autoimmune encephalomyelitis: the antigen specificity of T lymphocytes determines the topography of lesions in the central and peripheral nervous system. Lab Invest 76: 355-364
28. Adelmann M, Wood J, Benzel I et al. (1995) The N-terminal domain of the myelin oligodendrocyte glycoprotein (MOG) induces acute demyelinating experimental autoimmune encephalomyelitis in the Lewis rat. J Neuroimmunol 63: 17-27
29. Hedrich HJ (ed) (1990) Genetic monitoring of inbred strains of rats. Gustav Fischer, Stuttgart New York
30. Frankel WN (1995) Taking stock of complex trait genetics in mice. Trends Genet 11: 471-477
31. Mapmaker reference: http://www.genome.wi.mit.edu/ftp/distribution/software
32. Lander E, Kruglyak L (1995) Genetic dissection of complex traits: guidelines for interpreting and reporting linkage results. Nat Genet 11: 241-247
33. Microsatellite maps: http://waldo.wi.mit.edu/rat/public/ http://www.jax.org/
34. Snell GD (1948) J Genet 49: 87-99
35. Wakeland E, Morel L, Achey K et al. (1997) Speed congenics: a classic technique in the fast lane (relatively speaking). Immunol Today 18: 472-477
36. Markel P, Shu P, Ebeling C et al. (1997) Theoretical and empirical issues for marker-assisted breeding of congenic mouse strains. Nat Genet 17: 280-284
37. Comparative maps http://www.ncbi.nlm.nih.gov/Homology/http://www.jax.org
38. Brunsberg U, Gustafsson K, Jansson L et al. (1994) Expression of a transgenic class II Ab gene confers susceptibility to collagen-induced arthritis. Eur J Immunol 24: 1698-1702
39. Williams RM, Moore MJ (1973) Linkage of susceptibility to experimental allergic encephalomyelitis to the major histocompatibility complex locus in the rat. J Exp Med 138: 775-783
40. Fritz RB, Skeen MJ, Jen Chou CH et al. (1985) Major histocompatibility complex-linked control of the murine immune response to myelin basic protein. J Immunol 134: 2328-2332
41. Jansson L, Olsson T, Höjeberg B, Holmdahl R (1991) Chronic experimental allergic encephalomyelitis induced by the 89-101 myelin basic protein peptide in mice with I-Ar. Eur J Immunol 21: 693-699

42. Lorentzen J, Andersson M, Issazadeh S et al. (1997) Genetic analysis of inflammation, cytokine mRNA expression and disease course of relapsing experimental autoimmune encephalomyelitis in DA rats. J Neuroimmunol 80: 31-37
43. Mustafa M, Vingsbo C, Olsson T, Issazadeh S, Ljungdahl Å, Holmdahl R (1994) Alleles of both MHC class I and II regions have protective roles in MBP 63-88 induced EAE in rats. J Immunol 153: 3337-3344.
44. Issazadeh S, Kjellen P, Olsson T et al. (1997) Major histocompatibility complex-controlled protective influences on experimental autoimmune encephalomyelitis are peptide specific. Eur J Immunol 27: 1584-1587
45. de Graaf KL, Weissert R, Kjellen P et al. (1999) Allelic variations in rat MHC class II binding of myelin basic protein peptides correlate with encephalitogenicity. (Manuscript submitted)
46. Pfeiffer C, Stein J, Southwood S et al. (1995) Altered peptide ligands can control CD4 T lymphocyte differentiation in vivo. J Exp Med 181: 1569-1574
47. Murray JS (1998) How the MHC selects Th1/Th2 immunity. Immunol Today 19: 157-163
48. Sundvall M, Jirholt J, Yang HT et al. (1995) Identification of murine loci associate with susceptibility to chronic experimental autoimmune encephalomyelitis. Nat Genet 10: 313-317
49. Encinas JA, Lees MB, Sobel RA et al. (1996) Genetic analysis of susceptibility to experimental autoimmune encephalomyelitis in a cross between SJL/J and B10 S mice. J Immunol 157: 2186-2192
50. Baker D, Rosenwasser OA, O´Neill JK, Turk JL (1995) Genetic analysis of experimental allergic encephalomyelitis in mice. J Immunol 155: 4046-4051
51. Butterfield RJ, Sudweeks JD, Blankenhorn EP et al. (1998) New genetic loci that control susceptibility and symptoms of experimental allergic encephalomyelitis in inbred mice. J Immunol 161: 1860-1867
52. Kjellen P, Issazadeh S, Olsson T, Holmdahl R (1998) Genetic influence on disease course and cytokine response in relapsing experimental allergic encephalomyelitis. International Immunol 10: 333-340
53. Dahlman I, Jacobsson L, Glaser A et al. (1999) A genome-wide linkage analysis of chronic relapsing experimental autoimmune encephalomyelitis in the rat identifies a major susceptibility locus on chromosome 9. J Immunol (*in press*)
54. Dahlman I, Lorentzen JC, de Graaf KL et al. (1998) Quantitative trait loci disposing for both experimental arthritis and encephalomyelitis in the DA rat; impact on severity of myelin oligodendrocyte glycoprotein-induced EAE and antibody isotype pattern. Eur J Immunol 28: 2188-2196
55. Becker KG, Simon RM, Bailey-Wilson JE et al. (1998) Clustering of non-major histocompatibility complex susceptibility candidate loci in human autoimmune diseases. Proc Natl Acad Sci USA 95: 9979-9984
56. Kuokkanen S, Sundvall M, Terwilliger JD et al. (1996) A putative vulnerability locus to multiple sclerosis maps to 5 p14-p12 in a region syntenic to the murine locus Eae2. Nat Genet 13: 477-480

Chapter 23

Genetics of Hereditary Neuropathies

G.L. Mancardi

Introduction

Hereditary motor and sensory neuropathies (HMSN) have been previously classified on the basis of clinical course, mode of inheritance and neuropathological findings [1]. In the past years, considerable advances in the knowledge of this heterogeneous group of disorders have been made due to genetic studies demonstrating that duplication, deletion or mutation of specific genes of the peripheral myelin are the most common causes of HMSN. The classification of HMSN and of related disorders, from now forward called with the eponymous Charcot-Marie-Tooth disease (CMT), is continuously evolving. CMT has a prevalence of 1 case every 2500 [1] and is divided into CMT1 and CMT2 according to neurophysiologic and neuropathologic findings. In fact, in CMT1 the motor nerve conduction velocity (MNCV) of the median nerve is < 38 m/s. Nerve biopsy shows hypertrophy of the nerve, demyelination and remyelination of the nerve fibers, and Schwann cell proliferation around the demyelinated fibers in an "onion bulb" fashion. In contrast, in CMT2 MNCV is only slightly lowered (> 38 m/s) and sural nerve biopsy shows signs of primary axonal involvement with minimal signs of myelin sufferance. CMT1 and CMT2 are usually inherited as dominant disorders and are clinically indistinguishable. A severe form of CMT called Dejerine-Sottas syndrome (DSS) has onset in early childhood and a progressive and disabling clinical course. DSS has been considered in the past as a recessive disorder, but it is now apparent that the majority of cases are due to a *de novo* point mutation of myelin genes. Recessive forms of CMT nevertheless exist and are genetically and clinically heterogeneous. Complex forms of CMT, in which the peripheral polyneuropathy is associated with other symptoms, have been known for a long time, but their pattern of inheritance and the genes involved are still under investigation. Finally, in CMT group also includes the hereditary neuropathy with liability to pressure palsies (HNPP), characterized by recurrent episodes of peripheral nerve palsies due to mechanical trauma along the course of the nerve. Table I summarizes the classification of CMT and related disorders, based on recent

---

Department of Neurological Sciences, University of Genoa, Via De Torri 5, Genoa, Italy. e-mail: neurolab@cisi.unige.it

Table 1. Charcot-Marie-Tooth and related disorders

	Locus	Gene	Mechanism
Charcot-Marie-Tooth type 1			
CMT1A	17p11.2	PMP22	Duplication/point mutation
CMT1B	1q22-23	P0	Point mutation
CMT1C	Unknown	Unknown	Unknown
CMTX	Xq13.1	Cx32	Point mutation
Charcot-Marie-Tooth type 2			
CMT2A	1p35-p36	Unknown	Unknown
CMT2B	3q13.22	Unknown	Unknown
CMT2C	Unknown	Unknown	Unknown
CMT2D	7p14	Unknown	Unknown
CMT2E	1q22-23	P0	Point mutation
Dejerine-Sottas syndrome (DSS)			
DSSA	17p11.2	PMP22	Point mutation
DSSB	1q22-23	P0	Point mutation
Congenital hypomyelination	1q22-23	P0	Point mutation
	10q21.1	EGR2	Point mutation
Hereditary neuropathy with liability to pressure palsies (HNPP)			
HNPPA	17p11.2	PMP22	Deletion/point mutation
HNPPB	Unknown	Unknown	Unknown
Recessive forms of Charcot-Marie-Tooth			
CMT4A	8q	Unknown	Unknown
CMT4B	11q23	Unknown	Unknown
CMT4C	Unknown	Unknown	Unknown
Complex forms of CMT	Unknown	Unknown	Unknown

genetic findings. Animal and in vitro models of inherited neuropathies are necessary in order to clarify the pathogenetic mechanism of demyelination.

CMT1

A motor sensory distal symmetrical polyneuropathy with absent deep tendon reflexes, pes cavus and muscle atrophy involving mainly the leg muscles are the main clinical features of CMT1, also known in the past with the term "peroneal muscular atrophy". The motor disturbances are prevalent, sensory symptoms are usually mild, and deep tendon reflexes can be diminished only in the upper limbs. Onset is usually in the first or second decade of life and progression is slow. Considerable phenotypic variability exists, but in the majority of cases the disease is not severely disabling. In some cases the disorder is mild and limited to the presence of pes cavus and slight weakness in the dorsal flexion of the feet. Usually CMT1 is inherited as an autosomal dominant disorder, but some pedigrees have an X-linked pattern of inheritance. According to the different genetic defects, four types of CMT1 (CMT1A, CMT1B, CMT1C, CMTX) have been identified.

CMT1A and CMT1B

Neurophysiologic studies show an important reduction of MNCV (<38 m/s). Neuropathologic examinations of the nerves display demyelination and remyelination with onion bulbs made of Schwann cells and their processes which, together with an increase of collagen, determine the well-known nerve hypertrophy. CMT1A was first found to be associated with a duplication in chromosome 17p11.2 of the size of approximately 1.5 Mb [2, 3]. The gene of 22 kDa peripheral myelin protein (PMP22) maps to this region. It was subsequently demonstrated that PMP22 is indeed the CMT1A gene [4] and that the presence of three copies of PMP22 causes the disorder. Duplication of the 17p11.2 region is responsible for 70%-85% of CMT1A, while point mutations of PMP22 cause disease only rarely (1%) [5]. PMP22, a protein of the peripheral nervous system (PNS), is localized in the compact part of myelin and accounts for 2%-5% of total myelin proteins. For many years it has been known that some other cases of CMT1 were linked to the Duffy locus on chromosome 1. The human myelin P0 protein, whose gene maps to chromosome 1q22-23, is an important component (almost 50%) of the compact peripheral myelin. Point mutations of the P0 gene are associated with CMT1B [6]. CMT1B is by far rarer than CMT1A, accounting for 4% of all CMT1 cases.

CMT1C

Rare cases of CMT with a typical CMT1 phenotype but not associated with either PMP22 nor P0 gene defects are classified as CMT1C, and are due to an unknown genetic disorder.

CMTX

Nearly 10% of all CMT1 cases are characterized by: the absence of male-to-male transmission; a more severe clinical course in males as compared to females; a relevant lowering of MNCV in males (between 18 and 40 m/s); and slight neurophysiologic changes in females (MNCV between 25 and 61 m/s). The neuropathology of CMTX differs from that of CMT1A and 1B and is mainly characterized by axonal changes, atrophy and clusters of axonal regeneration, and minimal signs of deremyelination. CMTX is due to point mutations in the connexin-32 (Cx32) gene, which encodes a major component of non-compact myelin at the nodes of Ranvier and Schmidt-Lanterman incisures [7]. Cx32 is a constituent of gap junction channels through which ions and nutrients are transferred across the myelin sheath. A large number of different Cx32 mutations have been detected in CMTX families.

CMT2

This form of CMT is an axonal polyneuropathy with only slight lowering of MNCV (>38 m/s in the median nerve) and signs of axonal atrophy and axonal regeneration at nerve biopsy. Is inherited as an autosomal dominant disorder. The onset is later than that of CMT1, in the second or third decade of life, but clinical findings and clinical course are usually indistinguishable from those of CMT1. CMT2 is genetically heterogeneous, and in some pedigrees linkage has been found with chromosome 1p35-p36 (CMT2A) [8], chromosome 3q13.22 (CMT2B) [9], or chromosome 7p14 (CMT2D) [10]. In some cases CMT2 is associated with paralysis of the diaphragm and vocal cords (CMT2C), but in these cases no convincing linkage has been found until now. Recently, patients with a CMT2 phenotype and a mutation in the P0 gene have been reported [11], indicating that changes in the myelin sheath can be followed by axonal sufferance and atrophy.

DSS

Dejerine-Sottas syndrome (DSS) is a hypertrophic, demyelinating polyneuropathy with early onset and severe disabling clinical course. MNCV is markedly lowered (usually below 12 m/s), and SNCV is usually not detectable at all. Nerve biopsy shows a profound loss of myelinated fibers and diffuse "onion bulb", hypertrophic changes of Schwann cells. Occasionally, onion bulbs around demyelinated axons are made only by concentric layers of basal lamina. The majority of DSS cases are sporadic; in the past it was thought that this disorder was autosomal recessive. Recently DSS has been associated with point mutations in either the PMP22 gene (DSSA) [12] or the P0 gene (DSSB) [13]. Since most of the mutations described so far in DSS are present in the heterozygous state, the disease is probably caused by dominantly acting genetic defects. There are even more severe cases, where the majority of axons are completely devoid of myelin or only surrounded by a thin myelin sheath, with a complete absence of fully developed myelinated fibers. These forms, now classified as "congenital hypomyelination", have onset in the first months of life, usually have a very disabling clinical course and, in some cases, lead to premature death. Point mutations in the P0 gene [14] or a dominant mutation of the early growth response 2 (EGR) gene [15] have been implicated.

HNPP

Hereditary neuropathy with liability to pressure palsies (HNPP) is inherited as an autosomal dominant trait and is characterized by recurring episodic peri-pheral nerve palsies due to mechanical compression of the nerve trunks. Electrophysiological studies show nerve conduction slowing more evident at common entrapment sites; a variable number of "sausage-like" thickenings of the myelin sheath (tomacula), with a near normal number of myelinated fibers is observed in the sural nerve. Some variability in clinical and neurophysiological phenotype has been observed, and patients with a progressive sensory-motor neuropathy and the typical genetic features of HNPP have been described [16]. HNPP is commonly associated with deletion on chromosome 17p11.2 involving the same interval that is duplicated in CMT1A [17], although cases without deletion have been described [18]. Since only one copy of the PMP22 gene is present in HNPP patients, a gene dosage mechanism has been proposed as the most likely pathogenetic factor causing the disease.

CMT4

Autosomal recessive CMT cases, also called CMT4, are uncommon, although several pedigrees have been described. The recessive forms of CMT usually have an early onset and a more severe clinical course compared to the dominant forms. CMT4A often has delayed milestones, marked muscular distal atrophy rapidly progressing to the proximal parts of the legs and arms, mild sensory loss, and skeletal deformities and scoliosis. Patients become severely disabled by the end of the first or second decade of life. MNCV is lowered (<38 m/s) and sural nerve biopsy shows diffuse hypomyelination of the nerve fibers. Linkage with chromosome 8q has been demonstrated in Tunisian families [19]. CMT4B is characterized by a severe decrease of MNCV and by the presence at nerve biopsy of myelin outfoldings, with redundant loops and foldings of myelin at internodes [20]. Outfolding of myelin is also observed in CMT1A, but in CMT4B this change is diffused to the majority of myelinated fibers [21]. A linkage to chromosome 11q23 has been demonstrated in a large family [22], but probably this disorder is heterogeneous because this linkage was excluded in other pedigrees. CMT4C is mainly an axonal disorder with preserved MNCV and absence of changes of myelinated fibers at nerve biopsy.

Complex Forms of CMT

CMT has been associated with other clinical features, such as optic atrophy [23], pigmentary retinal degeneration [24], deafness [25], optic atrophy and deafness [26], deafness and mental retardation [27], and deafness, mental retardation and the absence of large myelinated fibers [28]. In all these cases the genetic defects are unknown.

CMT and Myelin Genes

In the PNS, myelin is made up by Schwann cells which initially surround and subsequently spiral axons, forming a multilamellar structure. Nerve myelin is composed of lipids and proteins specific for the compact and non-compact regions. Compact myelin proteins are PMP22, P0 and MBP, whereas non-compact myelin, present near the nodes of Ranvier and at the incisures of Schmidt-Lanterman, contains myelin-associated glycoprotein (MAG), Cx32, E-cadherin, and α6β4 integrin [29].

PMP22 is a 22 kDa glycoprotein expressed in the compact region of peripheral myelin, accounting for only 2%-5% of total myelin proteins. The function of PMP22 is not completely known but it probably has a dual role, as a protein favoring Schwann cell growth and also as a structural adhesive component of the myelin sheath. Duplication of PMP22 gene causes CMT1A, while deletion of PMP22 gene causes HNPP. The 1.5 Mb duplication or deletion in chromosome 17p11.2 containing the PMP22 gene is caused by unequal crossing over during meiosis, facilitated by the presence of two homologous regions with a low copy number of repeat sequences (CMT1A-REP repeat) that flank the proximal and distal duplication or deletion breakpoint regions [30]. Points mutations of PMP22 can also result in CMT1A, but the phenotypic manifestation of some missense or nonsense mutations in the PMP22 gene can also be HNPP or DSS. It is still unclear why different mutations in the same gene cause different syndromes. Because the PMP22 gene is present in three copies in CMT1A and in one copy in HNPP, gene dosage is proposed as the mechanism underlying both disorders. In fact in CMT1A and HNPP, an increase or an underexpression of PMP22 or its mRNA has been demonstrated [31-33], but the results are not straightforward in all cases. Probably the overexpression or underexpression of PMP22 determines a gain or loss of function of the PMP22 protein, which in turns destabilizes the myelin sheath structure, leading demyelination or the formation of tomacula.

P0 is a glycoprotein belonging to the immunoglobulin superfamily, accounting for 50%-60% of the total proteins of the peripheral nerve myelin. It is organized in tetramers which form a functional unit interacting with that of the opposing membrane, settling the tight apposition of Schwann cell membranes and maintaining the compactness of myelin [29]. Mutations of the P0 gene can determine different phenotypes, from mild cases of CMT1B to severe cases of DSS. More than 30 different mutations have been described so far, and according to the site of mutation different clinical symptoms can appear. In some cases there is probably only a partial loss of function of the protein and the clinical phenotype is mild; in other cases the loss of function can be complete or the mutated P0 gene can have an additive toxic effect, also involving the interaction with other proteins of the myelin sheath. In rare cases the dominant negative effect of the mutated P0 can be so intense to determine an almost complete impairment of the process of myelination.

Cx32, a member of the connexin family, is a constituent of gap junctions which links close surfaces of the Schwann cell membrane in non-compacted myelin of the paranodal region or in the Schmidt-Lanterman incisures. The connexins are organized in hexamers (connexons), which associate with connexons of the opposing cell membrane, thereby forming a pore through which small molecules and ions can diffuse directly across the myelin sheath and reducing the diffusion distance by 1000-fold as compared to the circumferential pathway [34]. Mutation of the Cx32 gene has been identified as the cause of a dominantly inherited form of CMTX. More than 90 different mutations have been identified, some of which are missense or nonsense mutations. However, similarly to other myelin proteins, there is no clear relationship between the location of mutation and the clinical phenotype.

Animal and In Vitro Models of CMT

Rats or mice overexpressing PMP22 have been generated by injecting fertilized rat or mouse oocytes with the PMP22 gene [35]. These animals showed the clinical and electrophysiologic signs of peripheral polyneuropathy and pathological evidence of chronic demyelination with onion bulb formation. In some cases, especially in rats overexpressing many copies of PMP22, the loss of myelin is particularly severe. PMP22 deficient mice, P0 deficient mice, and Cx32 null mice have also been generated [36]. The peripheral neuropathies obtained, although different from the human cases in some respect, are of main importance to study the pathophysiological mechanisms of CMT. More recently, in vitro models for the study of CMT have been developed utilizing long-term organotypic cultures of dorsal root ganglia from transgenic rat models of CMT1A. Signs of demyelination became apparent after 3-4 weeks of culture, suggesting that this in vitro model system is suitable for studying the biologic role of myelin proteins and the pathogenetic mechanisms underlying this group of inherited disorders.

References

1. Dyck PJ, Chance PF, Lebo RV, Carney JA (1993) Hereditary motor and sensory neuropathies. Peripheral neuropathy, 3rd edn. WB Saunders, Philadelphia, pp 1094-1136
2. Lupski JR, Montes de Oca-Luna R, Slaugenhaupt S et al. (1991) DNA duplication associated with Charcot-Marie-Tooth disease type 1A. Cell 66: 219-232
3. Raeymaekers P, Timmerman V, Nelis E et al. (1991) Duplication in chromosome 17p11.2 in Charcot–Marie-Tooth neuropathy type 1A. Neuromusc Dis 1: 93-97
4. Timmerman V, Nelis E, Van Hul W et al. (1992) The peripheral myelin protein gene PMP-22 is contained within the Charcot-Marie-Tooth disease type 1A duplication. Nat Genet 1: 171-175
5. Crespi V, Fabrizi G, Mandich P et al. (1999) Proposal of guide-lines for diagnosis of Charcot-Marie-Tooth disease and related neuropathies. It J Neurol Sci *(in press)*
6. Hayasaka K, Himoro M, Sato W et al. (1993) Charcot-Marie-Tooth neuropathy type 1B associated with mutations in the myelin protein zero gene. Nat Genet 5: 31-34
7. Bergoffen J, Scherer SS, Wang S et al. (1993) Connexin mutations in X-linked Charcot–Marie-Tooth disease. Science 262: 2039-2042
8. Ben Othmane K, Middleton LT, Loprest LJ et al. (1993) Localization of a gene (CMT2A) for autosomal dominant Charcot-Marie-Tooth disease type 2 to chromosome 1p and evidence of genetic heterogeneity. Genomics 17: 370-375
9. Kwon JM, Elliot JL, Yee WC et al. (1995) Assignment of a second Charcot-Marie-Tooth type II locus to chromosome 3q. Am J Hum Genet 57: 853-858

10. Ionasescu V, Searby C, Sheffield VC et al. (1996) Autosomal dominant Charcot-Marie-Tooth axonal neuropathy mapped on chromosome 7p (CMT2D). Hum Mol Genet 5: 1373-1375
11. Marrosu MG, Vaccargiu S, Marrosu G et al. (1998) Charcot-Marie-Tooth disease type 2 associated with mutation of the myelin protein zero gene. Neurology 50: 1397-1401
12. Roa BB, Dyck PJ, Marks HG et al. (1993) Dejerine-Sottas syndrome associated with point mutation in the peripheral myelin protein 22 (PMP22) gene. Nat Genet 5: 269-273
13. Hayasaka K, Himoro M, Sawaishy Y et al. (1993) De novo mutation in the myelin P0 gene in Dejerine-Sottas disease. Nat Genet 5: 266-268
14. Warner LE, Hilz MJ, Appel SH et al. (1996) Clinical phenotypes of different MPZ (P0) mutations may include Charcot-Marie-Tooth type 1B, Dejerine-Sottas, and congenital hypomyelination. Neuron 17: 451-460
15. Warner LE, Mancias P, Butler IJ et al. (1998) Mutations in the early growth response 2 (EGR2) gene are associated with hereditary myelinopathies. Nat Genet 18: 382-384
16. Mancardi GL, Mandich P, Nassani S et al. (1995) Progressive sensory-motor polyneuropathy with tomaculous changes is associated to 17p11.2 deletion. J Neurol Sci 131: 30-34
17. Chance PF, Alderson MK, Leppig KA et al. (1993) DNA deletion associated with hereditary neuropathy with liability to pressure palsies. Cell 72: 143-151
18. Pareyson D, Taroni F (1996) Deletion of the PMP 22 gene and hereditary neuropathy with liability to pressure palsies. Curr Opin Neurol 9: 348-354
19. Ben Othmane K, Hentati F, Lennon F et al. (1993) Linkage of a locus (CMT 4A) for autosomal recessive Charcot-Marie-Tooth disease to chromosome 8q. Hum Mol Genet 2: 1625-1628
20. Ohnishi A, Yoshiyuki M, Ikeda M et al. (1989) Autosomal recessive motor and sensory neuropathy with excessive myelin outfolding. Muscle Nerve 12: 568-575
21. Schenone A, Abbruzzese M, Uccelli A et al. (1994) Hereditary motor and sensory neuropathy with myelin outfolding: clinical, genetic and neuropathological study of three cases. J Neurol Sci 122: 20-27
22. Bolino A, Brancolini V, Bono F et al. (1996) Localization of a gene responsible for autosomal recessive demyelinating neuropathy with focally folded myelin sheaths to chromosome 11q23 by homozygosity mapping and haplotype sharing. Hum Mol Genet 5: 1051-1054
23. McLeod JG, Low PA, Morgan JA (1978) Charcot-Marie -Tooth disease with Leber optic atrophy. Neurology 28: 179-184
24. Massion-Verniory L, Dumont E, Potuin A (1946) Retinite pigmentaire familiale complique d'une amyotrophie neurale. Rev Neurol 78: 561-571
25. Satya-Murti S, Cacace AT, Hanson PA (1979) Abnormal auditory evoked potentials in hereditary motor-sensory neuropathy. Ann Neurol 5: 445-448
26. Rosenberg RN, Chutorian A (1967) Familial optico-acoustic nerve degeneration and polyneuropathy. Neurology 17: 827-832
27. Cowchock FS, Duckett SW, Streletz LJ et al. (1985) X-linked motor-sensory neuropathy type II with deafness and mental retardation: a new disorder. Am J Med Genet 20: 307-315
28. Mancardi GL, Di Rocco M, Schenone A et al. (1992) Hereditary motor and sensory neuropathy with deafness, mental retardation and absence of large myelinated fibers. J Neurol Sci 110: 121-130
29. Scherer SS (1997) The biology and pathobiology of Schwann cells. Curr Opin Neurol 10: 386-397

30. Pentao L, Wise CA, Chinault AC et al. (1992) Charcot-Marie-Tooth type 1A duplication appears to arise from recombination at repeat sequences flanking the 1.5 Mb monomer unit. Nat Genet 2: 292-300
31. Vallat JM, Sindou P, Preux PMP et al. (1996) Ultrustructural PMP-22 expression in inherited demyelinating neuropathies. Ann Neurol 39: 813-817
32. Hanemann CO, Stoll G, D'Urso D et al. (1994) Peripheral myelin protein-22 expression in Charcot-Marie-Tooth disease type 1a sural nerve biopsies. J Neurosci Res 37: 654-659
33. Schenone A, Nobbio L, Mandich P et al. (1997) Underexpression of messenger RNA for peripheral myelin protein 22 in hereditary neuropathy with liability to pressure palsies. Neurology 48: 445-449.
34. Scherer SS, Yi-Tian Xu, Nelles R et al. (1998) Connexin 32-null mice develop demyelinating peripheral neuropathy. Glia 24: 8-20
35. Sereda M, Griffiths I, Puhlhofer A et al. (1996) A transgenic rat model of Charcot-Marie-Tooth disease. Neuron 16: 1049-1060
36. Martini R, Zielasek J, Toyka KV (1998) Inherited demyelinating neuropathies: from gene to disease. Curr Opin Neurol 11: 545-556

Chapter 24

Immunopathogenetic Role of Anti-Neural Antibodies in Demyelinating Dysimmune Neuropathies

E. NOBILE-ORAZIO, M. CARPO

Introduction

Since the original reports by Latov et al. [1] of a patient with a demyelinating neuropathy and IgM antibodies to the myelin-associated glycoprotein [2], antibodies to several other neural antigens have been reported in patients with different forms of dysimmune neuropathies (Table 1). The identification of these reactivities in patients with definite neuropathy syndromes did not only introduce a useful tool in the diagnosis of neuropathies but also cast new light on the immunopathogenetic mechanisms involved. In this chapter, we review evidence supporting possible pathogenetic roles for some of the frequently reported antibody reactivities in demyelinating neuropathies.

Anti-MAG IgM Antibodies

Myelin-associated glycoprotein (MAG), a minor constituent of myelin membranes of the central nervous system (CNS) and peripheral nervous system (PNS), is of the approximate molecular mass of 100 kDa [3, 4]. MAG is mostly concentrated in the periaxonal region and other uncompacted regions of PNS myelin including Schmidt-Lanterman incisures, lateral loops and outer mesaxon, where it probably plays an important role in maintaining glial-axonal interactions [4].

High titers of anti-MAG IgM antibodies have been reported in approximately 50% of patients with neuropathy associated with IgM monoclonal gammopathy [2, 5-8]. In these patients IgM reacting with MAG have the same ligth chain of the M-protein. This reactivity is directed against the same carbohydrate moiety of MAG that serves as the epitope for HNK-1/L2 monoclonal antibody [9]. This epitope is also present in other glycoconjugates in nerve including the myelin glycoproteins P0 and PMP22 [10, 11], and the glycosphingolipids sulfoglucuronyl paragloboside (SGPG) and sulfoglucuronyl lactosaminyl paragloboside (SGLPG) [12]. IgM reactivity with MAG is often associated with IgM monoclonal gammopathy of undetermined significance (MGUS), but may be also found in

---

Giorgio Spagnol Service of Clinical Neuroimmunology, Centro Dino Ferrari, Institute of Clinical Neurology, University of Milan, IRCCS Ospedale Maggiore Policlinico, Via F. Sforza 35 - 20122 Milan, Italy. e-mail: eduardo.nobile@unimi.it

Table 1. Anti-neural antibodies in dysimmune neuropathies

Antigens	Isotype	Clinical syndrome	Antibody frequency (%)	Clinical impairment	Nerve pathology	References
MAG/SGPG/P0	IgM	PN+IgM	50	S>>M	Demyelination	[1, 8]
Sulfatide	IgM	PN axonal	?	S, S>M, SM	Axonal Degeneration	[50]
		PN+IgM	6	SM	Demyelination	[8]
GM1	IgM	MMN±IgM	20–80	M	Focal demyelination	[58]
		LMND+IgM	~5	M	MND	[57]
	IgG	GBS	20–30	M>>S	Axonal Degeneration	[81]
GQ1b	IgG	MFS	>90	MFS	Normal/Demyelination	[87]
Disialogangliosides (GD1b/GQ1b)	IgM	PN+IgM	2	S>>M	Demyelination	[97, 99]
GD1a	IgM	PN+IgM	2	M	Demyelination	[105]
	IgG	GBS	>5	M>>S	Axonal Degeneration	[103]
GM2	IgM	PN±IgM	?	M>S	Demyelination	[106]
	IgG+IgM	GBS	6	SM	Demyelination	[107]
Chondroitin sulfate C	IgM	PN+IgM	1	SM	Axonal Degeneration	[108]
β-tubulin	IgM	CIDP	57	M>>S	Demyelination	[109]
Hu (neuronal nucleus)	IgG	SSN/PEM	?	S>>M	Axonal Degeneration	[110]

MAG, Myelin-associated glycoprotein; *SGPG*, Sulfoglucuronyl paragloboside; *SM*, Sensorimotor neuropathy; *M*, Motor; *S*, Sensory; *PN +IgM*, Neuropathy associated with IgM monoclonal gammopathy; *MMN*, Multifocal motor neuropathy; *LMND*, Lower motor neuron disease; *GBS*, Guillain-Barré syndrome; *MFS*, Miller Fisher syndrome; *CIDP*, Chronic inflammatory demyelinating polyneuropathy; *SSN/PEM*, Subacute sensory neuronopathy-paraneoplastic encephalomyelitis.

patients with Waldenström macroglobulinemia (MW) and, occasionally, with lymphoma [7, 8, 13].

There is considerable evidence that the neuropathy in these patients is an autoimmune disease caused by binding of anti-MAG IgM antibodies to one or more glycoconjugates in nerve bearing the HNK-1/L2 epitope [2, 14]:

1. Even if anti-MAG antibodies may be found at low titers in neurological controls and probably represent a common constituent of the human antibody repertoire [7], high titers of these antibodies (>1/6400 by immunoblot in our laboratory) are almost invariably found in cases of neuropathy associated with IgM monoclonal gammopathy. We recently reviewed the clinical reports of over 700 patients with different forms of neuropathy tested in our laboratory for anti-MAG antibodies and found that 95% of patients with high titers of anti-MAG IgM had neuropathy associated with IgM monoclonal gammopathy. Furthermore, most of the patients with high antibody titers but no neuropathy symptoms had subclinical neuropathy [7, 8] that later became clinically manifest [15]. In other patients with neuropathy, the finding of anti-MAG antibodies led to the discovery of a previously unrecognized or otherwise undetectable IgM M-protein [7, 16]. Even if the lack of correlation between anti-MAG antibody titers and the severity of neuropathy among different individuals may argue against a possible pathogenetic role of these antibodies [17], this discrepancy is not unusual in autoimmune diseases and may reflect different fine antibody specificities [18, 19] or affinities [20] for the antigen in vitro than in vivo.

2. The neuropathy in patients with high anti-MAG IgM antibodies is quite homogeneous. The majority of affected patients are men presenting their first neuropathy symptoms in the sixth or seventh decade of life [2, 14]. The neuropathy usually runs a slowly progressive course [21, 22] with predominantly deep sensory involvement, gait ataxia and postural tremor in the upper limbs [2, 8, 14]. Motor impairment is usually less prominent and often appears later. Electrophysiological [8, 23] and morphological [24, 25] studies are consistent with a demyelinating neuropathy with the presence, by ultrastructural examination, of widely spaced myelin lamellae [24], found in over 90% of examined patients [26].

3. Deposits of IgM [6, 24, 27] and complement [28, 29] around the myelin sheaths of peripheral nerves, i.e. the presumed target organ of the immune response, were detected by direct immunofluorescence or immunohistochemistry in over 95% of patients with neuropathy associated with anti-MAG IgM monoclonal gammopathy [8].

4. With only few exceptions [30-32], therapeutical reduction of anti-MAG IgM levels in these patients, although difficult to achieve even using several immunosuppressive agents (e.g. prednisone, plasma exchange, cytotoxic agents, high-dose intravenous immunoglobulin (IVIg) and fludarabine) was associated with clinical improvement [1, 33-35].

5. Experimental demyelination of nerve has been induced in animals by intraneural or systemic injection of serum containing anti-MAG IgM antibodies. In

earlier studies, injections of anti-MAG IgM antibodies in feline nerves caused demyelination only when added with fresh complement [36, 37], suggesting that antibodies caused demyelination by complement fixation. Demyelination of nerve with prominent widening of myelin lamellae was also obtained, however, in chickens by systemic injection of purified anti-MAG IgM without complement [38]. These results led to the hypothesis that antibodies may inhibit myelin turnover or alter the normal periodicity of myelin by separating or impeding the fusion of apposing leaflets of intraperiod lines. Such a mechanism may explain the slow progression of neuropathy in these patients. It was not, however, investigated in this study whether human IgM was able to activate chicken complement. More recently Monaco et al. [39] induced demyelination with widely spaced myelin lamellae in rabbit nerve by intraneural injection of either anti-MAG IgM or the terminal complement complex (TCC). They also demonstrated that in nerves treated with anti-MAG IgM, the abnormalities were concurrent with activation of the rabbit's own complement to the formation of TCC. This finding, together with the lack of effect of anti-MAG IgM in C6-deficient rabbits and the correlation in nerve biopsies between the number of fibers with widened myelin lamellae and that of fibers showing TCC deposits [29], seems to confirm the possible effector role of complement in the demyelination induced by anti-MAG IgM.

Although the pathogenetic mechanisms of the neuropathy in patients with anti-MAG IgM have been, at least in part, elucidated, little is known about what causes IgM monoclonal gammopathy and why its immune reactivity is frequently directed at the same or closely related MAG epitopes. Low levels of anti-MAG IgM antibodies are found in approximately 20% of controls with and without circulating M-protein, including some normal subjects [7]. B cells capable of secreting anti-MAG antibodies are present at birth [40], indicating that these antibodies may be a common constituent of the human antibody repertoire [41]. It is therefore possible that anti-MAG M-proteins derive from monoclonal expansion of normally occurring anti-MAG IgM secreting clones, as has been postulated for other M-proteins with autoantibody activity [41, 42]. The reason for this expansion is not known. Monoclonal expansion may be caused by transformations or mutations of anti-MAG-secreting B cell clones that disrupt normal regulatory interactions or by activation of regulatory T cells, possibly induced by self or foreign antigens bearing similar carbohydrate epitopes [43]. Anti-MAG M-protein secretion in vitro is subjected to T cell regulation [44, 45] and anti-idiotypic antibodies to the M-protein have been demonstrated in some patients [46, 47], suggesting that anti-MAG M-protein secretion, although abnormally increased, is somehow regulated. This is indicated by the fact that in most patients, anti-MAG M-protein levels are stable for years and rapidly return to their original level when treatment directed at their reduction is suspended. Whether these antibodies may also have a regulatory role on the immune system [48] is not known. However, this possibility may prompt some cautions on the risks of therapies directed at reducing M-protein levels in these patients.

Anti-Sulfatide IgM Antibodies

Sulfatide (galatosylceramide-3-O-sulfate) is the major acidic glycosphingolipid in central and peripheral nerve myelin where its concentration is about 100-times higher than that of any ganglioside [49].

After the initial report of Pestronk et al. [50], more than 50 patients with neuropathy and high anti-sulfatide IgM antibodies have been reported, half of whom had IgM monoclonal gammopathy [8, 51-55]. Even if only few studies have so far been devoted to this reactivity, some data favor a possible pathogenetic role of these antibodies in the neuropathy.

As in the case of anti-MAG IgM antibodies, low titers of anti-sulfatide antibodies (up to 1/8,000 by ELISA in our laboratory) may be found in several neurological diseases and in controls, while high titers (>1/8,000) are almost invariably associated with neuropathy (see below). Chronic progressive, predominantly sensory *axonal* neuropathy often presenting with painful paresthesias has been initially associated with this reactivity [50-53], while subsequently reported patients had sensorimotor *demyelinating* neuropathy with prominent limb weakness and gait ataxia [8, 54, 55]. In a review of the clinical and electrophysiological features of almost 600 patients referred to our laboratory for anti-sulfatide antibody testing, we found high antibody titers (>1/8,000 by ELISA) in only 11 patients, all with chronic demyelinating sensorimotor neuropathy. In 7 of these cases, there was associated IgM monoclonal gammopathy (6% of patients with neuropathy associated with IgM monoclonal gammopathy).

Morphological studies on sural nerve biopsy in some of these patients showed abnormally spaced myelin lamellae with myelin deposits of the M-protein and complement by direct immunofluorescence [8, 55]. The high concentration of sulfatide in myelin as well as the close correlation between the number of fibers with abnormally spaced myelin lamellae and that of fibers with deposits of the terminal cytolytic complex of complement in one patient (patient 6 in [29]), support the hypothesis that, at least in some patients, myelin can be the target for these antibodies. Myelin deposits of IgM were not found in other patients, but in these patients IgM bound to dorsal root ganglia [51, 52], possibly revealing a different site of attack for the antibodies. It is unclear however whether these differences reflect different antibody specificities, the concomitant presence in some patients of other serum IgM reactivities [8, 50-53], or the use of different methods or reference values for these antibodies.

Even if the number of patients so far reported with this reactivity is relatively small, the clinical improvement observed in some patients concomitant to the therapeutical reduction of serum IgM levels (unpublished observations) as well as preliminary experimental studies showing that demyelination of nerve can been induced in animals by intraneural injection of human anti-sulfatide IgM antibodies (S. Monaco, personal communication) support the hypothesis that these antibodies, although infrequent, may have a pathogenetic role in human neuropathies.

Anti-Ganglioside Antibodies

Gangliosides are sialic acid-containing glycosphingolipids that are particularly abundant in neural membranes in the CNS and PNS where they represent about 10% of the total lipid content [56]. IgM reactivity with the ganglioside GM1 was initially [57] reported in a patient with lower motor neuron disease (MND); it was subsequently associated with multifocal motor neuropathy (MMN) [58] in a proportion of patients ranging from 20% [59] to over 80% [60], according to different laboratories and technique used. These antibodies are not specific for MMN but may be also found in Guillain-Barré syndrome (GBS) (see below), in other chronic neuropathies [61, 62], and occasionally, in MND [63-65]. However, a recent meta-analysis confirmed that anti-GM1 IgM can help distinguish MMN from other lower motor neuron syndromes [66]. We reached similar conclusions by reviewing a series of over 700 patients with different neuropathies or motor neuron syndromes referred to our laboratory since 1991 for anti-GM1 IgM antibody testing. High anti-GM1 IgM titers (>1/320 by ELISA) were highly predictive for dysimmune neuropathies, diagnosed in over 80% of positive patients, but were not specifically associated with MMN, where they were found in 20% of patients. These results support the opinion that anti-GM1 IgM may help to individuate patients with motor neuropathy syndromes susceptible to immune therapies.

The pathogenetic role of these antibodies remains unclear [64, 67, 68]. As already mentioned, high levels of serum anti-GM1 IgM antibodies are frequently found in MMN patients, some of whom improved after therapeutic reduction of antibody levels [58, 69]. There is however a consistent proportion of MMN patients, some of whom also respond to immunological treatments [70-73], not bearing high anti-ganglioside antibodies. On the other hand, increased anti-GM1 levels have also been found in other dysimmune neuropathies and, occasionally in MND. No definite correlation between the different clinical syndromes and the fine specificity of these antibodies has been so far convincingly demonstrated [64, 68], leaving it unclear how similar antibodies may cause different diseases. IgM deposits have been detected at the level of the nodes of Ranvier in the sural nerve of an MND patient with multifocal motor conduction block (CB). High anti-GM1 titers [74] and focal CB have been experimentally induced in vivo [75, 76] and in vitro [77] by intraneural injection or exposure to sera from patients with high anti-GM1 antibodies and MMN but not lower MND [76]. The latter results were not, however, confirmed using the purified anti-GM1 antibodies [78]. More recently, a similar blocking effect on mouse distal motor nerve conduction has been induced in vitro with sera from MMN patients with *and* without high anti-GM1 antibodies [79], suggesting that either current tests for anti-GM1 antibodies are not sensitive enough or that serum components or antibodies other than anti-GM1 antibodies might be responsible for CB in MMN. These antibodies may therefore represent a reliable marker of an ongoing immune response, whose primary antigenic target is not known but probably differs in the different diseases. Although probably not primarily pathogenetic, anti-GM1 IgM may help individuate patients susceptible to immune therapies.

More recently, high titers of anti-GM1 IgG antibodies have been reported in 20%-30% of patients with GBS [80-84]. A close association between these antibodies and an antecedent *Campylobacter jejuni* (CJ) infection has also been reported [80-85]. In our series of 113 patients with GBS, we found anti-GM1 IgG antibodies in only 12%. These patients represented, however, more than 80% of all anti-GM1 IgG-positive patients, indicating that these antibodies are highly predictive for GBS. The initially reported association of this reactivity with a severe, predominantly motor, axonal form of GBS with poor prognosis [81, 82, 84] was not subsequently confirmed [80, 83] and seems to be more related to an antecedent CJ infection than to anti-GM1 antibodies [85]. Also in these patients the spontaneous or therapeutical reduction of antibody titers most often correlated with clinical improvement [81, 84], supporting a possible pathogenetic role of anti-GM1 IgG antibodies in GBS even if further studies are necessary to confirm this hypothesis.

High titers of IgG antibodies to GQ1b have been reported in up to 100% of patients with Miller Fisher syndrome (MFS) [86-89] as well as in some patients with other acute diseases characterized by oculomotor impairment including GBS with ophthalmoplegia, Bickerstaff's brain stem encephalitis, or acute ophthalmoparesis [87, 89-91]. Even if the incidence of MFS and other diseases associated with anti-GQ1b IgG antibodies is apparently low (5% of GBS) [92], the high specificity of these antibodies and their almost constant presence in MFS indicate that they are a useful test in the diagnosis of MFS. The possible pathogenetic relevance of these antibodies is supported by the fact that their close association with diseases characterized by oculomotor impairment correlates with the abundant expression of GQ1b in the extramedullary portion of the oculomotor nerves [87]. This finding, together with the decrease of anti-GQ1b antibodies during clinical improvement [86, 87, 93, 94] and the recent observations that these antibodies may interfere with [95] or block the neuromuscular transmission in vitro [96] all point to a pathogenetic role of these antibodies.

High titers of IgM antibodies to GQ1b and other disialosyl gangliosides have been reported in some patients with chronic, predominantly sensory or sensorimotor, mostly demyelinating neuropathies associated with IgM monoclonal gammopathy [89, 97-102]. In all these patients, anti-GQ1b IgM antibodies also bound to other gangliosides containing a disialosyl residue including GD3, GD1b and GT1b. However, IgM deposits were not detected in sural nerve biopsies leaving unclear the possible pathogenetic role of these antibodies in the neuropathy. Nonetheless, the frequent correlation of this uncommon reactivity (2% in our series of over 100 patients with neuropathy associated with IgM monoclonal gammopathy) with definite clinical syndrome makes it unlikely that this association is merely casual.

High titers of serum IgG antibodies to the ganglioside GD1a have been reported in a small proportion of patients with GBS [103, 104], while high titers of anti-GD1a IgM antibodies have been reported in patients with a chronic, mostly demyelinating, motor neuropathy associated with IgM monoclonal gammopathy [104, 105]. We found moderately high anti-GD1a titers in patients with MND and other immunological diseases [104], a finding that is probably devoid of pathogenetic or diagnostic relevance. On the other hand, high antibody titers (>1/5120 by ELISA) always correlated with dysimmune neuropathies including GBS for IgG antibodies (3% of GBS patients), and chronic demyelinating motor neuropathy associated with IgM monoclonal gammopathy for IgM (3% of these patients). As also reported for anti-GM1 antibodies, anti-GD1a IgG antibodies have been initially associated with a severe axonal form of GBS [103], but this association was not always observed [104]. Even if in both GBS and motor neuropathy patients the therapeutical reduction of anti-GD1a antibodies was associated with clinical improvement [103, 104] supporting their pathogenetic role in the neuropathy, in none of the patients so far reported were deposits of antibodies detected in sural nerve biopsy so that the pathogenetic role of this reactivity remains to be elucidated.

Acknowledgements: This work was made possible by the financial support of Associazione Amici Centro Dino Ferrari and by grants from Telethon, Italy (No. 674), Associazione Italiana Sclerosi Multipla and IRCCS Ospedale Maggiore Policlinico, Milan, Italy.

References

1. Latov N, Sherman WH, Nemni R et al. (1980) Plasma cell dyscrasia and peripheral neuropathy with a monoclonal antibody to peripheral nerve myelin. N Eng J Med 303: 618-621
2. Latov N, Hays AP, Sherman WH (1988) Peripheral neuropathy and anti-MAG antibodies. Crit Rev Neurobiol 3: 301-332
3. Quarles RH (1989) Myelin-associated glycoprotein in demyelinating disorders. Crit Rev Neurobiol 5: 1-28
4. Trapp BD (1990) Myelin-associated glycoprotein. Location and potential functions. Ann N Y Acad Sci 605: 29-43
5. Murray N, Page N, Steck AJ (1986) The human anti-myelin-associated glycoprotein IgM system. Ann Neurol 19: 473-478
6. Nobile-Orazio E, Marmiroli P, Baldini L et al. (1987) Peripheral neuropathy in macroglobulinemia: Incidence and antigen specificity of M-proteins. Neurology 37: 1506-1514
7. Nobile-Orazio E, Francomano E, Daverio R et al. (1989) Anti-myelin-associated glycoprotein IgM antibody titers in neuropathy associated with macroglobulinemia. Ann Neurol 26: 543-550
8. Nobile-Orazio E, Manfredini E, Carpo M et al. (1994) Frequency and clinical correlates of anti-neural IgM antibodies in neuropathy associated with IgM monoclonal gammopathy. Ann Neurol 36: 416-424
9. McGarry RC, Helfand SL, Quarles RH, Roder JC (1983) Recognition of myelin-associated glycoprotein by the monoclonal antibody HNK-1. Nature 306: 376-378
10. Bollensen E, Steck AJ, Schachner M (1988) Reactivity with the peripheral myelin glycoprotein P0 in serum from patients with monoclonal IgM gammopathy and polyneuropathy. Neurology 38: 1266-1270
11. Snipes GJ, Suter U, Shooter EM (1993) Human peripheral myelin protein-22 carries the L2/HNK1 carbohydrate adhesion epitope. J Neurochem 61: 1961-1964
12. Chou DKH, Ilyas AA, Evans JE et al. (1986) Structure of sulfated glucuronyl glycolipids in the nervous system reacting with HNK-1 antibody and some IgM paraproteins in neuropathy. J Biol Chem 261: 11717-11725

13. Baldini L, Nobile-Orazio E, Guffanti A et al. (1994) Peripheral neuropathy in IgM monoclonal gammopathy and Waldenstrom macroglobulinemia: A frequent complication in males with low MAG-reactive serum monoclonal component. Am J Hematol 45: 25-31
14. Nobile-Orazio E (1998) Neuropathies associated with anti-MAG antibodies and IgM monoclonal gammopathies. In: Latov N, Wokke JHJ, Kelly JJ (eds) Immunological and infectious diseases of the peripheral nerve. Cambridge University, Cambridge, pp 169-189
15. Meucci N, Baldini L, Barbieri S et al. (1997) Anti-MAG IgM antibodies predict the development of neuropathy in asymptomatic patients with IgM monoclonal gammopathy. Neurology 48(Suppl 2): A52
16. Nobile-Orazio E, Latov N, Hays AP et al. (1984) Neuropathy and anti-MAG antibodies without detectable serum M-protein. Neurology 34: 218-221
17. Gosselin S, Kyle RA, Dyck PJ (1991) Neuropathy associated with monoclonal gammopathies of undetermined significance. Ann Neurol 30: 54-61
18. Ilyas AA, Chou DKH, Jungalwala FB et al. (1990) Variability in the structural requirement for binding of human monoclonal anti-myelin-associated glycoprotein immunoglobulin M antibodies and HNK-1 to sphingolipid antigens. J Neurochem 55: 594-601
19. Lieberman F, Marton LS, Stefansson K (1985) Pattern of reactivity of IgM from the sera of eight patients with IgM monoclonal gammopathy and neuropathy with components of neural tissues: evidence for interaction with more than one epitope. Acta Neuropathol (Berl) 68: 196-200
20. Ogino M, Tatum AH, Latov N (1994) Affinity of human anti-MAG antibodies in neuropathy. J Neuroimmunol 52: 41-46
21. Smith IS (1994) The natural history of chronic demyelinating neuropathy associated with benign IgM paraproteinemia. Brain 117: 949-957
22. Nobile-Orazio E, Meucci N, Baldini L et al. (1997) Long-term prognosis of neuropathy associated with anti-MAG IgM. J Neurol 244(Suppl. 3): S41
23. Kaku DA, England JD, Sumner AJ (1994) Distal accentuation of conduction slowing in polyneuropathy associated with antibodies to myelin-associated glycoprotein and sulphated glucuronyl paragloboside. Brain 117: 941-947
24. Meier C, Vandevelde M, Steck A, Zurbriggen A (1983) Demyelinating polyneuropathy associated with monoclonal IgM paraproteinemia. Histological, ultrastructural and immunocytochemical studies. J Neurol Sci 63: 353-367
25. Nemni R, Galassi G, Latov N et al. (1983) Polyneuropathy in nonmalignant IgM plasma cell dyscrasia: a morphological study. Ann Neurol 14: 43-54
26. Vital A, Vital C, Julien J et al. (1989) Polyneuropathy associated with IgM monoclonal gammopathy. Immunological and pathological study in 31 patients. Acta Neuropathol 79: 160-167
27. Takatsu M, Hays AP, Latov N et al. (1985) Immunofluorescence study of patients with neuropathy and IgM M-proteins. Ann Neurol 18: 173-181
28. Hays AP, Lee SSL, Latov N (1988) Immune reactive C3d on the surface of myelin sheaths in neuropathy. J Neuroimmunol 18: 231-244
29. Monaco S, Bonetti B, Ferrari S et al. (1990) Complement-mediated demyelination in patients with IgM monoclonal gammopathy and polyneuropathy. New Eng J Med 322: 649-652
30. Ernerudh JH, Vrethem M, Andersen O et al. (1992) Immunochemical and clinical effects of immunosuppressive treatment in monoclonal IgM neuropathy. J Neurol Neurosurg Psychiatry 55: 930-934

31. Melmed C, Frail D, Duncan I et al. (1983) Peripheral neuropathy with IgM monoclonal immunoglobulin directed against myelin-associated glycoprotein. Neurology 33: 1397-1405
32. Stefansson K, Marton L, Antel JP et al. (1983) Neuropathy accompanying IgM lambda monoclonal gammopathy. Acta Neuropathol (Berl) 59: 255-261
33. Blume G, Pestronk A, Goodnough LT (1995) Anti-MAG antibodies associated polyneuropathies: Improvement following immunotherapy with monthly plasma exchange and IV cyclophosphamide. Neurology 45: 1577-1580
34. Kelly JJ, Adelman LS, Berkman E, Bhan I (1988) Polyneuropathies associated with IgM monoclonal gammopathies. Arch Neurol 45: 1355-1359.
35. Nobile-Orazio E, Baldini L, Barbieri S et al. (1988) Treatment of patients with neuropathy and anti-MAG IgM M-proteins. Ann Neurol 24: 93-97
36. Hays AP, Latov N, Takatsu M, Shermann WH (1987) Experimental demyelination of nerve induced by serum of patients with neuropathy and anti-MAG IgM M-proteins. Neurology 37: 242-256
37. Willison HJ, Trapp BD, Bacher JD et al. (1988) Demyelination induced by intraneural injection of human anti-myelin-associated glycoprotein antibodies. Muscle Nerve 11: 1169-1176
38. Tatum AH (1993) Experimental paraprotein neuropathy; demyelination by passive transfer of human anti-myelin-associated glycoprotein. Ann Neurol 33: 502-506
39. Monaco S, Ferrari S, Bonetti B et al. (1995) Experimental induction of myelin changes by anti-MAG antibodies and terminal complement complex. J Neuropathol Exp Neurol 54: 96-104
40. Lee KW, Inghirami G, Sadiq SA et al. (1990) B-cells that secrete anti-MAG or anti-GM1 antibodies are present at birth and anti-MAG secreting B-cells are CD5+. Neurology 40(Suppl 1): 367
41. Dighiero G, Lymberi P, Guilbert B et al. (1986) Natural autoantibodies constitute a substantial part of normal circulating immunoglobulins. Ann N Y Acad Sci 475: 135-145
42. Avrameas S, Guilbert B, Dighiero B (1981) Natural antibodies against tubulin, actin, myoglobulin, thyroglobulin, fetuin and transferrin are present in normal human sera and monoclonal immunoglobulins from multiple myeloma and Waldenstrom macroglobulinemia may express similar antibody specificities. Ann Immunol 132C: 103-113
43. Latov N (1987) Waldenström's macroglobulinemia and nonmalignant IgM monoclonal gammopathies. In: Kelly JJ, Kyle RA, Latov N (eds) Polyneuropathies associated with plasma cell dyscrasia. Martinus Nijhoff, Boston, pp 51-72
44. Latov N, Godfrey M, Thomas Y et al. (1985) Neuropathy and anti-myelin-associated glycoprotein IgM M-proteins: T cell regulation of M-protein secretion in vitro. Ann Neurol 18: 182-188
45. Spatz L, Latov N (1986) Secretion of anti-myelin associated glycoprotein antibodies by B-cells from patients with neuropathy and nonmalignant monoclonal gammopathy. Cell Immunol 1: 434-440
46. Nobile-Orazio E, McIntosh C, Latov N (1985) Anti-MAG antibody and antibody complexes: Detection by radioimmunoassay. Neurology 35: 988-992
47. Page N, Murray N, Perruisseau G, Steck AJ (1985). A monoclonal anti-idiotypic antibody against a human monoclonal IgM with specificity for myelin-associated glycoprotein. J Immunol 134: 3094-3099
48. Tanaka M, Nishizawa M, Inuzuka T et al. (1985) Human natural killer cell activity is reduced by treatment of anti-myelin-associated glycoprotein (MAG) monoclonal mouse IgM antibody and complement. J Neuroimmunol 10: 115-127

49. Svennerholm L, Fredman P (1990) Antibody detection in Guillain Barre' syndrome. Ann Neurol 27(Suppl): 36-40
50. Pestronk A, Li F, Griffin J, Feldman EL, Cornblath D, Trotter J, Zhu S, Yee WC, Phillips D, Peeples DM, Winslow B (1991) Polyneuropathy syndromes associated with serum antibodies to sulfatide and myelin-associated glycoprotein. Neurology 41: 357-362
51. Quattrini A, Corbo M, Dhaliwal SK et al. (1992) Anti-sulfatide antibodies in neurological disease: Binding to rat dorsal root ganglia neurons. J Neurol Sci 112: 152-159
52. Nemni R, Fazio R, Quattrini A et al. (1993) Antibodies to sulfatide and to chondroitin sulfate C in patients with chronic sensory neuropathy. J Neuroimmunol 43: 79-86
53. van den Berg LH, Lankamp CLAM, de Jager AEJ et al. (1993) Anti-sulphatide antibodies in peripheral neuropathy. J Neurol Neurosurg Psychiatry 56: 1164-1168
54. Lopate G, Parks BJ, Goldstein JM et al. (1997) Polyneuropathies associated with high titre antisulphatide antibodies: Characteristics of patients with and without monoclonal proteins. J Neurol Neurosurg Psychiatry 62: 581-585
55. Ferrari S, Morbin M, Nobile-Orazio E et al. (1998) Antisulfatide polyneuropathy: Antibody-mediated complement attack on peripheral myelin. Acta Neuropathol (in press)
56. Wiegandt H (1985) Gangliosides. In: Wiegandt H (ed) Glycolipids. Elsevier, Amsterdam, pp 199-260
57. Freddo L, Yu RK, Latov N et al. (1986) Gangliosides GM1 and GD1b are antigens for IgM M-protein in a patient with motor neuron disease. Neurology 36: 454-458
58. Pestronk A, Cornblath DR, Ilyas A et al. (1988) A treatable multifocal motor neuropathy with antibodies to GM1 ganglioside. Ann Neurol 24: 73-78
59. Lange DJ, Trojaburg W, Latov N et al. (1992) Multifocal motor neuropathy with conduction block: Is it a distinct clinical entity? Neurology 42: 497-505
60. Pestronk A, Choksi R (1997) Multifocal motor neuropathy. Serum IgM anti-GM1 ganglioside antibodies in most patients detected using covalent linkage of GM1 to ELISA plates. Neurology 49: 1289-1292
61. Lamb NL, Patten BM (1991) Clinical correlations of anti-GM1 antibodies in amyotrophic lateral sclerosis and neuropathies. Muscle Nerve 14: 1021-1027
62. Sadiq SA, Thomas FP, Kilidireas K et al. (1990) The spectrum of neurological disease associated with anti-GM1 antibodies. Neurology 40: 1067-1072
63. Kinsella LJ, Lange DJ, Trojaburg W et al. (1994) Clinical and electrophysiologic correlates of elevated anti-GM1 antibody titers. Neurology 44: 1278-1282
64. Kornberg AJ, Pestronk A (1994) The clinical and diagnostic role of anti-GM1 antibody testing. Muscle Nerve 17: 100-104
65. Nobile-Orazio E, Legname G, Daverio R et al. (1990) Motor neuron disease in a patient with a monoclonal IgMk directed against GM1, GD1b and high-molecular-weight neural-specific glycoproteins. Ann Neurol 28: 190-194
66. van Schaik IN, Bossuyt PMM, Brand A, Vermeulen M (1995) Diagnostic value of GM1 antibodies in motor neuron disorders and neuropathies: A meta analysis. Neurology 45: 1570-1577
67. Nobile-Orazio E (1996) Multifocal motor neuropathy. J Neurol Neurosurg Psychiatry 60: 599-603
68. Willison HJ (1994) Antiglycolipid antibodies in peripheral neuropathy: Fact or fiction. J Neurol Neurosurg Psychiatry 57: 1303-1307
69. Feldman EL, Bromberg MB, Albers JW, Pestronk A (1991) Immunosuppressive treatment in multifocal motor neuropathy. Ann Neurol 30: 397-401
70. Azulay JP, Blin O, Pouget J et al. (1994) Intravenous immunoglobulin treatment in patients with motor neuron syndromes associated with anti-GM1 antibodies. Neurology 44: 429-432

71. Bouche P, Moulonguet A, Younes-Chennoufi AB et al. (1995) Multifocal motor neuropathy with conduction block: a study of 24 patients. J Neurol Neurosurg Psychiatry 59: 38-44
72. Chaudhry V, Corse A, Cornblath DR et al. (1993) Multifocal motor neuropathy: Response to human immune globulin. Ann Neurol 33: 237-242
73. Nobile-Orazio E, Meucci N, Barbieri S et al. (1993) High dose intravenous immunoglobulin therapy in multifocal motor neuropathy. Neurology 43: 537-544
74. Santoro M, Thomas FP, Fink ME et al. (1990) IgM deposits at the node of Ranvier in a patient with amyotrophic lateral sclerosis, anti-GM1 antibodies and multifocal conduction block. Ann Neurol 28: 373-377
75. Santoro M, Uncini A, Corbo M et al. (1992) Experimental conduction block induced by serum from a patient with anti-GM1 antibodies. Ann Neurol 31: 385-390
76. Uncini A, Santoro M, Corbo M et al. (1993) Conduction abnormalities induced by sera of patients with multifocal motor neuropathy and anti-GM1 antibodies. Muscle Nerve 16: 610-615
77. Arasaki K, Kusunoki S, Kudo N, Kanazawa I (1993) Acute conduction block in vitro following exposure to anti-ganglioside sera. Muscle Nerve 16: 587-593
78. Harvey GK, Toyka KV, Zielasek J et al. (1995) Failure of anti-GM1 IgG or IgM to induce conduction block following intraneural transfer. Muscle Nerve 18: 388-394
79. Roberts M, Willison HJ, Vincent A, Newsom-Davis J (1995) Multifocal motor neuropathy human sera block distal motor nerve conduction in mice. Ann Neurol 38: 111-118
80. Enders U, Karch H, Toyka KV et al. (1993) The spectrum of immune responses to *Campylobacter jejuni* and glycoconjugates in Guillain-Barré syndrome and in other neuroimmunological disorders. Ann Neurol 34: 136-144
81. Yuki N, Yoshino H, Sato S Miyatake T (1990) Acute axonal polyneuropathy associated with anti-GM1 antibodies following Campylobacter enteritis. Neurology 40: 1900-1902
82. Walsh FS, Cronin M, Koblar S, Doherty P et al. (1991) Association between glycoconjugate antibodies and Campylobacter infection in patients with Guillain Barre' syndrome. J Neuroimmunol 34: 43-51
83. Vriesendorp FJ, Mishu B, Blaser MJ, Koski CL (1993) Serum antibodies to GM1, GD1b, peripheral nerve myelin, and *Campylobacter jejuni* in patients with Guillain-Barre' syndrome and controls: Correlation and prognosis. Ann Neurol 34: 130-135
84. Nobile-Orazio E, Carpo M, Meucci N et al. (1992) Guillain Barre' syndrome associated with high titers of anti-GM1 antibodies. J Neurol Sci 109: 200-206
85. Rees JH, Gregson NA, Hughes RAC (1995) Anti-ganglioside GM1 antibodies in Guillain Barré syndrome and their relationship to *Campylobacter jejuni* infection. Ann Neurol 38: 809-816
86. Chiba A, Kusunoki S, Shimizu T, Kanazawa I (1992) Serum IgG antibody to ganglioside GQ1b is a possible marker of Miller Fisher syndrome. Ann Neurol 31: 677-679
87. Chiba A, Kusuno ki S, Obata H et al. (1993) Serum anti-GQ1b IgG antibody is associated with ophthalmoplegia in Miller Fisher syndrome and Guillain-Barré syndrome: Clinical and immunohistochemical studies. Neurology 43: 1911-1917
88. Yuki N, Sato S, Tsuji S et al. (1993) Frequent presence of anti-GQ1b antibody in Fisher's syndrome. Neurology 43: 414-417
89. Carpo M, Pedotti R, Lolli F et al. (1998) Clinical correlate and fine specificity of anti-GQ1b antibodies in peripheral neuropathy. J Neurol Sci 155: 186-191
90. Yuki N, Sato S, Tsuji S et al. (1993) An immunological abnormality common to Bickerstaff's encephalitis and Fisher's syndrome. J Neurol Sci 118: 83-87

91. Yuki N (1996) Acute paresis of extraocular muscles associated with IgG anti-GQ1b antibody. Ann Neurol 39: 668-672
92. Ropper AH (1992) The Guillain Barré syndrome. N Eng J Med 326: 1130-1136
93. Yuki N, Miyatake T, Ohsawa T (1993) Beneficial effect of plasmapheresis on Fisher's syndrome. Muscle Nerve 16: 1267-1268
94. Yuki N (1995) Successful plasmapheresis in Bickerstaff's brain stem encephalitis associated with anti-GQ1b antibody. J Neurol Sci 131: 108-110
95. Roberts M, Willison H, Vincent A, Newsom-Davis J (1994) Serum factor in Miller-Fisher variant of Guillain-Barré syndrome and neurotransmitter release. Lancet 343: 454-455
96. Buchwald B, Wiishaupt A, Toyka KV, Dudel J (1995) Immunoglobulin G from a patient with Miller-Fisher syndrome rapidly and reversibly decrease evoked quantal release at the neuromuscular junction in mice. Neurosci Lett 201: 163-166
97. Ilyas AA, Quarles RH, Dalakas MC et al. (1985) Monoclonal IgM in a patient with paraproteinemic polyneuropathy binds to a carbohydrate containing disialosyl groups. Ann Neurol 18: 655-659
98. Obi T, Kusunoki S, Takatsu M et al. (1992) IgM M-protein in a patient with sensory dominant neuropathy binds preferentially to polysialogangliosides. Acta Neurol Scand 86: 215-218
99. Daune GC, Farrer RG, Dalakas MC, Quarles RH (1992) Sensory neuropathy associated with monoclonal immunoglobulin M to GD1b ganglioside. Ann Neurol 31: 683-685
100. Willison HJ, Paterson G, Veitch J et al. (1993) Peripheral neuropathy associated with monoclonal IgM anti-Pr2 cold agglutinins. J Neurol Neurosurg Psychiatry 56: 1178-1183
101. Oka N, Kusaka H, Kusunoki S et al. (1996) IgM M-protein with antibody activity against gangliosides with disialosyl residue in sensory neuropathy binds to sensory neurons. Muscle Nerve 19: 528-530
102. Yuki N, Miyatani N, Sato S et al. (1992) Acute relapsing sensory neuropathy associated with IgM antibody against B-series gangliosides containing GalNacβ1-4 (Gal3-2αNeuAc8-2αNeuAc)β1 configuration. Neurology 42: 686-689
103. Yuki N, Yamada M, Sato S et al. (1993) Association of IgG anti-GD1a antibody with severe Guillain-Barre' syndrome. Muscle Nerve 16: 642-647
104. Carpo M, Nobile-Orazio E, Meucci N et al. (1996) Anti-GD1a ganglioside antibodies in peripheral motor syndromes. Ann Neurol 39: 539-543
105. Bollensen E, Schipper HI, Steck AJ (1989) Motor neuropathy with activity of monoclonal IgM antibody to GD1a ganglioside. J Neurol 236: 353-355
106. Ilyas AA, Li SC, Chou DKH et al. (1988) Gangliosides GM2, IVGalNAcGM1b, and IVGalNAcGD1a as antigens for monoclonal immunoglobulin M in neuropathy associated with gammopathy. J Biol Chem 263: 4369-4373
107. Irie S, Saito T, Nakamura K et al. (1996) Association of anti-GM2 antibodies in Guillain-Barré syndrome with acute cytomegalovirus infection. J Neuroimmuno l68: 19-26
108. Sherman WH, Latov N, Hays AP et al. (1983) Monoclonal IgM-k antibody precipitating with chondroitin sulphate C from patients with axonal polyneuropathy and epidermolysis. Neurology 33: 192-201
109. Connolly AM, Pestronk A, Trotter JL et al. (1993) High titer selective serum anti-β-tubulin antibodies in chronic inflammatory demyelinating polyneuropathy. Neurology 43: 557-562
110. Graus F, Cordon-Cardo C, Posner JB (1985) Neuronal antinuclear antibody in sensory neuronopathy from lung cancer. Neurology 35: 538-543

Chapter 25

Treatment of Inflammatory Demyelinating Polyneuropathy

G. COMI, L. ROVERI

Introduction

The management of acute and chronic inflammatory demyelinating polyneuropathy (Guillain-Barrè syndrome (GBS) and chronic inflammatory demyelinating polyradiculoneuropathy (CIDP)) will be the main topic of this chapter. A few comments will also be made about treatment of the demyelinating form of paraproteinaemic demyelinating polyneuropathy (PDN) and of multifocal motor neuropathy (MMN). We will briefly describe the main characteristics of these neuropaties and examines case series and trials which evaluated the principal therapeutic strategies for GBS, CIDP, PDN and MMN, such as intravenous immunoglobulin therapy (IVIg), steroid treatment, plasma exchange, and immunosuppressor administration. Controlled trials demonstrated that IVIg, steroid treatment and plasma exchange are effective in GBS and CIDP. For PDN, the therapeutic strategies are the same as for idiopathic CIDP, but usually the clinical response is poorer. For MMN, IVIg are definitely the first choice treatment.

Guillain-Barrè Syndrome

Guillain-Barrè syndrome (GBS) is an inflammatory demyelinating disorder of the peripheral nervous system resulting from an abnormal immune response directed against components of peripheral nerve. The syndrome is characterised by rapidly evolving symmetrical limb weakness, loss of tendon reflexes, absent or mild sensory signs, and variable autonomic dysfunctions. The condition occurs world wide and has became the leading cause of acute neuromuscular paralysis affecting patients of all ages and both sexes with an incidence rate of 0.4-2.4 cases per 100 000. A quarter of all patients require artificial ventilation, 10% die as a result of complications of the disease, and 10% are left with such severe disability that they cannot walk unaided a year later [1-3]. Mortality is about 5%-8% [4], the most common causes of death being sepsis, pulmonary embolism, myocardial

---

Department of Clinical Neurophysiology, University of Milan, San Raffaele Scientific Institute, Via Olgettina 60 - 20132 Milan, Italy. e-mail: g.comi@hsr.it

infarction and arrhythmias due to autonomic neuropathy. Although GBS has an appreciable morbidity and mortality, about 80% of patients make a good recovery. The most reliable indicators for significant residual disability at 12 months from onset include older age, severely reduced compound muscle action potentials (CMAPs), prolonged ventilation (>1 month), and rapid progression or quadriparesis in less than 1 week [1, 4]. Prognosis is better in children [5]. Utmost vigilance and anticipation of potential complications are necessary to optimise the chances of a favourable outcome. An epidemiological survey of 79 GBS patients showed that 62% had made a complete recovery 1 year later, 17% were unable to run, and 9% were unable to walk unaided. In this series 8 patients (8%) died and 3 patients remained bedridden or ventilator-dependent at 1 year; all 13 patients were over 60 years of age [6].

Approximately two-thirds of all cases are preceded by an infection, which is most commonly mild and affects the upper respiratory or gastrointestinal tracts. A large number of organisms have been described in association with subsequent GBS, but the most common are *Campylobacter jejuni* [7-10], *Mycoplasma pneumoniae* [11], cytomegalovirus [12-15], Epstain-Barr virus [13, 16], HIV [17, 18], and HVC [19]. Most surveys show a slight peak in late adolescence and young adulthood, coinciding with an increased risk of infections with cytomegalovirus and *Campylobacter jejuni*, and a second peak in the elderly [14, 20, 21]. In addition to infections, a variety of other antecedent events such as surgery and malignancies, particularly Hodgkin's disease and other lymphomas, have been put forward as possible triggers. However, the link with GBS is not established and remains anecdotal. Several case reports or small series have linked GBS to vaccination on the grounds of a mere temporal association, but no causal relation has been established. Most currently used vaccines do not seem to be associated with any increased risk [22-26]. Nevertheless, there is no doubt that rabies vaccines carry an increased risk of inducing GBS, probably because of contamination with myelin antigens [27, 28]. There have been several reports of GBS occurring in association with certain drugs, although the infrequency of the association and the lack of pathological material in such cases shed doubt on any real pathogeni involvement. Drugs that have been reported in association with GBS include streptokinase [29], captopril [30], danazol [31], and intravenous heroine [32].

Although the pathogenesis of GBS remains incompletely defined, there is increasing support for the concept that GBS results from an aberrant organospecific immune response rather than a direct effect of the infecting agent [33]. The more attractive hypothesis is that an infectious agent induces an immune response that is cell-mediated or humoral, or a combination of the two [34, 35]. The salient pathological findings are lymphocytic infiltrates in spinal roots and peripheral nerves, with subsequent macrophage-mediated segmental stripping of myelin [36, 37].

The onset of neuropathic symptoms may occur acutely (within days) or subacutely (up to 4 weeks) and reaches a plateau, with subsequent resolution of paralysis. GBS typically begins with paraesthesiae at the tips of the fingers and toes. Sensory symptoms may then gradually ascend and are often out of proportion to any signs that may be elicited. Pain is a rather common presenting complaint and usually occurs in the shoulders, thighs or lumbar region. Radiation of lower back pain into the buttocks and legs may give rise to an initial diagnosis of lumbosacral nerve root compression, but weakness is much more prominent and areflexia is not limited to the ankle jerks. The pain is probably due to a radiculoneuritis and may be severe but responsive to conventional analgesics. For these reasons GBS may initially be mistaken for an acute musculoskeletal problem. One of the 2 features required for the diagnosis is progressive weakness of all 4 limbs. This is characteristically ascending and relatively symmetrical. It may be predominantly proximal, predominantly distal or equally proximal and distal. Hyporeflexia or areflexia is the other feature required for the diagnosis of GBS, although reflexes may be preserved in the first few days of the illness. The diagnosis of GBS is a clinical one and relies on history of symmetrical weakness and areflexia. The two most helpful investigations in GBS are cerebrospinal fluid (CSF) examination and nerve conduction studies. Examination of CSF reveals the phenomenon of albumino-cytological dissociation [38], whereby the CSF protein is elevated without a concomitant CSF pleocytosis. This occurs in about 90% of patients, although the protein is often normal in the first week. A raised CSF white cell count (>10/µl) is unusual and is suggestive of HIV infection in which GBS may occur as part of a seroconversion illness. Nerve conduction studies are the most sensitive and specific, and abnormalities in these tests occur earlier than elevated CSF protein. The characteristic electrophysiological feature is conduction block whereby there is a significant decrement in the CMAP amplitude on proximal nerve stimulation compared with distal nerve stimulation. In addition, F waves that reflect proximal nerve root conduction may be delayed or absent even when all other conduction parameters are normal.

Other forms of GBS occur but are rare and include an acute motor-sensory axonal neuropathy leading to axonal degeneration [37, 39-43]. Another variant form of GBS is the Miller Fisher syndrome (MFS) which has distinct immunological and pathological features. About 90% of MFS cases have a characteristic pattern of antibodies to GQ_{1b} ganglioside [44, 45]. The antibodies recognise epitopes that are expressed specifically in the nodal regions of oculomotor nerves, and also in dorsal-roots ganglion cells and cerebellar neurones [46, 47]. This pattern corresponds with the clinical feature of ophthalmoplegia, ataxia and areflexia.

Treatment of GBS is aimed at suppressing the immune response, thereby preventing further damage to the peripheral nerves, allowing remyelination and hence restoring nerve conduction.

General Treatment

Patients with GBS need to be admitted to hospital for careful observation in order to anticipate potential complications that may occur in the acute phase of the disease. Particular attention must be paid to: cardiorespiratory function by means of online cardiac monitoring and serial assessment of the ventilatory reserve; prevention of thromboembolic complications; appropriate bowel care, pain and infection management; and prevention of complication due to prolonged immobility. During the plateau phase, physiotherapy is essential to prevent contractures and to facilitate recovery of motor function. Psychological support is important to help patient and family to cope with this frightening and frustrating illness.

Plasma Exchange

In the mid-1980s, three large clinical trials, randomised and controlled but unmasked in design, independently demonstrated significant beneficial effect on the rate of recovery from several therapeutic plasma exchange (PEs) when begun in the first 2 weeks of disease. In the North American trial of 245 patients with severe GBS, 122 patients were randomly assigned to PE (5 exchanges of 50 ml/kg body weight each, given over 7-14 days), and 123 were assigned to conventional treatment (Table 1). On average, patients treated by PE improved more rapidly, could be weaned from assisted ventilation earlier, and reached ambulation 1 month earlier. PE was ineffective when started later than 2 weeks from onset of symptoms [48]. These results were corroborated by a French study [49] in which PE appeared to halt progression of GBS in addition to hastening recovery. In a follow-up study of the same French population, 71% of patients treated by PE recovered full motor strength as opposed to 52% of controls [50]. The main criticism to these trials was the lack of a sham PE control group and thus the lack of blindness. Within 1-2 weeks of initial improvement after PE, secondary worsening may be seen in 10% of patients [51]. These limited relapses may be due to persistent active disease or to antibody rebound; additional treatment by PE lead to renewed improvement [52]. More recently the French Cooperative Group on PE in GBS presented another large study of more than 550 patients that addressed the question of the appropriate number of PE treatments [53]. The recommendations derived from these studies are to use 2 PE treatments for mild GBS, and 4-5 for severe GBS, starting as soon as possible. PE is reasonably safe, but not totally free of risk, particularly in haemodynamically unstable GBS patients. Such risks, the high cost, and the limited availability of PE facilities prompted the search for alternative treatments.

Immunoglobulins

Based on some encouraging results of small studies [54, 55], a multicentre study was performed in the Netherlands comparing intravenous immunoglobulins (IVIg) and plasma exchange on 150 GBS patients. IVIg was given at a dose of 0.4 g/kg body weight for 5 consecutive days, while PE treatment followed the conventional schedule. At 4 weeks, significantly more patients showed functional improvement with intravenous IgG [56]. In this study, the PE group did worse than the PE group in the North American trial. However, some factors may have favoured the IVIg group: at entry, 15 patients assigned to PE versus 3 patients assigned to IVIg needed ventilation. This study was criticised in that the two groups were not equally matched and the study lacked masking. Therefore, the

Table 1. Treatment of GBS: overview of clinical trials

Clinical trials	Outcome	Treatments			$p <$
North American [48]		PE	Controls		
	Improved 1 grade at 1 month (%)	59	39		0.01
	Improved at 6 months (%)	97	87		0.01
	Days to improve 1 grade	19	40		0.001
	Days to walk unaided	53	85		0.001
French [49, 50]		PE	Controls		
	Days to walk unaided	70	111		0.01
	Days to weaning	18	31		0.005
Dutch [56]		PE	IVIg		
	Improved 1 grade at 1 month (%)	34	53		0.02
	Days to improve 1grade	41	27		0.05
	Days to walk unaided	69	55		0.07
Dutch [57]		PE	IVIg	PE+IVIg	
	Improved 1 grade at 1 month (%)	57.9	55.4	59.5	ns
	Days to walk unaided	49	51	40	ns
	Days to weaning	29	26	18	ns
GBS Steroid Trial Group [65]		Steroids	Controls		
	Improved at 1 month (%)	54	51		ns
	Days to walk unaided	38	50		ns
	Days to weaning	18	27		ns
Dutch [66]		IVIg	IVIg+ Steroids		
	Improved 1 grade at 1 month (%)	53	76		ns
	Days to improve 1g	27	20		ns
	Days to walk unaided	55	27		ns
	Days to weaning	15	6		ns

PE, plasma exchange; *IVIg*, intravenous immunoglobulins; *ns*, not significant.

two treatments were assessed again in a large multicentre randomised double-blind trial [57]. PE was compared with IVIg (0.4 g/kg body weight for 5 days) and with a combined treatment of PE (5 times over 10-14 days), followed by IVIg (0.4 g/kg body weight for 5 days) in 379 adult patients with severe GBS. The primary outcome criterion was the change of disability grade at 4 weeks from randomisation. The functional disability was assessed by means of a seven-point disability scale. The PE group improved by 0.92 grades, the IVIg group by 0.82 grades, and the combined treatment group by 1.11 grades. The three groups did not differ significantly in this outcome criterion, nor did they differ in any of the secondary-outcome measures (time to recover unaided walking, time to discontinue ventilation, recovery from disability during 48 weeks). Improvement of at least 1 point in the disability scale was observed in 57.9% of patients assigned to PE, in 55.4% of patients treated with IVIg, and in 59.5% of patients randomised to receive PE and IVIg. Both PE [58] and IVIg [59] are effective in paediatric patients. Although there are no clear guidelines as to the treatment of choice there may be a preference for IVIg as PE is technically more difficult to perform in children because of the problems of venous access. Furthermore, a recent retrospective study of 15 children suggested that IVIg might be superior to PE in reducing the number of days to walk independently [60].

Limited relapses may be observed in about 10% of patients treated with IVIg. Two independent studies drew attention to a high relapse rate after treatment with IVIg [61, 62]. In a retrospective study on 54 patients with GBS, evidence was not found of increased relapses in patient with GBS treated with IVIg as opposed to those treated with PE [63]. Moreover, a recent study examined the risk factors for treatment-related relapses occurring in GBS patients [64]. Their findings were that treatment modality (PE, IVIg either alone or in combination with high dose methylprednisolone) seems not to have any influence on the risk of clinical relapse. On the other hand, patients with fluctuation showed a trend to have the fluctuation after a protracted disease course, suggesting that treatment-related clinical fluctuations are due to a more prolonged immune attack. A second interesting issue is the severity of disease that requires intervention and whether patients should be treated while still ambulant, since it is possible that early treatment may be associated with higher relapse rate. Although the time from onset of weakness to treatment was similar in both groups, in order to be randomised the patients had to be unable to walk 10 m independently. In this respect it is still unclear whether ambulant patients tend to relapse more often.

Corticosteroids

Despite initial enthusiasm in the 1960s for the use of steroids in GBS, corticosteroids proved to be of no benefit in GBS. In a large double-blind, placebo-controlled multicentre trial of methylprednisolone (500 mg intravenously for 5 days within 2 weeks of onset) vs. placebo, the groups did not differ significantly in any of the outcome measures [65]. However, use of PE in the placebo group could have obscured a possible beneficial effect of methylprednisolone. Sixty-six patients

(53%) in the methylprednisolone treated group and 77 patients (65%) in the placebo group received PE, a difference that is almost significant ($p= 0.08$). PE was significantly more often applied in the placebo group compared to the methylprednisolone group ($p=0.01$). A pilot study suggested a beneficial interaction between IVIg and steroids [66]. In this study 19 of 25 patients (76%) improved by one or more functional grades after 4 weeks compared with 53% of historical controls treated with IVIg alone. This preliminary observation is now being tested in an international randomised clinical trial.

Interferon Beta

New insights in GBS pathophysiology have emphasised the role of cellular immune reactions, and the role of proinflammatory and anti-inflammatory cytokines, especially tumour necrosis factor (TNF)-α and transforming growth factor (TGF). Interferon-β has been tried in peripheral nervous system inflammatory disease [67] and proposed in GBS [68]. Créange et al. [69] described some beneficial effects in case report of a patient fulfilling the diagnostic criteria of GBS with axonal features, generally characterised by a slower recovery than the demyelinating form of GBS. The patient received 4 PE (50 ml/kg on alternate days) and then 4 days later was started on 6 mUI interferon-β-1a on alternate days for 2 weeks until he was able to walk 10 m without aid. The most rapid improvement of disability occurred during the period when the patient was on interferon-β-1a, and there was a subsequent persistent disability after interferon-β-1a interruption. Further studies are needed to establish the potential usefulness ofinterferon-β in GBS.

CIDP

Chronic inflammatory demyelinating polyradiculoneuropathy (CIDP) is an acquired chronic disorder of the peripheral nervous system. The aetiology of CIDP is unknown, but immunologic mechanisms are clearly involved [70]. Nerve biopsy revealed inflammatory infiltration, mainly in the endoneurium or less frequently around the epineurial capillaries. Electron microscopy demonstrated macrophages engaged in the phagocytosis of myelin [71]. Humoral and cell-mediated responses against a variety of myelin-derived autoantigens have been detected in some CIDP patients [72-76], however these findings are substantially inconsistent and a common target epitope has never been found. Studies on T cell activation by measuring the levels of a variety of cytokines in serum or in CSF failed to show consistent patterns. A stronger argument for involvement of the immune system in the pathogenesis of CIDP has come from passive transfer experiments. The infusion in healthy recipient animals of hyperimmune serum from animals with experimental allergic neuritis [77] or from patients [78, 79] induced a chronic relapsing inflammatory neuritis. More recently, repeated transfer of P2 protein-reactive lymphocytes produced a chronic neuritis in Lewis rats

[80]. Taken together, these studies suggest that both humoral and cellular immune mechanisms are involved in the pathogenesis of CIDP and constitute the rational for the use of immunoactive treatments.

Reliable incidence data for CIDP are not available, but clinical experience suggests that it is lower than that of GBS; because of the usually prolonged course of the disease, however, the prevalence is not negligible. A slight male predominance has been found in most of the clinical series. Mean age at onset varies between 31 and 47 years in different studies, the variability being essentially explained by selection bias. The disease affects all age groups. The occurrence of familial cases is exceptional and no significant relationship with HLA antigens has been consistently found [81]. As a consequence, a genetic predisposition to CIDP is improbable. Preceding infections as a precipitating factor have been suggested in some studies [71, 82], however their role is definitely less relevant in CIDP than in GBS.

CIDP is characterised by sensory loss and weakness, areflexia, elevated CSF protein, and electrodiagnostic evidence of multifocal demyelination with or without superimposed axonal degeneration. Distribution of sensorimotor deficits is usually symmetrical, but asymmetrical distributions can be observed, particularly in the early phases of the disease. Most patients present with a combination of weakness, numbness and paraesthesiae; cranial nerve involvement at onset is rare. Weakness affects both proximal and distal muscles of the legs and arms, with only a slight predominance of distal over proximal segments and of lower over upper limbs. This is a quite uncommon finding in polyneuropathies. Facial muscles are affected in about 20% of patients, while extraocular muscle impairment is rare. Sensory loss is common and includes impaired touch and vibratory sensations, with lesser involvement of pain and temperature sensitivities. About 10% of patients have pure motor syndrome (lower motor neuron variant) and another 10% have pure sensory syndrome (sensory ataxic variant) at presentation; a strong predominance of motor versus sensory impairment can persist for many years. A minority of patients has sensory-motor deficits in the distribution of specific nerve territories, suggestive of mononeuritis multiplex. Pain is usually considered infrequent in CIDP; however in a recent study, pain, usually in the form of distal burning dysaesthesia or muscle aching and cramps, was reported by 42% of patients [83].

CIDP is characterised by a high clinical heterogeneity that can make the diagnosis quite difficult because there are no specific tests for its diagnosis. The essence of the diagnosis is that it is a chronic acquired demyelinating disease and that other potential causes of demyelinating neuropathy have been excluded. This means that the diagnosis is based on anamnestic data, clinical examination and laboratory tests: electrophysiologic studies, cerebrospinal fluid examination and nerve biopsy. These tests should demonstrate the presence of demyelination and inflammation and exclude other causes. The anamnesis should demonstrate the chronicity of the disorder: Dyck et al. [84] in 1975 required progression of 6 months, Barhon et al. [85] 2 months, Hughes [86] 4 weeks, Cornblath et al. [87] 2 months, and Dyck et al. [88] in 1993 8 weeks. This time interval is important to differentiate CIDP from GBS. Eight weeks are probably enough for this purpose. Several sets of diagnostic criteria have been developed, using a combination of clinical and instrumental variables. In general, criteria for inclusion in clinical trials must be more specific and criteria for clinical activity must be more sensitive, because patients with atypical clinical findings may respond to immunoactive treatments as well as patients with typical clinical patterns. The more widely used diagnostic criteria for research protocols are those published in 1991 by the Ad Hoc Subcommittee of the American Academy of Neurology [89]. Moreover, some authors have proposed to classify patients in 3 classes according to the results of 3 instrumental tests (neurophysiology, CSF examination, and nerve biopsy) as definite (3 tests abnormal), probable (2 tests abnormal) and possible (1 test abnormal). It is noteworthy that only 30%-60% of the patients included in the largest clinical series could be classified as definite CIDP.

Three types of disease courses are classically described: monophasic, relapsing and progressive. The classification is based on retrospective data and has no prognostic implications for the successive type of course. No data are available on the natural history of the disease because almost all patients receive treatment, which can substantially modify the disease evolution. About one-third of patients have a monophasic course with full recovery or residual disability; about 40% of these patients require persisting treatments to prevent deterioration. About 25% have a progressive course (slow or stepwise) and 35% a relapsing course. In patients with relapsing courses the interval between relapses is about 10 months [85, 90], although relapses may occur even 31 years after the initial attack [71]. Outcome was very poor in early reports: in the series of Dyck et al. [84] 11% of patients died of the disease, 6% died from other diseases, 11% were bedridden or wheelchair-bound, and 8% were ambulatory but unable to work. However, such poor outcomes were probably due to inadequate treatment and biased selection. In a recent long-term follow-up, only 4% of the patients died and a further 9% had moderate or severe disability [91]. The outcome of CIDP in children seems to be even better when the disease has a monophasic course, while cases with a progressive course from onset may have considerable long-term morbidity with persistent disability [92].

Table 2. Disorders (generally autoimmune) that, when occurring in association with CIDP, form a condition sometimes termed "CIDP plus"

HIV infection
Thyreotoxicosis
Benign monoclonal gammopathy
Inflammatory bowel disease
Chronic active hepatitis
Hereditary motor-sensory neuropathy
Hodgkin's disease
Central-peripheral inflammatory demyelination

CIDP may occur in association with other conditions (Table 2) most of which have a presumed autoimmune pathogenesis. There is a large debate if these cases should be classified as "CIDP plus" or should be maintained separate from idiopathic CIDP. This association is not surprising because it is well known that autoimmune diseases tend to occur in the setting of other autoimmune disorders.

General Treatment

During severe acute relapse or in the advanced phases of progressive courses, CIDP patients may require ventilatory assistance. Physiotherapy is important to preserve and improve muscle strength and to prevent tendon retraction. Bracing and other aids may be useful in cases of severe weakness. As in other chronic disabling diseases, psychological support is essential.

Corticosteroids

As far back as 1958, Austin [93] documented the efficacy of steroid treatment in CIDP. Since then, many uncontrolled studies reported the positive effects of steroid treatment in CIDP. These results were confirmed by the controlled trial of Dyck and co-workers [94] in which 28 patients were randomised to receive 120 mg prednisone every other day tapered in 3 months, or no treatment. Prednisone was shown to cause a small but significant improvement in neurological disability and some instrumental tests. Patients with a recurrent course responded as well as patients with a progressive course. Improvement after steroid treatment occurred in 86% of the patients. In two other large, open studies, clinical improvement was observed in 65% and 95% of the treated patients [82, 85]. The clinical improvement usually started after some weeks (1.9 months in Barohn et al.'s study [85]), but could be delayed until 3.5 months [95]. The peak of improvement may be reached after 6-12 months, indicating the need to continue treatment for years, with tapering of the dose. Unfortunately relapse after discontinuation of the steroid treatment occurs in most patients, 70% in Barohn et al.'s study [85]. This observation raises problem that chronic steroid treatment may induce serious side effects such as osteoporosis, diabetes and hypertension. About 50% of CIDP patients do not respond or have unsatisfactory responses to steroid treatment or become refractory after an initial positive response. In a minority of these patients, increasing the dose results in the reappearance of the response, which is frequently only transitory.

Plasma Exchange

Many small open studies indicated the efficacy of plasma exchange (PE) in CIDP [96-98]. These results were subsequently confirmed in a double-blind-cross-over controlled trial [99] in which patients with static or worsening disease were randomly assigned to PE ($n=15$) or to sham exchange ($n=14$). After 3 weeks, a significant improvement of nerve conduction parameters was observed in the PE group. Moreover, the neurologic disability score improved in 5 patients of this group to a greater degree than for any patient receiving sham exchange. Improvement was observed both in patients with progressive and relapsing courses. However, in patients who responded to treatment, the improvement generally began to fade 10-14 days after the treatment was stopped. This rapid deterioration occurring within a few days of stopping has been confirmed in another study [100]. Rebound phenomenon may be caused by an overshooting synthesis of antibodies or other pathogenic factors, or by alterations of immunoregulatory mechanisms [101, 102]. Most patients in the Dyck et al.'s study [99] had long-lasting disease and severe disability. Interestingly, in a more recent double-blind sham-controlled cross-over trial in patients with a disease of short duration and never treated before, a positive effect of PE was observed in 80% of patients [100]. Clinical improvement occurred independently of the type of disease course, and was clearly associated with a reduction of conduction block in most nerves. Eight of the 12 patients with clinical improvement after PE showed relapse; six of these patients, with an apparently stable disease at randomisation, deteriorated rapidly over a few days and became more severely paralysed than they had been at entry into the study. These patients subsequently improved and stabilised with further PE, but corticosteroids or immunosuppressors were added to PE to maintain the disease stability in all patients. Long-term follow-up was achieved in 16 patients: 13 patients had no or minor disability and 3 patients remained with moderate distal weakness.

Comparing the controlled study of Dyck et al. [99] and Hahn et al. [100], it seems that the better results obtained by Hahn et al. can be explained by the more vigorous PE scheduling (10 vs. 6 treatments) and by the earlier treatment. In patients with long-lasting disease, irreversible nerve damage (secondary axonal degeneration) may occur, reducing the probability and extent of recovery.

Immunoglobulins

Intravenous immunoglobulins (IVIg), initially used as replacement therapy in primary and secondary antibody deficiency syndromes [103, 104], have recently acquired an important role in the treatment of some dysimmune pathologies, including some diseases of the central and peripheral nervous systems [105, 106].

Compared to other immunoacting treatments, IVIg are characterised by a good safety profile. The main risk consists of the transmission of infectious agents that can only be excluded if the manufacturing process is optimal.

IVIg display a quite wide range of effects on the immune system, including anti-idiotypic suppression, down-regulation of B cell and T cell activation, blockade of Fc receptors on phagocytic cells, neutralization of superantigen a complement-mediated effects, downregulation of cytokine production, and neutralization of cytokines. Which of these mechanisms, alone or in variable combination, plays a major role in the treatment of CIDP is still a matter of debate.

Some open studies [81, 87, 107-111] (Table 3) suggested an efficacy of IVIg in CIDP. In the largest trial [81], 52 patients were treated with IVIg in a dosage of 0.4

Table 3. IVIg treatment of chronic inflammatory demyelinating polyneuropathy: results of uncontrolled studies

Reference	Patients	
	Treated (n)	Improved (n)
van Doorn et al. [81]	52	30
Cornblath et al. [87]	15	6
Hoang-Xuan et al. [107]	6	4
Faed et al. [108]	9	9
Hodkinson et al. [109]	8	3
Nemni et al. [110]	9	6
Vermeulen et al. [111]	17	13
Total	116	71

g/kg body weight per day for 5 consecutive days; when necessary, treatment was repeated. Twenty patients did not improve; 2 patients had a short-lasting improvement with subsequent infusions having no effects; 9 patients had a complete remission, and 21 needed intermittent infusions to maintain the improvement. Response to treatment was associated with short duration of disease in a progressive phase, symmetric weakness distribution, and relevant nerve conduction velocity slowing. In another study [109], IVIg resulted effective also in some patients not responsive to steroids or PE.

Four controlled studies on the effects of IVIg in CIDP have been published [81, 111-113]: van Doorn et al. [81] performed a placebo-controlled, cross-over study in 7 patients who had all shown a favourable response to IVIg in a previous open study. The disability of all patients improved after IVIg and did not change after placebo. Clinical changes were paralleled by neurophysiological changes. The clinical improvement occurred in all patients within one week of treatment, but all patients deteriorated within 3-11 weeks after treatment discontinuation. The same group of researchers [111] was not able to confirm the beneficial effects of IVIg treatment in a group of 28 newly diagnosed CIDP patients. Fifteen patients were randomised to IVIg and 13 to placebo: clinical improvement occurred in 4 patients of the first group and in 3 patients of the placebo group. Mean nerve conduction velocity and amplitude ratio of the evoked responses tended towards amelioration in the active treatment group while they were stable or deteriorated in the placebo group, but the neurophysiological changes were not significant. All patients who did not improve during the double-blind phase of the trial were treated with IVIg in a successive open phase: 6 of 10 patients in the previous placebo group improved, while none of the 11 patients of the previous IVIg group improved. The third controlled study [112] was an observer-blind, cross-over trial with a washout period of 6 weeks comparing IVIg and PE in 20 patients with progressive or static neuropathy. Thirteen received both treatments; the other patients did not worsen sufficiently after the first treatment to receive the second treatment. Both treatments resulted equally effective. As pointed out by the authors, the total dose of IVIg (1.8 g/kg over 6 weeks) was low and better results might have been found with higher doses. The fourth study is a double-blind, placebo-controlled, cross-over study [113]. Thirty patients, many of whom had failed on other treatments, were randomised to receive IVIg 0.4 g/kg for 5 days or placebo, and after 4 weeks were crossed-over to the alternate treatment. A significant difference in favour of IVIg was observed in all the clinical parameters and in most neurophysiological parameters. Nineteen patients improved on IVIg, but 10 of them relapsed after 3-22 weeks (median, 10 weeks). All 10 patients have been maintained and stabilised with IVIg pulse therapy of ≤1 g per kg body weight, given as a single infusion prior to the expected relapse.

In a series of 44 consecutive patients with CIDP, the response rates among PE, IVIg and steroids were similar but functional improvement was greatest with PE [83]. An improvement was observed in 39% of the patients for at least 2 months with an initial therapy. Among those who failed to respond, more than one-third improved with a second treatment.

Neuropathies Associated with Monoclonal Gammopathy

The association between neuropathies and monoclonal gammopathies was realised in the early 1970s, but the frequency of such an association is still undetermined. Monoclonal gammopathy, predominantly IgG, has a high prevalence in the elderly population and approximately one-third of these patients has related neuropathy [114]. Kelly et al. [115] reported that monoclonal gammopathy was present in 10% of patients with idiopathic peripheral neuropathy. However, prevalence studies on consecutive patients with benign monoclonal gammopathy showed figures ranging from 5% [116] to 70% [117]. Peripheral neuropathy is more frequently associated with IgM monoclonal gammopathy [118]. In the majority of patients with neuropathy and associated monoclonal protein, no underlying disease is found (monoclonal gammopathy of unknown significance (MGUS)).

Autoantibodies against peripheral nerve constituents have been demonstrated in many patients. In about half the patients with IgM monoclonal gammopathy ,the M protein reacts with myelin-associated glycoprotein (MAG) and many syndromes have been described [119]. Many open trials with steroids [120-122], immunosuppressive agents [116, 123, 124], or PE alone or in variable combination [125-127], have been conducted in recent years with widely variable results, so that no conclusion can be derived. The only exception is a double-blind study on the efficacy of PE vs. sham exchange [128] involving 39 patients with stable or worsening neuropathy and MGUS. A significant improvement of clinical and neurophysiological parameters was observed in the PE group, while no changes occurred in the sham group. Patients with IgG or IgA gammopathy had a better response to PE than did those with IgM gammopathy. However, only some patients seemed to respond to PE [127, 129], and PE did not produce additional benefits when added to the chlorambucil treatment in patients with IgM gammopathy [130]. In another open study [122], the response to PE was only slightly better than that to steroids, and again clinical improvement occurred mostly in patients with IgG and IgA gammopathies. There are few data on the efficacy of IVIg in MGUS neuropathy, coming from small, open, uncontrolled studies. We treated 5 patients with IgG or IgA gammopathy, and found clinical and neurophysiological improvement in 3 patients. Cook et al. [131] treated 2 patients with IgM gammopathy, who had a steadily progressive course over 3 years in spite of treatments with steroids and immunosuppressive agents. Both patients had rapid clinical improvement 5-10 days after the first immunoglobulin infusion, lasting 3-6 weeks. Retreatment determined improvement after each consecutive infusion. A positive response to IVIg has been also reported by Simmons et al. [127] in patients with IgA gammopathy. Léger et al. [132] treated 4 patients with IgG gammopathy and 13 patients with IgM gammopathy with monthly infusions for 6-24 months. Two of the patients with IgG and 6 of the patients with IgM had persistent clinical improvements. Interestingly, some of the responders had not previously responded to other immunosuppressive agents. Ellie et al. [129] treated 17 patients with IgM gammopathy and anti-MAG reactivity; a clear-cut improvement was observed in 24% of the cases and a transient, mostly subjective, improvement in another 35%.

In an open prospective study, Notermans et al. [133] analysed the effect of intermittent cyclophosphamide and prednisone treatment. Eleven patients had IgM-MGUS and five IgG-MGUS. During a follow-up period of three years, eight patients improved and six patients stabilised. There was no difference in response between patient with IgM-MGUS and IgG-MGUS, nor in patients with or without anti-MAG antibodies.

Finally, Blume et al. [134] combined plasma exchange and cyclophosphamide to treat four patients with anti-MAG antibody-associated progressive polyneuropathy. They found improvement in all patients in the 5–24 months after treatment with improvement persisting for 1–2 years. Improvement after immunoglobulin infusion was not associated with changes in serum IgM.

Published data have demonstrated that some patients with paraproteinaemic neuropathy respond to prednisone, PE, IVIg, and immunosuppressive agents. Treatment responses are unpredictable in single patients even though usually patients with IgG and IgA gammopathy have a higher probability to respond positively. The efficacy of treatment when present is usually transitory, but persistent with retreatment.

Multifocal Motor Neuropathy

Multifocal motor neuropathy (MMN) with persistent conduction block is a rare, recently recognised neurological disorder characterised by an insidious onset and usually slow progression of weakness in the territory of one or more nerves, nerve branches or roots in a highly variable combination from patient to patient. Weakness usually starts in the upper limb, distally and asymmetrically; the spread of the weakness to other homolateral and controlateral segments of the upper and lower limbs is quite common in the course of the disease. Muscular atrophy is frequently observed, but it can be absent, even for a long time in the affected muscles, in spite of severe weakness. Sensory deficits are absent or mild even in the territory of a nerve with severe involvement of the motor fibres. Spinal fluid examination is usually normal. A completely confident diagnosis rests on electrodiagnostic studies showing a conduction block, confined to motoraxons, in a short segment of the nerve, outside the typical entrapment sites. Motor nerve biopsy demonstrated endoneurial oedema, lymphocytic inflammation, mild loss of myelinated fibres and rare onion bulbs [135]. Mild pathological abnormalities have also been found in sural sensory fibres, mostly thinly myelinated large calibre fibres [136]. The role of antibodies to GM1 ganglioside in the pathogenesis of MMN remains controversial. The frequent occurrence of high titres of anti-GM1 antibodies seemed to provide useful diagnostic information. However, in different series of MMN patients, figures ranging from 20% [137] to 80% [138] have been reported, with most series ranging from 40% to 60%. Furthermore, sera from patients with Guillain-Barrè syndrome, amyotrophic lateral sclerosis (ALS), CIDP, spinal muscular atrophy and other neurological disorders may also have increased anti-GM1 antibodies titres, even though titres are usually lower in MMN.

Steroid treatment resulted ineffective in 23 of 25 patients reported in the literature [139-142]. The deleterious effect of steroids in MMN patients has been emphasised in a study by Donaghy et al. [143]. These authors reported an increase in weakness within one month of starting oral prednisolone in 4 patients with pure motor demyelinating neuropathy, 2 of which had a prompt improvement following high dose IVIg. More recently, a study by van den Berg et al. [144] confirmed that steroid therapy is not effective in patients with MMN. Plasma exchange (PE) also had no effects on MMN [139-142]. The absence of response to steroid treatment and plasma exchange is sometimes useful for differentiating MMN from CIDP. We were able to find in the literature 14 patients treated with cyclophosphamide; in 9 of them treatment was of some efficacy [139-142].

The efficacy of IVIg has been tested in many small open studies (Table 4) [145-150]. The dose and schedule of treatment were quite variable, ranging from 1.0 to 2.4 mg/kg over a period of 2-5 days, every 2-8 weeks. Globally, 63 patients were treated and 58 had some benefit. Improvement usually started 2-10 days after treatment and lasted 6-10 weeks. The range of improvement was variable from patient to patient and in the same patient strength ameliorated in some segments but not in others. The response to treatment was usually better in muscles with no or mild atrophy. The study of Bouche et al. [145] is of particular interest because 19 patients were followed for at least 2 years. Treatment was 2 g/kg over 5 days, monthly for 6 months, then every 2-3 months. Improvement or stabilisation occurred in 18 cases, almost exclusively in patients without amyotrophy. In patients with anti-GM1 antibodies, the improvement of strength was associated with a decrease of antibody titre. Improvement of muscle power was usually, but not always, associated with a decrease of conduction block in the corresponding

Table 4. IVIg treatment of multifocal motor neuropathy: results of uncontrolled studies

Reference	Patients	
	Treated (n)	Improved (n)
Donaghy et al. [143]	4	2
Bouche et al. [145]	19	17
Chaudry et al. [146]	9	9
Comi et al. [147]	5	5
Kaji et al. [148]	2	2
Nobile-Orazio et al.[149]	5	4
Parry [150]	1	1
Azulay et al.[151]	18	12
Total	63	52

nerve; this discrepancy can be explained by conduction block arising proximally in the motor fibres. The short-term efficacy of IVIg has been confirmed by two small, double-blind placebo-controlled cross-over studies [151, 152].

More recently, Azulay et al. [153] analysed the long-term efficacy of IVIg in MMN. Eighteen MMN patients treated with high dose IVIg were followed for 9-48 months. Clinical benefits were seen in 12 patients (67%) but most patients needed repeated courses of IVIg to maintain the improvement.

There are consistent data supporting the short-term efficacy of IVIg in patients with MMN with persistent conduction block; early treatment is advisable because the response is quite poor or absent in atrophic muscles. The response seems to persist with treatment in most patients, but doses and frequency of administration have to be adjusted on single cases. Due to the slow progression of the disease in many cases, at present it is impossible to say whether IVIg treatment is able to persistently stop the disease progression.

Conclusions

All the substantial controlled trials so far undertaken have indicated that PE and IVIg are equally effective treatment of GBS. However, GBS is a heterogeneous condition and it is possible that some patients respond differently to PE and to IVIg. Large studies are necessary to address this critical point. In the meantime, PE and IVIg remain the main stay of treatment. It is reasonable to use either of the two treatments unless there are specific contraindication to use one of the two. The most important contraindications for PE are cardiovascular and haemodynamic instability and difficulty to find venous access. IVIg should be avoided in patients with previous allergic reaction to IVIg, in patients with known IgA deficiency in whom anaphylaxis is more common, and in the presence of renal failure which may be exacerbated. If a patient deteriorates during either PE of IVIg treatment, the dilemma for the physician is whether the patient should be switched to an alternative treatment. The follow-up of the Dutch PE and IVIg trial indicated that about half of patients who continued to deteriorate during the first week of treatment ameliorated during the second week. Therefore, the physician should wait at least 2 weeks before deciding to change treatment. Another interesting issue to be addressed in further studies is to determine the best treatment for relapsing patients. There is still uncertainty as to whether ambulant patients tend to relapse more often. Interestingly but not unexpectedly, patients who have an acute axonal motor GBS do not relapse. This may be because the axons once destroyed cannot regenerate quickly enough to be attacked again by the original immune effector mechanisms that may have already subsided.

Controlled trials have demonstrated that IVIg, PE and corticosteroids are effective in CIDP. Uncontrolled studies also suggest that cytotoxic drugs are effective, but they are a second choice treatment because of the serious short-term and long-term side effects. IVIg and PE are equally effective and equally expensive. Most patients respond to the same extent to both treatments, but some patients respond only to IVIg or PE; because both treatments are active within 1-4 weeks, patients can be quickly shifted to the second treatment if a response is not observed after the first treatment. We prefer to start with IVIg because PE is more invasive. If after IVIg treatment secondary deterioration occurs following an initial clinical improvement, treatment must be repeated and the treatment interval has to be determined, taking into consideration the high inter-patient variability of the duration of the response. When stabilisation is obtained, both the dose and frequency of IVIg infusion have to be reduced. In some cases, low doses of steroids need to be added to IVIg treatment to consolidate the responses. If no response is seen to IVIg, a trial of PE should be undertaken and the opportunity to combine immunosuppressive drug treatment has to be considered. As a guideline for the PE schedule, 10 exchanges 2-3 times a week is usually sufficient, with subsequent tapering of PE frequency. Corticosteroids are the first immunosuppressive agents that should be added to PE, if necessary. Occasionally, patients may require cyclophosphamide, azathioprine, cyclosporin A or other cytotoxic agents.

For MGUS, the therapeutic strategies are the same as for idiopathic CIDP, but usually the clinical response is poorer. For MMN, IVIg are definitely the first choice treatment.

Acknowledgements: We thank the project INCAT BIOMED 2 contract number BMH4-CT96-0324 for supporting the preparation of this article.

References

1. Winer JB, Hughes RAC, Osmond C (1988) A prospective study of acute idiopathic neuropathy. I. Clinical features and their prognostic value. J Neurol Neurosurg Psychiatry 51: 605-612

2. Alters M (1990) The epidemiology of Guillain-Barrè syndrome. Ann Neurol 27(Suppl): S7-12
3. Arnason BGW, Soliven B (1993) Acute inflammatory demyelinating polyradiculoneuropathy. In: Dyck PJ, Thomas PK, Griffin JW et al. (eds) Peripheral neuropathy, 3rd edn. WB Saunders, Philadelphia, pp 1437-1497
4. Ng KKP, Howard RS, Fish DR et al. (1995) Management and outcome of severe Guillain-Barrè syndrome. QJM 88: 243-250 (abstract)
5. Bos AP, van der Mechè FGA, Witsenburg M, van der Voort E (1987) Experiences with Guillain-Barrè syndrome in a paediatric intensive care unit. Intens Care Med 13: 328-331
6. Rees JH, Thompson RD, Hughes RAC (1996) An epidemiological study of Guillain-Barrè syndrome. J Neurol Neurosurg Psychiatry 61: 215 (abstract)
7. Rees JH, Gregson NA, Griffiths PL et al. (1993) *Campylobacter jejuni* and Guillain-Barrè syndrome. QJM 86: 623-624
8. Mishu B, Ilya AA, Koski CL et al. (1993) Serologic evidence of previous *Campylobacter jejuni* infection in patients with the Guillain-Barrè syndrome. Ann Intern Med 118: 947-953
9. Rees JH, Soudain SE, Gregson NA, Hughes RAC (1995) *Campylobacter jejuni* infection and Guillain-Barrè syndrome. N Engl J Med 333: 1374-1379
10. Jacobs BC, van Doorn PA, Schmitz PIM et al. (1996) *Campylobacter jejuni* infection and anti-GM$_1$ antibodies in Guillain-Barrè syndrome. Ann Neurol 40: 181-187
11. Steele JC, Gladstone RM, Thanasophon S et al. (1969) *Mycoplasma pneumoniae* as a determinant of the Guillain-Barrè syndrome. Lancet 2(7623): 710-713
12. Dowling PC, Cook SB (1981) Role of infection in Guillain-Barrè syndrome: laboratory confirmation of herpes virus in 41 cases. Ann Neurol 9(suppl): 44-45
13. Winer JB, Hughes RAC, Anderson MJ et al. (1988) A prospective study of acute idiopathic neuropathy: II antecedent events. J Neurol Neurosurg Psychiatry 11: 613-618
14. Boucquey D, Sindic CJM, Lamy M et al. (1991) Clinical and serological studies in a series of 45 patients with Guillain-Barrè syndrome. J Neurol Sci 104: 56-63
15. Visser LH, van der Mechè FGA, Meulstee J et al. (1996) Cytomegalovirus infection and Guillain-Barrè syndrome: the clinical, electrophysiologic, and prognostic features. Neurology 47: 668-673
16. Gautier-Smith PC (1965) Neurological complication of glandular fever (infectious mononucleosis). Brain 88: 323-334
17. Berger JR, Difini JA, Swerdloff MA et al. (1987) HIV seropositivity in Guillain-Barrè syndrome. Ann Neurol 22: 393-394
18. Gross FJ, Mindel JS (1991) Pseudotumor cerebri and Guillain-Barrè syndrome associated with human immunodeficiency virus infection. Neurology 41: 1845-1846
19. Lacaille F, Zylberberg H, Hagegè H et al. (1998) Hepatitis C associated with Guillain-Barrè syndrome. Liver 18: 49-51
20. Kaplan JE, Schonberger LB, Hurwitz ES, Katona P (1983) Guillain-Barrè syndrome in the United States, 1978-1981: additional observation from the national surveillance system. Neurology 33: 633-637
21. Blaser MJ (1997) Epidemiologic and clinical features of *Campylobacter jejuni* infections. J Infect Dis 176(suppl 2): S103-105
22. Roscelli JD, Bass JW, Pang L (1991) Guillain-Barrè syndrome and influenza vaccination in the US Army, 1980-1988. Am J Epidemiol 133: 952-955
23. Hughes RAC, Rees J, Smeeton N, Winer J (1996) Vaccines and Guillain-Barrè syndrome. BMJ 312: 1475-1476
24. De Silveira CM, Salisbury DM, De Quadros CA (1997) Measles vaccination and Guillain-Barrè syndrome. Lancet 349: 14-15

25. Kinnunen E, Junttila O, Haukka J, Hovi T (1998) Nationwide oral poliovirus vaccination campaign and the incidence of Guillain-Barrè syndrome. Am J Epidemiol 147: 69-73
26. Salisbury DM (1998) Association between oral poliovaccine and Guillain-Barrè syndrome? Lancet 351: 79-80
27. Hemachudha T, Griffin DE, Chen WW, Johnson RT (1988) Immunologic studies of rabies vaccination-induced Guillain-Barrè syndrome. Neurology 38: 375-278
28. Knittel TH, Ramadori G, Majet WJ et al. (1989) Guillain-Barrè syndrome and human diploid cell rabies vaccine. Lancet 1(8649): 1334-1335
29. Arrowsmith JB, Milstain JB, Kuritsky JN et al. (1985) Streptokinase and Guillain-Barrè syndrome. Ann Intern Med 103: 302 (letter)
30. Chakraborty TK, Ruddell WS (1987) Guillain-Barrè neuropathy during treatment with captopril. Postgrad Med J 63: 221-222
31. Hory B, Blanc D, Boillot A et al. (1985) Guillain-Barrè syndrome following danazol and corticosteriod therapy for hereditary angioedema. Am J Med 79: 111-115
32. Loizou LA, Boddie HG (1978) Polyradiculoneuropathy associated with heroin abuse. J Neurol Neurosurg Psychiatry 41: 855-857
33. Giovannoni G, Hartung H-P (1996) The immunopathogenesis of multiple sclerosis and Guillain-Barrè syndrome. Curr Opin Neurol 9: 165-177
34. Hartung H-P, Pollard JD, Harvey GK, Toyka KV (1995) Immunopathogenesis and treatment of the Guillain-Barrè syndrome-Part I. Muscle Nerve 18: 137-153
35. Hartung H-P, Pollard JD, Harvey GK, Toyka KV (1995) Immunopathogenesis and treatment of the Guillain-Barrè syndrome-Part II. Muscle Nerve 18: 154-164
36. Honavar M, Tharakan JHJ, Hughes RAC et al. (1991) A clinico-pathological study of the Guillain-Barrè syndrome: nine cases and literature review. Brain 114: 1245-1269
37. Griffin JW, Li CY, Ho TW et al. (1995) Guillain-Barrè syndrome in northern China: the spectrum of neuropathological changes in clinically defined cases. Brain 118: 577-595
38. Guillain G, Barrè JA, Strohl A (1916) Sur un syndrome de radiculonevritè avec hyperalbuminose du liquide céphalo-rachidien sans réaction cellulaire: remarques surles caractères cliniques et graphiques des réflexes tendineux. Bull Soc Med Hop Paris 40: 1462-1470
39. Feasby TE, Gilbert JJ, Brown WF et al. (1986) An acute axonal form of Guillain-Barrè polyneuropathy. Brain 109: 1115-1116
40. Yuki N, Yoshino H, Sato S et al. (1992) Severe acute axonal form of Guillain-Barrè syndrome associated with IgG anti-GD_{1a} antibodies. Muscle Nerve 15: 899-903
41. Brown WF, Feasby TE, Hahn AF (1993) Electrophysiological changes in the acute "axonal" form of Guillain-Barrè syndrome. Muscle Nerve 16: 200-205
42. Hafer-Macko C, Hsieh S-T, Li CY et al. (1996) Acute motor axonal neuropathy: an antibody-mediated attack on axolemma. Ann Neurol 40: 635-644
43. Griffin JW, Li CY, Macko C et al. (1996) Early nodal changes in the acute motor axonal neuropathy pattern of the Guillain-Barrè syndrome. J Neurocytol 25: 33-51
44. Chiba A, Kusonoki S, Shimizu T, Kanazawa I (1992) Serum IgG antibody to ganglioside GQ1b is a possible marker of Miller-Fisher syndrome. Ann Neurol 31: 677-679
45. Willison HJ, Veitch J, Paterson G, Kennedy PGE (1993) Miller-Fisher syndrome is associated with serum antibodies to GB1b ganglioside. J Neurol Neurosurg Psychiatry 56: 204-206
46. Chiba A, Kusonoki S, Obata H et al. (1993) Serum anti-GQ1b IgG antibody is associated with ophthalmoplegia in Miller-Fisher syndrome and Guillain-Barrè syndrome: clinical and immunohistochemical studies. Neurology 43: 1911-1917

47. Kornberg AJ, Pestronk A, Blume GM et al. (1996) Selective staining of the cerebellar molecular layer by serum IgG in Miller-Fisher and related syndrome. Neurology 47: 1317-1320
48. McKann GM, Griffin JW, Cornblath DR et al. (1988) Plasmapheresis and Guillain-Barrè syndrome: analysis of prognostic factors and the effect of plasmapheresis. Ann Neurol 23: 347-353
49. French Cooperative Group on Plasma Exchange in Guillain-Barrè Syndrome (1987) Efficiency of plasma exchange in Guillain-Barrè syndrome: role of replacement fluids. Ann Neurol 22: 753-761
50. French Cooperative Group on Plasma Exchange in Guillain-Barrè Syndrome (1992) Plasma exchange in Guillain-Barrè syndrome: one-year follow-up. Ann Neurol 32: 94-97
51. Kleyweg RP, van der Meché FGA (1991) Treatment related fluctuation in Guillain-Barrè syndrome after high-dose immunoglobulins or plasma exchange. J Neurol Neurosurg Psychiatry 54: 957-960
52. Rudnicki S, Vriesendorp F, Koski CI, Mayer RF (1992) Electrophysiologic studies in the Guillain-Barrè syndrome: effects of plasma exchange and antibody rebound. Muscle Nerve 15: 57-62
53. French Cooperative Group on Plasma Exchange in Guillain-Barrè Syndrome (1997) Appropriate numbers of plasma exchanges in the Guillain-Barrè syndrome. Ann Neurol 41: 298-306
54. Vermeulen M, van der Meché FGA, Speelman JD (1985) Plasma exchange and gamma-globulin infusion in chronic inflammatory polyneuropathy. J Neurol Sci 70: 317-326
55. Curro Dossi B, Tezzon F (1987) High-dose intravenous gammaglobunin for chronic inflammatory demyelinating polyneuropathy. Ital J Neurol Sci 8: 321-326
56. van der Meché FGA, Schmitz PIM, Dutch Guillain-Barrè Syndrome Study Group (1992) A randomised trial comparing intravenous immunoglobulin and plasma exchange in Guillain-Barrè syndrome. N Engl J Med 326: 1123-1129
57. Plasma Exchange/Sandoglobulin Guillain-Barrè Syndrome Trial Group (1997) Randomised trial of plasma exchange, intravenous immunoglobulin, and combined treatments in Guillain-Barrè syndrome. Lancet 349: 225-230
58. Epstein MA, Sladky JT (1990) The role of plasmapheresis in childhood Guillain-Barrè syndrome. Ann Neurol 28: 65-69
59. Shahar E, Murphy EG, Roifman CM (1990) Benefit of intravenously administered immune serum globulin in patients with Guillain-Barrè syndrome. J Pediatric 116: 141-144
60. Vajsar J, Sloane A, Wood E et al. (1994) Plasmapheresis versus intravenous immunoglobulin treatment in childhood Guillain-Barrè syndrome. Arch Pediatr Adolesc Med 148: 1210-1212
61. Irani DN, Cornblath DR, Chaudhry V et al. (1993) Relapse in Guillain-Barrè syndrome after treatment with human immune globulin. Neurology 43: 872-875
62. Castro LH, Ropper AH (1993) Human immune globulin infusion in Guillain-Barrè syndrome: worsening during and after treatment. Neurology 43: 1034-1036
63. Romano JG, Rotta FT, Potter P et al. (1998) Relapses in the Guillain-Barrè syndrome after treatment with intravenous immune globulin or plasma exchange. Muscle Nerve 21: 1327-1330
64. Visser LH, van der Meché FGA, Meulstee J et al. (1998) Risk factors for treatment related clinical fluctuations in Guillain-Barrè syndrome. J Neurol Neurosurg Psychiatry 64: 242-244
65. Guillain-Barrè Syndrome Steroid Trial Group (1993) Double-blind trial of intravenous methylprednisolone in Guillain-Barrè syndrome. Lancet 341: 586-589

66. Dutch Guillain-Barrè Study Group (1994) Treatment of Guillain-Barrè syndrome with high-dose methylprednisolone: a pilot study. Ann Neurol 35: 749-752
67. Choudhary PP, Thompson N, Hughes RAC (1995) Improvement following interferon beta in chronic inflammatory demyelinating polyradiculoneuropathy. J Neurol 242: 252-253
68. Hughes RAC (1998) Modulating the immune response in demyelinating diseases. J Neurol Neurosurg Psychiatry 64(2): 148
69. Créange A, Lerat H, Meyrignac C, Degos J-D et al. (1998) Treatment of Guillain-Barrè syndrome with interferon-β. Lancet 352: 368-369
70. Hartung H-P, Reiners K, Toyka KV, Pollard JD (1994) Guillain-Barrè syndrome and CIDP. In: Hohlfeld R (ed) Immunology of neuromuscular disease. Kluwer, Dordrecht, pp 33-104
71. Prineas JW, McLeod JG (1976) Chronic relapsing polyneuritis. J Neurol Sci 27: 427-458
72. Koski CL, Humphrey R, Shin ML (1985) Anti-peripheral myelin antibody in patients with demyelinating neuropathy: quantitative and kinetic determination of serum antibody by complement component 1 fixation. Proc Natl Acad Sci USA 82: 905-909
73. van Doorn PA, Brand A, Vermeulen M (1987) Clinical significance of antibodies against peripheral nerve tissue in inflammatory polineuropathy. Neurology 37: 1798-1802
74. Fredman P, Vedeler CA, Nyland H ET AL. (1991) Antibodies in antisera from patients with inflammatory demyelinating poliradiculo-neuropathy react with ganglioside LM1 and sulfatide of periferal nerve myelin. J Neurol 238: 76-79
75. Conolly AM, Pestronk A, Trotter JL et al. (1993) High titer selective serum anti tubulin antibodies in chronic inflammatory demyelinating polyneuropathy. Neurology 43: 557-562
76. Simone IL, Annunziata P, Maimone D (1993) Serum and CSF antibodies to ganglioside GM1 in Guillain-Barrè syndrome and chronic inflammatory polyneuropathy. J Neurol Sci 114: 49-55
77. Saida K, Sumner AJ, Saida T (1980) Antiserum mediated demyelinization: relationship between remyelinization and functional recovery. Ann Neurol 8: 12-24
78. Saida T, Saida K, Lisak RP et al. (1982) In vivo demyelinating activity of sera from patients with Guillain-Barrè syndrome. Ann Neurol 11: 69-75
79. Heininger K, Liebert UG, Toyka KV, Haneveld FT et al. (1984) Chronic inflammatory demyelinating polyradiculoneuropathy. Reduction of nerve conduction velocities in monkeys by systemic passive transfer of immunoglobulin G. J Neurol Sci 66: 1-14
80. Lassmann H, Fierz W, Neuchrist C et al. (1991) Chronic relapsing esperimental allergic neuritis induced by repeated transfer of P2-protein reactive T cell lines. Brain 144: 429-442
81. van Doorn PA, Vermeulen M, Brand A et al. (1991) Intravenous immunoglobulin treatment in patients with chronic inflammatory demyelinating polyneuropathy. Clinical and laboratory characteristics associated with improvement. Arch Neurol 48: 217-220
82. McCombe PA, Pollard JD, McLeod JG (1987) Chronic inflammatory demyelinating polyradiculoneuropathy: a clinical and electrophysiological study of 92 cases. Brain 110: 1617-1630
83. Gorson KC, Allam G, Ropper AH (1997) Chronic inflammatory demyelinating polyneuropathy: Clinical feature and response to treatment in 67 consecutive patients with and without a monoclonal gammopathy. Neurology 48: 321-328
84. Dyck PJ, Lais AC, Ohta M et al. (1975) Chronic inflammatory polyradiculoneuropathy. Mayo Clin Proc 50: 621-637

85. Barohn RJ, Kisser JT, Warmolts JR, Mendell JR (1989) Chronic inflammatory demyelinating polyradiculoneuropathy. Clinical characteristics, course, and recommendations for diagnostic criteria. Arch Neurol 46: 878-884
86. Hughes RAC (1990) Guillain-Barrè syndrome. Springer, Berlin Heilderberg New York
87. Cornblath DR, Chaudry V, Griffin JW (1991) Treatment of chronic inflammatory demyelinating polyneuropathy with intravenous immunoglobulin. Ann Neurol 30: 104-106
88. Dyck PJ, Prineas J, Pollard J (1993) Chronic inflammatory demyelinatingpolyradiculoneuropathy. In: Dyck PJ, Thomas PK, Griffin JW, Low A, Poduslo JF (eds) Peripheral neuropathy, 3rd edn. WB Saunders, Philadelphia, pp 1498-1517
89. Ad Hoc Subcommittee of the American Academy of Neurology (1991) AIDS Task Force: Research criteria for diagnosis of chronic inflammatory demyelinating polyneuropathy (CIDP). Neurology 41: 617-618
90. Albers J, Kelly J (1989) Aquired inflammatory demyelinating polyneuropathies: clinical and electrodiagnostic features. Muscle Nerve 12: 435-451
91. Simmons Z, Albers JW, Bromberg MB, Feldman EL (1995) Long term follow-up of patients with chronic inflammatory demyelinating polyneuropathy without and with monoclonal gammopathy. Brain 118: 359-368
92. Nevo Y, Pestronk A, Kornberg AJ, Connolly AM et al. (1996) Childwood chronic inflammatory demyelinating neuropathies: clinical course and long term follow-up. Neurology: 47: 98-102
93. Austin JH (1958) Recurrent polyneuropathies and their corticosteroid treatment. Brain 81: 157-192
94. Dyck PJ, O'Brien PC, Oviatt KF, Dinapoli RP et al. (1982) Prednisone improves chronic inflammatory demyelinating polyradiculoneuropathy more than no treatment. Ann Neurol 11: 136-141
95. de Vivo DC, Engel WK (1970) Remarcable recovery of steroid-positive recurrent polineuropathy. J Neurol Neurosurg Psychiatry 33: 62-69
96. Levy RL, Newkirk R, Ochoa J (1979) Treating chronic relapsing Guillain-Barrè syndrome by plasma exchange (letter). Lancet 2 (8136): 259-260
97. Server AC, Lefkowith J, Braine H, McKhann GM (1979) Treatment of chronic relapsing inflammatory polyradiculoneuropathy by plasma exchange. Ann Neurol 6: 258-261
98. Toyka KV, Augspach R, Wietholter H et al. (1982) Plasma exchange in chronic inflammatory polyneuropathy: evidence suggestive of pathogenic humoralfactor. Muscle Nerve 5: 479-484
99. Dyck PJ, Daube J, O'Brien P et al. (1986) Plasma exchange in chronic inflammatory demyelinating polyneuropathy. N Engl J Med 314: 461-465
100. Hahn Af, Bolton CF, Pillay N, Chalk C et al. (1996) Plasma-exchange therapy in chronic inflammatory demyelinating polyneuropathy. A double-blind sham controlled cross-over study. Brain 119: 1055-1066
101. Heininger K, Gibbels K, Besinger UA, Borberg H et al. (1990) Role of therapeutic plasmapheresis in chronic inflammatory demyelinating polyradiculoneuropathy. In: Rock G (ed) Apheresis. Wiley-Liss, New York, pp 275-281
102. Thornton CA, Griggs RC (1994) Plasma exchange and intravenous immunoglobulin treatment of neuromuscular disease. Ann Neurol 35: 260-268
103. Fanaroff AA, Korones SB, Wright LL et al. (1994) A controlled trial of intravenous immune globulin to reduce nosocomial infections in very-low-birth-weight infants. National Institute of Child Health and Human Development Neonatal Research Network. N Engl J Med 330: 1007-1013

104. Schwarzt SA (1993) Clinical use of immune serum globulin as replacement therapy in patients with primary immunodeficiency syndromes. Clin Rev Allergy 10: 1-12
105. Ronda N, Hurez V, Kazatchine MD (1993) Intravenous immunoglobulin therapy of autoimmune and systemic inflammatory diseases. Vox Sang 65: 65-72
106. Sullivan KM, Kopecky KJ, Jocom J et al. (1990) Immunomodulatory and antimicrobial efficacy of intravenous immunoglobulin in bone marrow transplantation. N Engl J Med 323: 705-712
107. Hoang-Xuan K, Leger JM, Ben Younes-Chennoufi A et al. (1993) Traitement des neuropathies dysimmunitaires par immunoglobulines polyvalentes intraveineuses. Etude ouverte de 16 cases. Rev Neurol 149: 385-392
108. Faed JM, Day B, Pollock M et al. (1989) High-dose intravenous human immunoglobulin in chronic inflammatory demyelinating polyneuropathy. Neurology 39: 422-425
109. Hodkinson SJ, Pollard JD, McLeod JG (1990) Cyclosporin A in the treatment of chronic demyelinating polyradiculoneuropathy. J Neurol Neurosurg Psychiatry 53: 327-330
110. Nemni R, Amadio S, Fazio R et al. (1994) Intravenous immunoglobulin treatment in patients with chronic inflammatory demyelinating neuropathy not responsive to other treatments. J Neurol Neurosurg Psychiatry 57(Suppl): 43-45
111. Vermeulen M, van Doorn PA, Brand A et al. (1993) Intravenous immunoglobulin treatment in patients with chronic inflammatory demyelinating polyneuropathy: a double-blind, placebo-controlled study. J Neurol Neurosurg Psychiatry 56: 36-39
112. Dyck PJ, Litchy WJ, Kratz KM et al. (1994) A plasma exchange versus immune globulin infusion trial in chronic inflammatory demyelinating polyradiculoneuropathy. Ann Neurol 36: 838-845
113. Hahn AF, Bolton CF, Pillay N ET AL. (1996) Intravenous immunoglobulin treatment in chronic inflammatory demyelinating polyneuropathy. A double-blind, placebo-controlled cross-over study. Brain 119: 1067-1077
114. Kahn SN, Riches PG, Kohn J (1980) Paraproteinemia in neurological disease: incidence, associations, and classification of monoclonal immunoglobulins. J Clin Pathol 33: 617-621
115. Kelly JJ, Kyle RA, O'Brien PC, Dyck PJ (1981) Prevalence of monoclonal protein in peripheralneuropathy. Neurology 31: 1480-1483
116. Krol V, Straaten MJ, Ackerstaff RG, De Maat CE (1985) Peripheral plyneuropathy and monoclonal gammopathy of undetermined significance. J Neurol Neurosurg Psychiatry 48(7): 706-708
117. Osby E, Noring L, Hast R, Kjellin KG, Knutsson E, Siden A (1982) Benign monoclonal gammopathy and peripheral neuropathy. Br J Haematol 51(4): 531-539
118. Vrethem M, Cruz M, Huang W et al. (1993) Clinical, neurophysiological and immunological evidence of polyneuropathy in patients with monoclonal gammopathies. J Neurol Sci 114: 193-199
119. Nobile-Orazio E, Manfredini E, Carpo M et al. (1994) Frequency and clinical correlates of anti-neural IgM antibodies in neuropathy associated with IgM monoclonal gammopathy. Ann Neurol 36: 416-424
120. Dalakas MC, Engel WK (1981) Polyneuropathy with monoclonal gammopathy: studies of 11 patients. Ann Neurol 10: 45-52
121. Kelly JJ, Adelman E, Berkman E, Bhan I (1988) Polyneuropathies associated with IgM monoclonal gammopathies. Arch Neurol 45: 1355-1359
122. Yeung KB, Thomas PK, King RHM et al. (1991) The clinical spectrum of peripheral neuropathies associated with benign monoclonal IgM, IgG and IgA paraproteinaemia. Comparative clinical, immunological and nerve biopsy findings. J Neurol 238: 383-391

123. Nobile-Orazio E, Baldini L, Barbieri S et al. (1988) Treatment of patients with neuropathy and anti-MAG IgM M-proteins. Ann Neurol 24: 93-97
124. Waterston JA, Brown MM, Ingram DA, Swash M (1992) Cyclosporin A therapy in paraprotein-associated neuropathy. Muscle Nerve 15: 445-448
125. Dellagi K, Chedru F, Clauvel JP, Brouet JC (1984) Neuropathie periphérique de la macroglobulinemie de Waldenstrom. Presse Med 13: 1199-1201
126. Sherman WH, Olarte MR, McKiernan G et al. (1984) Plasma exchange treatment of peripheral neuropathy associated with plasma cell dyscrasia. J Neurol Neurosurg Psychiatry 47: 813-819
127. Simmons Z, Bromberg MB, Feldman EL, Blaivas M (1993) Polyneuropathy associated with IgA monoclonal gammopathy of undetermined significance. Muscle Nerve 16: 77-83
128. Dyck PJ, Low PA, Windebank AJ et al. (1991) Plasma exchange in polyneuropathy associated with monoclonal gammopathy of undetermined significance. N Engl J Med 325: 1482-1486
129. Ellie E, Vital A, Steck A et al. (1996) Neuropathy associated with 'benign' anti-myelin-associated glycoprotein IgM gammopathy: clinical, immunological, neurophysiological, pathological findings and response to treatment in 33 cases. J Neurol 243: 34-43
130. Oksenehendler E, Chevret S, Léger JM et al. (1995) Plasma exchange and chlorambucil in polyneuropathy associated with monoclonal IgM gammopathy. J Neurol Neurosurg Psychiatry 59: 243-247
131. Cook D, Dalakas M, Galdi A et al. (1990) High-dose intravenous immunoglobulin in the treatment of demyelinating neuropathy associated with monoclonal gammopathy. Neurology 40: 212-214
132. Léger JM, Younes-Chennoufi AB, Chassande B et al. (1994) Human immunoglobulin treatment of multifocal motor neuropathy and polyneuropathy associated with monoclonal gammopathy. J Neurol Neurosurg Psychiatry 57(suppl): 46-49
133. Notermans NC, Lokhorst HM, Franssen H et al. (1996) Intermittent cyclophosphamide and prednisone treatment of polyneuropathy associated with monoclonal gammopathy of undetermined significance. Neurology 47(5): 1227-1233
134. Blume G, Pestronk A, Goodnough LT (1995) Anti-MAG antibody-associated polyneuropathies: improvement following immunotherapy with monthly plasma-exchange and IV cyclophosphamide. Neurology 45(8): 1577-1580
135. Kaji R, Hirota N, Oka N et al. (1994) Anti-GM1 antibodies and impaired blood-nerve barrier may interfere with remyelination in multifocal motor neuropathy. Muscle Nerve 17: 108-110
136. Corse AM, Chaudry V, Crawford TO et al. (1996) Sensory nerve pathology in multifocal motor neuropathy. Ann Neurol 39: 319-325
137. Lange DJ, Trojaborg W, Latov N et al. (1992) Multifocal motor neuropathy with conduction block: is it a distinct clinical entity? Neurology 43(3 Pt 1): 497-505
138. Kornberg AJ, Pestronk A (1994) The clinical and diagnostic role of anti-GM1 antibody testing. Muscle Nerve 17(1): 100-104
139. Feldman EL, Bromberg MB, Alberts JW, Pestronk A (1991) Immunosuppressive treatment in multifocal motor neuropathy. Ann Neurol 30: 397-401
140. Krarup C, Stewart JD, Sumner AJ et al. (1990) A syndrome of asymmetric limb weakness with motor conduction block. Neurology 40: 118-127
141. Parry GJ, Clarke S (1988) Multifocal acquired demyelinating neuropathy masquerading as motor neurondisease. Muscle Nerve 11: 103-107
142. Pestronk A, Cornblath DR, Ilyas AA et al. (1988) A treatable multifocal motor neuropathy with antibodies to GM1 ganglioside. Ann Neurol 24: 73-78

143. Donaghy M, Mills KR, Boniface SJ et al. (1994) Pure motor demyelinating neuropathy: deterioration after steroid treatment and improvement with intravenous immunoglobulin. J Neurol Neurosurg Psychiatry 57(7): 778-783
144. van den Berg LH, Lokhorst H, Wokke JH (1997) Pulsed high-dose dexamethasone is not effective in patients with multifocal neuropathy. Neurology 48(4): 1135
145. Bouche P, Moulonguet A, Younes-Chennoufi AB et al. (1995) Multifocal motor neuropathy with conduction block: a study of 24 patients. J Neurol Neurosurg Psychiatry 59: 38-44
146. Chaudry V, Corse AM, Cornblath DR et al. (1993) Multifocal motor neuropathy: response to human immune globulin. Ann Neurol 33: 237-242
147. Comi G, Amadio S, Galardi G et al. (1994) Clinical and neurophysiological assessment of immunoglobulin therapy in five patients with multifocal motor neuropathy. J Neurol Neurosurg Psychiatry 57(suppl): 35-37
148. Kaji R, Shibasaki H, Kimura J (1992) Multifocal demyelinating motor neuropathy: cranial nerve involvement and immunoglobulin therapy. Neurology 42:506-509
149. Nobile-Orazio E, Meucci N, Barbieri S et al. (1993) High-dose intravenousimmunoglobulin therapy in multifocal motor neuropathy. Neurology 43: 537-544
150. Parry GJ (1996) AAEM case report # 30: multifocal motor neuropathy. Muscle Nerve 19: 269-276
151. Azulay JP, Blin O, Pouget J et al. (1994) Intravenous immunoglobulin treatment in patients with motor neuron syndromes associated with anti-GM1 antibodies: a double-blind, placebo-controlled study. Neurology 44: 429-432
152. van den Berg LH, Kerkoff H, Oey PL et al. (1995) Treatment of multifocal motor neuropathy with high dose intravenous immunoglobulins: a double-blind, placebo-controlled study. J Neurol Neurosurg Psychiatry 59: 248-252
153. Azulay JP, Rihet P, Pouget J et al. (1997) Long term follow-up of multifocal motor neuropathy with conduction block under treatment. J Neurol Neurosurg Psychiatry 62(4): 391-394

Subject Index

Adjuvant 70
Allergic disorder 73
α-β cristallin 163
Antigen 2
  antigen-based immunotherapy 33
  altered peptide ligand 132, 214, 215
  CNS antigen 84, 164
  cryptic 2
  degeneracy of recognition 176, 177
  determinant spreading 163
  mimicry 91
  myelin 162-166, 170-179, 274-281
  presentation 2, 132
  presentation in the CNS 89-96
  processing 2, 132
  self antigen 2, 26, 132
  shared epitope 11
  superantigen 132
  TCR peptide 213
  viral 2
Antigen presenting cell 1, 28, 29, 89-96
Apoptosis 85
Astrocyte 48, 90-96, 101, 152, 157
Autoimmune thyroiditis 31, 72

B cell 16-22, 60, 69, 73
  allelic exclusion 21, 22
  development 16
  Ig assembly 16-21
Bence Jones protein 18
Blood brain barrier 47-53, 83-87, 90, 149-159, 195-203, 236
Bystander damage 162

CD1 166
CD11a 92
CD28 171
CD30 71, 76
CD31 42, 49
CD34 53, 63
CD40 4, 92, 157, 239
CD40 ligand 4, 156, 157, 239

CD52 217, 218
CD54 92-94, 153
CD80 92, 153, 156, 218
CD86 92, 153, 156
CD106 93, 153
CD144 48
Cerebrospinal fluid (CSF) 90, 157
Charcot-Marie-Tooth (CMT) 265-268
Chemokine 3, 43, 58-64
Chemokine receptor 60-62, 71, 157
  Th1/Th2 63-64
Chronic inflammatory demyelinating poliradiculoneuropathy (CIDP) 293-299
Connexin 270, 271
Copolimer1 215
CTLA-4 3, 119, 122, 153, 156, 218

Dejerine-Sottas (DSS) 268
Dendritic cell 2, 62, 89-90
Delayed-type hypersensitivity (DTH) 69, 74, 90, 187
Demyelination 239, 240, 265-268, 274-281, 287-303

Experimental allergic encephalomyelitis (EAE) 31, 86, 90, 119, 151-153, 163, 164, 189, 196, 210-222, 233-239, 246-261
Endoplasmic reticulum 16, 17
  retention 20
Endothelium 47, 157

Ganglioside 279-281, 301
GATA3 72
Gene therapy 221
Genomic screening 116-127, 251-257
GFAP 109
Glia 101-111
  cell progenitor 101
  limitans 150
  trophic factor 101-111
GM-CSF 63, 69
Golgi complex 17

Subject Index

Graft rejection 72
Guillain-Barrè (GBS) 287-289

Heat shock protein 164, 165
Helicobacter pylori 73
Hereditary neuropathy with liability to pressure palsy (HNPP) 265, 266, 268, 269
HIV 2, 90, 288

ICE 86
IFN-β 203, 220, 293
IFN-γ 33, 3, 27, 63, 69, 91, 139-140, 162, 174, 187, 220, 237, 259
IFN-γ receptor 28, 73
IL-1 141-143, 162
IL-1 receptor antagonist 138, 141-143
IL-2 3, 91, 140, 174, 187
IL-4 33, 3, 27, 69, 72, 140, 174, 187, 221
IL-5 69, 174
IL-10 3, 29, 141
IL-12 33, 3, 27, 69, 138, 187, 237
IL-12 receptor 30-31, 73
IL-13 69
Immune surveillance 83
Immunoglobulin
  allelic exclusion 21, 22
  assembly 16-22
  function 16-22
  quality control 21
Immunotherapy 210-222, 279-281, 290-293, 297-299
Inflammatory bowel disease 31, 72
iNOS 86
Insulin dependent diabetes mellitus (IDDM) 29-31, 72, 116, 132
Integrin 39-41, 150-152, 156, 157, 218, 219
Intravenous Ig 290, 291, 297

Junction 47-53, 149, 150
  adherens 49
  interendothelial 48
  tight 50, 149, 150

LAG-3 71, 76
Leukocyte adhesion deficiency syndrome (LAD) 39
Linkage analysis 119, 133-135, 252-254

Magnetic resonance imaging (MRI) 155, 195-203, 239
  clinical correlate 201, 202
  diffusion weighted 200
  gadolinium enhanced 196-198

magnetization transfer imaging 198, 199
magnetic resonance spectroscopy 199
Matrix metalloproteinase (MMP) 157
MHC-binding peptide 7-13, 172
  promiscuos peptide 12
MHC molecule 7, 10, 164, 130-135
  congenic rat 251
  disease association 10, 133-135, 247-249, 257-261
  expression on CNS antigen presenting cell 84, 93, 94
  in EAE 257-260
  in MS 130-135
  peptide-MHC interaction 8
  polymorphism 7
Microglia 84, 90-96, 102, 157, 190
Migration 38-43, 47-53, 58-64, 83, 152-155
  leukocyte 42, 43, 83, 84, 218
Miller Fisher syndrome (MFS) 289
Monocyte 61, 62, 157
Multifocal motor neuropathy (MMN) 300-302
Multiple sclerosis 11, 12, 31, 72, 75, 116-127, 130-135, 137-144, 155, 162-166, 170-179, 185-192, 195-203, 210-222, 246-249
Myasthenia gravis 31
Myelination 101-111, 269-271
Myelin protein 101-111
  CNPase 166, 179
  MAG 179, 274-281, 299
  MBP 109, 162, 163, 170-177, 188-189, 214, 234, 247, 250, 258
  MOG 109, 177, 178, 189, 221, 234, 247, 250, 258
  myelin antigen 162-166
  P0 267, 271
  PLP 109, 162, 163, 175, 178, 188-189, 214, 234, 247, 250, 258
  PMP 22 267, 271

Neuron 84, 94, 104, 107
NF-AT 189, 190
NFkB 157
NOD mouse 29-31

Oligodendrocyte 84, 94, 101, 152, 157, 166

Plasma exchange 290, 296

Rheumatoid arthritis 11, 32

Schwann cell 102, 268
Selectin 38, 39, 150-152, 156

Subject Index

Spinal cord development 108
STAT 29, 72
Sulfatide 279-281
Systemic lupus erithematosus 72, 76

TGF-β 76, 111, 141, 216, 259, 261
Th1/Th2 3, 27-31, 63, 68-77, 90-96, 173, 174, 186, 187, 210-222, 237, 261

Theiler's murine encephalomyelitis virus (TMEV) 239, 240
TNF receptor 32, 86
TNF-α 3, 27, 32, 63, 69, 76, 140, 152, 162, 174, 187, 220, 237
Tolerance 2, 26, 217
  oral 216

Uveoretinitis 31